T0214026

Topics in Safety, Risk, Reliability and Quality

Volume 39

More information about this series at https://link.springer.com/bookseries/6653

Teddy Steven Cotter

Engineering Managerial Economic Decision and Risk Analysis

Economic Decision-Making and Risk Analysis

 Springer

Teddy Steven Cotter
Department of Engineering Management &
Systems Engineering
Old Dominion University
Norfolk, VA, USA

ISSN 1566-0443   ISSN 2215-0285   (electronic)
Topics in Safety, Risk, Reliability and Quality
ISBN 978-3-030-87769-9   ISBN 978-3-030-87767-5   (eBook)
https://doi.org/10.1007/978-3-030-87767-5

This Springer imprint is published by the registered company Springer Nature Switzerland AG
The registered company address is: Gewerbestrasse 11, 6330 Cham, Switzerland

Preface

This is not another book on engineering economics. Rather, this book is directed to the engineering manager or the undergraduate student preparing to become an engineering manager, who is, or will become, actively engaged in the management of economic risk trade-off decisions for engineering investments within an organizational system. In today's global economy, this may mean managing the economic risks of engineering investments across national boundaries in international organizations, government, or service organizations. As such, this is an applied book. The book's goal is to provide an easy-to-understand, up-to-date, and coherent treatment of the management of the economic risk trade-offs of engineering investments. This book accomplishes this goal by cumulatively sequencing knowledge content from foundational economic and accounting concepts to cost estimating to the traditional engineering economics knowledge culminating in fundamental engineering managerial economic decision making incorporating risk into engineering management economic decisions.

From years of working in industry and teaching an introductory engineering economics course, the author found that most practicing engineering managers and students preparing for careers in engineering have little or no prior coursework or training in accounting, micro- or business economics, finance, or cost estimation from the engineering perspective. Accordingly, this book places these preparatory subjects in the first four chapters of Part 1, Economic Context of Engineering Management Decisions, so that when the student reaches fundamental engineering economic concepts in Part 2, he or she understands cash flow accounting and the economic and financial framework within which he or she must operate in organizational settings. Next, this book presents the fundamentals of engineering managerial economic analysis in Part 2, Economic Analysis of Engineering Activities and Projects. Finally, this book synthesizes engineering managerial economics into engineering managerial decision and risk analysis in Part 3, Managing Engineering Investments, by tying together prior concepts for the selection of risk-adjusted minimum attractive rate of return, capital budgeting decisions, and benefit–cost decisions in public and regulated sectors. The book culminates with making engineering economic managerial investment risk decisions for uncertain future cash flows. The final chapter sets forth probability and statistics fundamentals necessary to model future variable cash flows and applies those fundamentals in sensitivity analysis and simulation modeling.

Features of This Book

Key design features of this book include:

1. The logical sequencing of engineering managerial economic knowledge from preparatory background to engineering economics fundamentals to engineering managerial economic decision and risk analysis of variable future cash flows.
2. A structured problem-solving approach is illustrated in manual solutions and corresponding spreadsheet solutions. The manual solutions illustrate fundamental engineering managerial economic decision and risk analysis theory and principles, and the corresponding spreadsheet solutions illustrate application in practice.
3. A cash flow accounting approach is taken throughout to engineering managerial economic analysis. Accrual accounting is introduced in chapter two and contrasted to cash flow accounting where appropriate throughout the book, so the engineering manager or student understands the relationship between the two.
4. Microsoft Excel® cash flow examples are included throughout the book to introduce the engineering manager or student to the use of spreadsheets in the analysis of the economics risk decisions in the management of engineering functions and projects. These examples may be used as templates to set up actual engineering managerial economic spreadsheet analyses.
5. A mix of private enterprise and public or regulated sectors example problems to provide the student with a robust exposure to engineering managerial economic problems.
6. Comprehensive coverage appropriate for the engineering manager performing or the student seeking to learn to perform engineering managerial economic analyses including:

   - Economic analyses with or without considering the time value of money. General and cost accounting, cost estimation, and economic value-added concepts.
   - Planning horizons.
   - Budgeting and the use of economic ratios.
   - Economic criteria metrics including present worth, equivalent uniform annual worth, future worth, internal rate of return, external rate of return, benefit–cost analyses, and cost-effectiveness. Other metrics such as undiscounted and discounted payback are also discussed.
   - Historical and IRS Publication 946 MACRS depreciation methods and their effects on cash flow accounting and economic criteria.

- Inflation-adjusted cash flows, indexes, and inflationary effects on before and after-tax economic metrics.
- Challenger–defender and graphical incremental analyses to select the best mutually exclusive alternative.
- Opportunity cost after-tax replacement analyses, optimal replacement interval, and Section 1031 like-kind property exchanges.
- Cost of capital MARR determination.
- Capital budgeting engineering managerial decisions.
- Public and regulated sector economic analyses using benefit–cost analyses.
- Introduction to economic decision risk analysis and management of engineering economic activities and projects including fundamentals of probability, structuring decisions, sensitivity analysis, and an introduction to simulation.
- End-of-chapter problems to facilitate self-study as well as homework assignments. Answers to problems are provided in a separate solution manual.

Course Coverage

This book is designed for two types of students and for any type of class. First is the student interested in preparing for the fundamentals of engineering examination. A course for this student may be taught as a traditional engineering economics course with material from the following chapters

- Chapter 2, General Accounting and Finance Fundamentals.
- Chapter 4, Cost Estimating Fundamentals.
- Chapter 5, Time Value of Money.
- Chapter 6, Measures of Investment Worth.
- Chapter 7, Depreciation Effects on Investment Worth.
- Chapter 9, Inflation Effects on Engineering Investments.
- Chapter 10, Incremental Analysis.
- Chapter 15, Introduction to Management Economic Decision Theory and Risk Analysis (with introductory probability and statistics).

Studying these chapters along with working solutions to the associated chapter problems will prepare the student for the economics portion of the fundamentals of engineering examination. The second type of student is the one who seeks proficiency in engineering managerial economics as a practicing engineering manager. A course for this student should sequentially follow the design of the book from Chap. 1 through 15.

The book and supporting lectures, homework, and homework solution manual are designed for use in a traditional classroom lecture setting, in a blended online-classroom setting, or as a self-contained, online asynchronous learning package. In the traditional classroom lecture setting, this book may be used as a stand-alone hardcopy or electronic book. In the blended online-classroom setting, the instructor has multiple options: (1) hardcopy book with chapters, example problems, homework problems, and examinations integrated into a learning management system; (2) electronic book with student tutorials and self-tests integrated into a learning management system along with homework solutions and examinations to support the traditional classroom lectures; and (3) self-contained online learning package with integrated electronic book, student tutorials, homework assignments and solutions, and examinations for use by distant learning students or supporting traditional classroom discussions. Instructor lecture notes, Microsoft PowerPoint® or Adobe PDF® slides, question and problem solutions along with Microsoft Excel® workbook solutions and templates, and test questions may be downloaded from an online source.

Norfolk, VA, USA   Teddy Steven Cotter, Ph.D.

Contents

Part II   Economic Analysis of Engineering Activities and Projects

Part I
Economic Context of Engineering Management Decision

Chapter 1
Managerial Economics of Engineering Organizations

Abstract This book is about engineering managerial decision making and the risks associated with those decisions. Decisions and their associated risks are driven by the context of the decision situation. For engineering managers, this context is the management of engineering and technical functions within for-profit and not-for-profit organizations. Chapter 1 establishes the context of engineering managerial decision making. First, the chapter sets forth definitions of the engineering manager and the four fundamental classes of engineering functions. Second, the basic classes of for-profit organizations within which engineering managers work are described. Third, the product life cycle and annual financial cycle context of capital expenditures and operating costs are defined. Fourth, the chapter describes engineering and technical function contributions to the stages of the product life cycle. Finally, the chapter concludes with a general description of the budgeting process of engineering operations and projects.

1.1 Introduction

All engineering activities and projects must be executed within an organizational context, either private industry or public governmental services. All organizational decisions are constrained by scarcity of resources. This constraint drives three fundamental decisions for organizational survival and growth.

1. What are the organization's objectives: (a) the customer base, (b) the product or service will be produced, and (c) the processes that will produce the product or service?
2. What resources are needed and available to produce the product or service: (a) monetary, (b) human resources, (c) knowledge and information resources, (d) materials, and (e) physical facilities and equipment?
3. How will resources be allocated: to which (a) products or services, (b) projects, (c) departments, and (d) in what amounts given varying conditions?

© The Author(s), under exclusive license to Springer Nature Switzerland AG 2022

T. S. Cotter, *Engineering Managerial Economic Decision and Risk Analysis*,
Topics in Safety, Risk, Reliability and Quality 39,
https://doi.org/10.1007/978-3-030-87767-5_1

This book is concerned with the third decision. More specifically, this book is concerned with the optimum allocation of money to engineering projects for the introduction of new products and services or the expansion of existing products and services. Although this book focuses on the third decision, it must be recognized that these three fundamental decisions cannot be made in isolation. Rather, they are iterative and interactive across annual budgets and the budget planning process.

1.2   Engineering and Engineering Management

Definition: *Engineeringmanagerial economics* is the body of knowledge and methods devoted to the systematic evaluation of the economic benefits resulting from ongoing engineering activities or proposed engineering projects in relation to the economic costs of those activities or projects.

To fully appreciate the scope of this definition of engineering managerial economics, we must decompose it into its components. The fundamental components are engineering, management, and economics. *Management* is the organizational function that *develops operational plans and goals, employs personnel for productive operations, and controls productive operations to achieve economic goals.* ABET, the Accreditation Board of Engineering and Technology, defines engineering as "*... the profession in which a knowledge of the mathematical and natural sciences gained by study, experience, and practice is applied with judgment to develop ways to utilize economically the materials and forces of nature for the benefit of mankind.*" From this definition, we can derive the properties of engineering as a profession. Property 1—Engineers gain knowledge of mathematics and natural sciences and apply that knowledge to design and potentially build and maintain materials, structures, or devices, processes, information, and systems. Thus, engineering is a transformation process taking as its inputs mathematical and natural science knowledge and combining and transforming them into, at minimum, designs for, and at best, outputs that can be used by other men and women to enhance (benefit) their lives in some manner. Property 2—From property one, the fundamental output of the engineering transformation process is a design. At minimum, engineering is a design process. The engineer may or may not build the designed material, structure, or device, information, process, or system. Others with sufficient technical knowledge can take the engineer's design and build the unit of intent. Likewise, the engineer may or may not maintain the intended material, structure, or device, information, process, or system. Again, others with sufficient technical knowledge can perform the necessary maintenance. Property 3—The engineering transformation process is constrained by the economics of the "... materials and forces of nature ..." or technology of the time that is available to the transformation process. Technology itself may be defined as "the application of scientific knowledge for practical purposes." Hence, the economics of engineering arises from the limits of knowledge of mathematical and natural sciences at the time it is applied. As technical knowledge increases over time, either the same beneficial outputs may be produced at lower economic cost per unit or outputs with greater benefits per unit may be produced at essentially the same economic cost per unit. Otherwise, mankind does not benefit from the engineering transformation process.

The engineering outputs of materials, structures, and devices can be roughly categorized as products. So, the first general engineering classification can be considered as the product engineer. Product engineers may only design a material, structure, or device, be the technical development interface between design and the production process, or perform both the design and development activities. Product engineers must deal with issues of cost, performance, producibility, quality, reliability, durability, serviceability, aesthetics, features, conformance to standards, and user expectations. Given the limits of technological knowledge, trade-offs among product characteristics are generally required to achieve some balance of features that will be beneficial to others. Once in production, product engineers may also seek to enhance certain product features without adversely impacting others. Typical knowledge and skills required of the product engineer include:

- Fundamental mathematical principles.
- Fundamental scientific principles relative to the product (specific technology).
- Computer-aided design and simulation.
- General manufacturing processes.
- Physical analysis methods.
- Statistical analysis methods and tools.
- Reliability analysis methods and tools.
- Test equipment, tools, and methods.
- Structured problem-solving skills.
- Root cause analysis.
- Project management skills.

The engineering output of a process is considered as *process engineering*. Process engineering focuses on the design, operation, control, and optimization of the production process. Production processes may be chemical, physical, or biological and may yield physical or service products. Like product engineers, process engineers also may only design the process, be the technical development interface between design and actual development of the process, or perform both the design and development activities depending on the scope and scale of the process. Process engineers must deal with issues of cost, performance quality, reliability, aesthetics, conformance to standards, and client expectations. Given the limits of socio-technical knowledge, trade-offs among process characteristics are generally required to achieve some balance of process features that will yield intended benefits in the product or service. Once released to the producers, the process engineer may also seek to enhance certain process features without adversely impacting others. Process engineers require the same general skills as the product engineer above with the exception that the process engineer requires fundamental scientific principles relative to process development and specific knowledge of the process type.

The engineering output of data, information, or knowledge falls under the general class of *information engineering*. Information engineering falls into two general categories: (1) data generation, distribution, analysis and synthesis and transformation into information or knowledge, and the use of information or knowledge in systems and (2) the design, development, implementation, and maintenance of physical and software information systems. Like product and process engineers, information engineers also may only design software or physical information systems, be the technical development interface between design and actual development, or perform both the design and development activities depending on the scope and scale of the information system. Information engineers must deal with issues of cost, delivery, and performance and measure software development in terms of quality, reliability, conformance to standards, and client expectations. Information engineers require knowledge and skills in:

- Data flow and data analysis.
- Database design.
- Information theory.
- Entity analysis and modeling.
- Cluster analysis to define the scope of design areas for proposed information systems.
- Ontology development.
- Knowledge engineering in intelligence applications.
- Expert systems.
- Artificial intelligence.
- Data mining.
- Decision support systems.
- Geographic information system (GIS).
- Project management skills.

The engineering output of complete systems with emergent properties and behaviors is the domain of *systems engineering*. Systems engineering is interdisciplinary in that it integrates and utilizes engineering knowledge from all engineering disciplines within a general systems framework to design, implement, and manage complex systems over their life cycles. Systems engineers may be involved with any or all process of the design, development, implementation, delivery, and management of systems. Systems engineers must deal with the broad issues of cost, delivery, and performance of complete systems. Systems engineering requires knowledge and skills in:

- Fundamental mathematical principles.
- Fundamental scientific principles.
- General engineering principles.
- System architecture.
- System dynamics.
- Systems analysis.
- Statistical analysis.
- Reliability analysis.

- Requirements analysis.
- Feasibility analysis.
- Logistics.
- Maintenance.
- System modeling and simulation.
- Optimization.
- Operations research.
- Decision making.
- Configuration management.
- Project management.

Engineering management combines technical expertise with managerial knowledge to coordinate operational performance of engineering disciplines within an organizational setting or coordinate and control technical project outputs. In addition to engineering technical expertise, engineering managers require additional knowledge and skills in (Farr and Merino 2010):

- Basic accounting and finance.
- Decision analysis.
- Engineering economics.
- Leadership and teams.
- Management theory and concepts.
- Modeling and simulation.
- Operations research.
- Project management.
- Quality and reliability management.
- Risk analysis and management.
- Strategic management.
- Systems engineering.

1.3 Types of Business Organizations

All engineering activities take place within the context of either business (for-profit and not-for-profit) or governmental (non-military and military) organizations. Accordingly, the engineer and engineering manager must understand the organizational type within which he or she works. There are three fundamental types of business organizations.

A *proprietorship* is an unincorporated business owned by one individual. This individual makes all initial investments in the business, personally owns all its assets, and is legally liable for all its debts. The proprietorship is not a legal entity; rather, the individual who owns the business is the legal entity.

Proprietorships have three distinguishing characteristics. First, a proprietorship is formed simply by a person setting up and conducting business. There are no legal or organizational requirements to set up a business as a proprietorship. Second, all revenues, debts, profits, and losses must be recognized and taxed as the owner's personal income. Third, the owner is legally accountable for all regulatory and legal debts. In addition to the personal legal liability, the major disadvantage of the proprietorship is that it cannot issue stocks and bonds to finance its operations. Within engineering, the typical form of proprietorship is as an engineering consulting firm.

A *partnership* is owned by two or more individuals, who jointly invest money, skills, and assets and who share in profits or losses in accordance with the terms set forth in their partnership agreement. In the absence of a written partnership agreement, the legal system will assume that a partnership exists where the individuals who participate in an enterprise agree to share the associated risks, returns, and losses proportionately. The partnership is not a legal entity; rather, the individuals entering the partnership are individual legal entities.

A partnership is a legal agreement among two or more individuals (the legal entities) to share the profits and liabilities of a business venture. Various partnership agreement structures are available: sharing profits and liabilities equally or in stated proportions to the individuals' contributions to the partnership; differing contributions of capital, assets, and expertise knowledge; or involvement with management and daily operations. Partnerships may be given favorable tax status in some states or municipalities. There are three categories of partnerships. In a *general partnership* (GP), the partners share profits and legal and financial liabilities equally. In a *limited partnership* (LP), one partner must be the general partner and bear full personal responsibility for the partnership's liabilities, and at least one other silent partner's liability is limited to the amount invested in the partnership. Generally, the silent partner cannot participate in the daily management of the partnership. A *limited liability partnership* (LLP) is a partnership in which each partner's liability is limited to his or her individual assets. If one partner is sued, the other partners' personal assets are not at risk. LLPs are a common organizational structure for accounting, architectural, engineering consulting, and law firms. Partnerships have many advantages. They are easy to form and have a low formation cost. Partnerships allow a mix of investments of assets, capital, expertise, and skills necessary for a particular business venture. Since the partners' personal assets back the business venture, partnerships can borrow money more easily than a proprietorship. Each partner pays only personal income tax on his or her proportion of the partnership's taxable income. Partnerships also have disadvantages. Each partner is proportionally liable for the venture's debts. In the case of bankruptcy, if any partner cannot meet his or her proportion of the debts, the remaining partners must cover the unresolved debts. The partnership has a limited life. It must be dissolved and reorganized if any one partner leaves the venture.

A *corporation* is a legal entity recognized under state or federal law. As a legal entity, it is separate and distinct from its owners. A corporation possesses most of the rights an individual possesses. It can enter into contracts, raise financing from the sale of stocks and bonds, borrow or loan money, own productive assets, and sue or be sued. A corporation is often referred to as a "legal person."

A group of shareholder owners create a corporation to pursue a common objective. The shareholder owners incorporate under the corporate laws in the jurisdiction of residence (state or federal government). The shareholder owners are each responsible only to pay their respective shares to the corporation's treasury, and generally, each receives one vote per share. Annually, each shareholder owner votes his or her shares to elect a board of directors that, in turn, hires and oversees management of the daily corporate activities. A corporation can be for-profit or not-for-profit. A corporation can have an infinite or limited life. If a corporation achieves its common objective, the shareholder owners can terminate the corporation and liquidate its assets. In this process, a liquidator is appointed, and the liquidator sells the corporation's assets, pays creditors, and distributes any remaining funds among the shareholder owners. The separation of the corporation as a legal entity from its shareholder owners provides four major advantages. First, a corporation can raise capital from a broader base of shareholder owners and investors through the sale of stocks and bonds. Second, he shareholder owners can easily transfer ownership by selling or trading their shares of stock. Third, the shareholder owners have only limited liability to the value of their respective shares of stock. Liabilities or legal judgments against the corporation are separate and cannot be reapportioned to the shareholder owners. Fourth, a corporation pays taxes at corporate rates, often favorable to the corporate structure, separate from the personal taxes paid by its shareholder owners. The disadvantages to the corporate form of business are that it is expensive to incorporate, and corporations are subject to numerous governmental laws and regulations.

The typical business organization starts out small and, if successful, grows. In the USA, sole proprietorships make up the majority of businesses by number (approximately 70%). Partnerships are second most numerous (approximately 23%) followed by corporations (7%). Proprietorships employ about 53% of the non-governmental workforce but generate only about 6% of total sales in the USA. Typically, a proprietorship employs ten or fewer people. Conversely, corporations employ about 38% of the workforce but generate about 73% of total sales, and partnerships employ about 9% of the workforce and generate about 21% of total sales. Accordingly, the majority of engineering and engineering management positions are found in corporations and partnerships.

This book addresses the economic decisions for the set of engineers and engineering managers employed in corporations, partnerships, state and local governmental agencies, and non-military federal governmental agencies. Economic decisions in the business sector tend to be driven by cash flow and return on investment criteria. Conversely, economic decisions in governmental agencies tend to be driven by some form of benefit–cost criteria.

1.4 Engineering Economic Decisions

As can be seen in the Engineering and Engineering Management section above, engineers and engineering managers contribute technical expertise to every aspect of organizational, product, process, and service design, implementation, maintenance, and responsible retirement and disposal. In these roles, engineers are responsible to consider the effective use of capital, in the form of investment or taxes, to acquire, implement, and use productive assets. In the design phase, engineers' primary task is to plan capital expenditure for equipment, facilities, and supporting infrastructure needed for the organization to produce its products or services. For the purchase of an asset, engineers must estimate all life cycle cash flows for capital expenditure and installation, revenues and expenses, and salvage and disposal. The life cycle of an asset is the minimum of its useful life, life to obsolescence, or project life (given the asset's life is greater than the project life). Inaccuracies in cash flow estimates can have serious impacts on realized project net profits or benefit–cost. Overinvestment in expensive assets can result in never recovering the cost of the asset for operations cash flows. Conversely, underinvestment in cheap assets can result in negative net profits (losses) due to excessive annual operating and maintenance expenses. Either can drive up the cost denominator of the B/C ratio in governmental projects.

> *Capital expenditure* is the purchase of an asset with a life greater than one fiscal year.
>
> An *asset* is land, physical facilities, equipment, money, monetary instruments, and intellectual property owned or controlled by a business organization for the purpose of producing products or services to provide a return on investment in the asset.
>
> A *fiscal year* is a 12-month time period that a business organization uses for the accounting purpose of preparing financial statements. The fiscal year may or may not be equal to the calendar year.

In the implementation and maintenance phases, engineers' primary task is **cost control** for the efficient operation and maintenance of productive assets. In the retirement and disposal phase, engineers' primary task is defining all activities necessary for responsible disposal and minimizes the project costs associated with disposal.

The configuration of facilities and operations are determined by the life cycle(s) of products and services produced and delivered. Hence, engineering decisions, and the corresponding cash flow implications, exist along three timescales. The short-term scale is less than one fiscal year. Short-term economic decisions are comparatively simple yes/no or quantity decisions with minimal cost impact and risk to support existing products and services: i.e., replace a burned-out pump this week, the quantity and type of maintenance supplies this month, or the purchase replacement PID controller. Short-term economic cash flows are treated as expenses in the period incurred. Intermediate-term economic decisions are tied to the budget planning period, which itself is determined by the organization's product or service life cycle or the time to transition assets to support innovative new products. For products with short life cycles or short transition periods, the budget planning period may be one to three years; intermediate life cycles or transition periods, three to seven years; and product or service long-term life cycles or transition periods, seven to ten or twelve years.

The *product life cycle* is the expected life of a product or service from the time of its conception to its termination and disposal.

Intermediate economic decisions to support production of current products and services within the budget planning period:

- Involve costs in the $10,000s or $100,000s.
- Are of intermediate risk—wrong outcomes can be fixed but at costs that may impact profitability.
- Have organizational and stakeholder impacts—the organization's financial status or reputation may be adversely impacted and cause a loss of stock value.
- Require formal organized analysis.
- Require comparison to a single criterion or a limited number of criteria with intermcdiate estimate variability.

Complex long-range economic decisions to develop future products and services for future budget planning periods:

- Involve costs in the millions or even billions of dollars for new production lines or facilities.
- Are of high risk—wrong outcomes may result in organizational failure.
- Have significant organizational and stakeholder impacts—loss of employment for employees, loss of cash flows and tax revenues for the community, and potential loss of a customer for its suppliers and a supplier for its customers.
- Require formal organized analysis.
- Require comparison to multiple criteria (economic, technical, legal, and societal) with long-range variability and uncertainty.

Current, intermediate, and complex long-term economic decisions are tied to the organization's long-term strategic planning and budgeting as illustrated by Fig. 1.1.

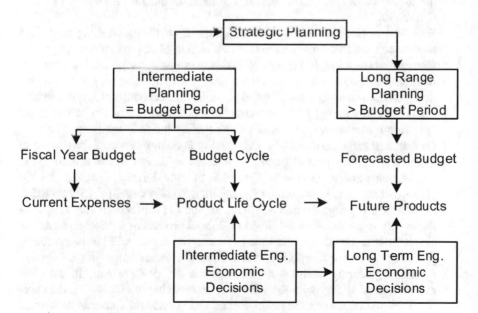

Fig. 1.1 Linkage between strategic planning and engineering economic analyses

In each of these timescales, engineering economic analysis is used to determine which projects (investments) are worthwhile, how to prioritize those projects, and the optimum or near-optimum effective designs given funding and other constraints. Fundamentally, there are five basic classes of projects for which engineering economic analysis is required.

Cost Reduction. Typically, new processes and products are implemented at competitive costs. As processes age, they incur increasing efficiency losses due to wear or obsolescence. If a new product is successful, competitors will enter with competing designs at lower price per unit for the same performance or with increased performance at the same price per unit. Both causal effects drive the need for continual cost reductions in labor, materials, overhead, process, and service costs to remain competitive. The primary metric for tracking cost reduction is total cost per unit.

Productivity, Quality, or Service Optimization. Whereas cost reduction seeks to reduce the total cost per unit for some level of performance, optimization seeks to maximize utilization of labor, materials, energy, consumables, and information for a stated level of process output (productivity), maximize product performance relative to quality standards or customer expectations, or maximize service responsiveness, professionalism, and attentiveness per customer contact. Optimization projects tend to fall into six basic categories.

1. *Process optimization.* In process optimization projects, the process engineer seeks to adjust the process parameters, without violating technical or economic constraints, with the goal of maximizing throughput or minimizing the utilization of labor, materials, energy, and consumables for a stated level of throughput.
2. *Workforce optimization.* In workforce optimization, the process engineer seeks to balance workforce requirements for a stated level of product throughput.
3. *Energy consumption.* In highly automated manufacturing processes, energy is the first or second highest cost element. Energy consumption may be minimized through building or equipment redesign, production scheduling to consume high levels of energy during low demand and low-cost times of the day or week, and renegotiation of energy contracts with electricity and fuel suppliers.
4. *Quality and reliability optimization.* Quality is conformance of the totality of product or service performance and features to design intent and customer requirements and expectations. Quality is measured along eight dimensions: (1) performance, (2) reliability, (3) durability, (4) serviceability, (5) aesthetics, (6) features, (7) organizational reputation, and (8) conformance to standards. Reliability is measured as the time to failure of product or service performance characteristic. The longer the time to failure, the higher will be the perceived product or service quality. Quality and reliability optimization projects involve formal engineering analyses of the negative and positive gaps in the eight dimensions of quality with the joint goals of reducing or eliminating the negative gaps and increasing the positive gaps. Quality and reliability improvement projects involve making changes in the total cost of quality, levels of customer satisfaction, or reduction in waste, rework, and scrap losses.

5. *Quality and reliability optimization.* Quality is conformance of the totality of product or service performance and features to design intent and customer requirements and expectations. Quality is measured along eight dimensions: (1) performance, (2) reliability, (3) durability, (4) serviceability, (5) aesthetics, (6) features, (7) organizational reputation, and (8) conformance to standards. Reliability is measured as the time to failure of product or service performance characteristic. The longer the time to failure, the higher will be the perceived product or service quality. Quality and reliability optimization projects involve formal engineering analyses of the negative and positive gaps in the eight dimensions of quality with the joint goals of reducing or eliminating the negative gaps and increasing the positive gaps. Quality and reliability improvement projects involve making changes in the total cost of quality, levels of customer satisfaction, or reduction in waste, rework, and scrap losses.

6. *Inventory carrying costs.* The total carrying, or holding, cost of maintaining inventory includes warehousing costs for space, wages, salaries, utilities, maintenance, and insurance plus the cost of the parts. Thus, maintaining raw materials, work-in-process, and finished product inventory for long periods of time comes at a cost. On the other hand, inventories provide organizations the ability to be responsive to customer demand while avoiding missed sales opportunities. Engineering projects to optimize inventory carrying costs involve balancing inventory levels throughout the production process against forecasted customer demand to minimize out-of-stock and overproduction occurrences.

Process and Equipment Selection. The construction of a new production facility requires the selection of productive processes and equipment. In the case of product extension to new markets, existing processes may be adapted from similar existing production facilities. For completely new innovative products, specialized equipment and processes may have to be designed. For either extreme and for mixed cases in between, equipment and process selection are a critical factor in the future competitive success. The selection process is driven by the process type, and equipment selection is driven by fourteen key factors. There are five fundamental types of production processes.

1. *Machining and assembly* production are additive processes. They are exemplified by production lines with minimal changeover and setup that produce similar or like products in long production runs. They are capital-intensive processes that run 24 h per day, 7 days per week to cover the high fixed equipment costs.

2. *Discrete processes* may be additive, nonadditive, or some mix thereof. They may require few or multiple changeovers and setups and produce a variety of products from highly similar structures to one-off designs.

3. *Job shops* are a collection of production areas with internally similar process but highly diverse intra-area processes. They may be additive, nonadditive, or some mix thereof. The product design determines the routing through the different production areas. Job shops are made up of a mix of highly automated and manual processes.

4. *Continuous processes* are nonadditive processes that produce gases, liquids, powders, or slurries. They are the counterpart of high-volume machining and assembly production in that they produce their products of multiple batches in long runs with minimal changeovers and setups. Likewise, they are capital intensive processes that run 24 h per day, 7 days per week to cover the high fixed equipment costs.
5. *Batch processes* produce specialty gases, liquids, powders, or slurries. Typically, they produce one to a few batches. They are the counterpart of discrete processes or job shop production processes.

For each process type, equipment selection is driven by four key factors.

1. *Socio-technical suitability*: Selected equipment must be technically compatible with the product to be produced and must consider the integration of people with the equipment.
2. *Initial installed cost*: Project budgets set constraints on the level and type of available equipment technology. The total initial installed cost is the sum of design, procurement, transportation, installation, commissioning, and scale-up costs necessary to place the asset in service fit for use.
3. *Equipment construction feasibility* due to technological, lead time, and schedule, physical installation, commissioning, and scale-up constraints.
4. Impact on *overall facility design elements* of size, location, interface, and interference of supporting mechanical, electrical, plumbing, and information architectures. Trade-offs among these elements are risk drivers of project change orders and schedule and cost overruns.

Process and Equipment Replacement. Typically, process or equipment replacement decisions involve large investments with moderate to high risk of failure. Replacement decisions are necessitated by:

1. The development of operational deficiencies in the existing defender process or equipment resulting in declining productivity, limited capacity, high or increasing setup cost, increasing energy consumption, increasing maintenance costs, or physical impairment.
2. Challenger replacement assets becoming available with advantages such as newer technological capabilities, lower per-unit production costs, higher throughput, lower energy costs, or lower maintenance costs.
3. Changing competitive environment or market demands. These can include (1) demand-driven user preference for new products with increasing technological capabilities, (2) supply-driven increasing functionality in competitor products and services, (3) changing industry specifications or government regulations, or (4) increased demand that cannot be met by existing equipment or processes.
4. 4.Obsolescence occurs when the existing defender equipment or process is no longer economically viable even though it remains in acceptable working order. Frequently, obsolescence occurs when a challenger replacement becomes available that has economic, productive, quality, or other technological advantages relative to the existing defender equipment or process.

The decision regarding the replacement of an existing defender asset with a challenger asset is based on incremental internal rate of return analysis of the difference between periodic net cash flows if the existing defender asset is kept versus the periodic net cash flows if the challenger asset is placed in service.

Product or Service Expansion. Investments in product or service expansion are made to increase revenues resulting from excess demand for existing products or services or projected demand for new products or services. The basic investment decision is whether to expand throughput by outsourcing production or service delivery through a contractor or build additional facilities. Typically, product or service expansion decisions involve large investment with high risk of failure.

1.5 Engineering Economic Principles

Engineering economics is based on fundamental principles that, if followed, assure maximizing investment benefits to investment costs.

1. *Money has time value.* All investments are made with rented borrowed money. The rent paid for borrowed money is in the form of *interest* on loans or *dividends* for stocks. From the lenders or investors' perspective, they prefer to receive a fixed sum of money sooner rather than later. The sooner a lender or investor receives interest or dividend payments, the sooner he or she can re-lend or reinvest the money in other interest or dividend-bearing assets. The longer a loan or investment remains outstanding, the greater the risk of not being repaid and the greater the cumulative interest for the number of loan or investment periods.

2. *All investments in productive assets must be economically justified.* Production or service assets must yield positive economic cash flows beyond the repayment of the principle and accrued interest payments. These positive economic cash flows are required for reinvestment toward the maintenance of existing production and service capacity and for new production or service capacity.

3. *All economic decisions must be based on differences in cash flows among the alternatives considered.* This principle assures that economic decisions are made based on comparison of differences in increases of cash flows due to differential benefits generated by the assets relative to decreases in cash flow required for differential investment in the assets.

4. *Use a common unit of measurement (metric) across alternative to make consequences commensurable.* Alternative consequences must be measured with a common measurement unit to be commensurable.

5. *Alternatives must be examined over a common planning horizon.* The planning period is the product or process life cycle or multiples thereof over which alternative consequences are relevant. A common planning horizon is required to make measurement of consequences commensurable.

6. *Only differences are relevant.* Prospective consequences that are common to all alternatives under consideration need not be considered, because they affect the outcomes of each alternative equally.

7. *Only feasible alternatives can be considered.* Decisions must lead only to courses of action that can be completed within physical laws and principles, economic laws and principles, political constraints, and legal requirements. Investment decisions must give weight to noneconomic consequences.

8. *Investment criteriamust include the time value of money due to capital acquisition and rationing.*

9. *Separable decisions should be made separately.* This principle requires careful evaluation of capital allocation to determine the number and types of economic decisions to be made.

10. *The relative degrees of uncertainty associated with future cash flows must be incorporated into the economic decision process.* Since engineering economic cash flow estimates occur in the future, there are probabilities that realized future cash flows will differ from the original estimates.

11. *The effectiveness of capital budgeting procedures is a function of their implementation across organizational levels.* Capital budgeting and allocation procedures must be clearly and succinctly specified, understood, and implemented at all organizational levels having responsibility for economic decisions.

12. *Post-decision audits are required to improve the quality of decisions over time.* The only way to judge and improve predictive ability is to audit the results of economic decision over their product, process, and systems life cycles and incorporate feedback on differences into future economic predictions.

1.6   Engineering Life Cycle Costs

The United States National Archives define products and services as (http://www.archives.gov/preservation/products/definitions/products-services.html):

> **Products** are tangible and discernible items that the organization produces, including digital file-based output.
>
> A **service** is the production of an essentially intangible benefit, either in its own right or as a significant element of a tangible product, which through some form of exchange satisfies an identified need.

Combining these definitions, products may be defined as:

> Tangible and discernible items or intangible service benefits that an organization produces to satisfy identified human needs or wants.

Key characteristics of products include:

- Tangible items or intangible services that provide benefits.
- Produced by organizations that can be either private or public.
- Intended to satisfy identified human needs or wants.

Three fundamental exchanges must take place for an organization to produce a product.

1. Securing the capital (financing).
2. Purchasing assets (investing) by which to produce products.
3. Generating returns (producing) from the sales of its products.

Figure 1.2 illustrates the flow from capital to products. Cash can be obtained from one of two sources to finance operations. First, organizations can borrow money from lenders. Cash to support ongoing operations is generally obtained as unsecured, short-term loans from a bank. A short-term loan is the acquisition of cash at a stated interest rate that must be repaid with interest by a stated due date, which is within a year from the loan date. For long-term financing of new facilities and equipment, organizations will issue notes, debentures, or bonds through a financial banker. A note or debenture is an unsecured debt, whereas a bond is secured by a mortgage on organizational property. A bond is a fixed income instrument that represents a loan made by an investor to a borrower. Bonds are issued by organizations, municipalities, states, and sovereign governments to finance projects and operations. Owners of bonds are the debtholders, or creditors, of the issuer. Bond details include the terms for variable or fixed interest payments made by the borrower and the end date when the principal of the loan is due to be paid to the bond owner.

Second, organizations can sell stock certificates to investors. A stock certificate is the physical piece of paper representing ownership in a company. Stock certificates state the number of shares owned, the date of issue, an identification number, usually a corporate seal and signatures. There are two primary types of stock certificates. Common stocks are shares of ownership of the organization. Common stockholders have voting rights on corporate issues, such as the board of directors and accepting takeover bids. Typically, common stockholders have one vote per share. Common stockholders may receive dividends out of net profit if the organization has cash

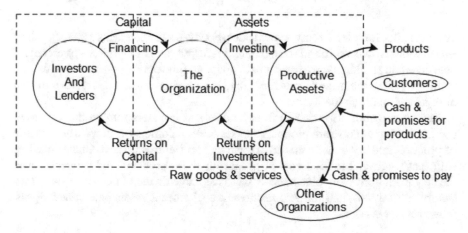

Fig. 1.2 Fundamental exchanges in the production of tangible products and services

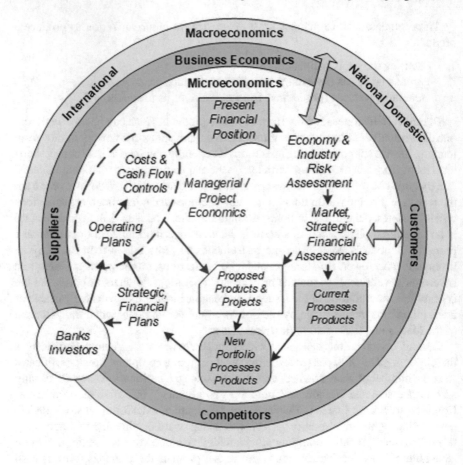

Fig. 1.3 Organizational annual budgeting finance cycle

available after long-term reinvestment. Conversely, preferred stocks are stock shares issued without ownership and voting rights. Preferred stockholders have "preference" over common stockholders in payment of dividend (preferred dividends must be paid before issuing common stock dividends) and in receiving cash from assets in the event of organizational liquidation.

The organization invests case from lenders and investors into productive assets used to convert inputs into products and services. Income cash flows from the sale of products and services to customers and clients (returns on investment) must be sufficient to cover the total cost of producing those products and services, repay the debt from borrowing plus interest, pay acceptable dividend rates to stockholders (returns on capital), and reinvest maintenance of current assets plus in new assets, products, and services.

Figure 1.3 illustrates the typical organizational annual budgeting finance cycle needed to support ongoing productive operations. First, we need a working definition of finance.

Finance—The **acquisition of capital** from lenders and investors, **budgeting** scarce resources effectively, and **investing funds** in portfolios of assets and projects that achieve market and financial objectives **yielding the maximum return/risk trade-offs with the goal of maximizing the wealth or value of the organization** to its stockholders.

The finance manager's role is to interface between the organization and sources of external financing to assure that the organization has sufficient cash flows to maintain ongoing productive operations and has access to cash to expand future productive capacity. The budgeting finance cycle begins with establishing the organization's current financial position from its income statement, balance sheet, statement of retained earnings, and statement of cash position. The finance manager evaluates the effects of economic, legal, political, and technical risks in the international and national domestic markets on the availability and interest cost of capital. He or she evaluates the competitive risks with suppliers, competitors, and customers in the organization's relevant business environment. He or she integrates the macroeconomic and industry environmental risks into the organization's risk models and develops market, strategic, and financial assessments. Risk assessments are integrated with market demand for current products and proposed products and competitive improvement projects into a new portfolio of processes and products. The organization's management sets financial objectives for the new portfolio and develops primary and alternate strategic and financial plans to achieve the objectives given assessed risks. The finance manager takes the new portfolio of processes and products and strategic and financial plans to banks and investment bankers to secure the financing necessary to achieve financial objectives. Financial managers at the banks and investment banks perform their respective risk assessments and decide on the amount of financing and interest rate for loans or the rate of return required for the sale of stocks. The risk-adjusted, weighted interest rate on loans and rate of return on stocks becomes the organization's minimum attractive rate of return (MARR) or the hurdle rate that the new portfolio must earn for the organization to remain viable.

$$\text{MARR} = \sum_i w_{Li} \times I_{Li} \times (1 - T_C) + \sum_j w_{Sj} \times \text{RoE}_{Sj} \qquad (1.1)$$

where $w_{Li} = Li / (\sum_i Li + \sum_j Sj)$, $w_{Sj} = Sj / (\sum_i Li + \sum_j Sj)$, $I_{Li} =$ interest rate charged on loan i, $TC =$ the organization's combined tax rate, and $\text{RoE}_{Sj} =$ return on equity interest rate for stock Sj. Given new financing, the organization then creates operating plans, develops it portfolio of processes and products, deploys its processes incurring cash flows and costs, and sells the portfolio of products generating revenues. Cash flows, costs, and revenues are summarized into the annual income statement, balance sheet, statement of retained earnings, and statement of cash position at the end of each fiscal year, and the budgeting finance is repeated.

Example 1.1

An organization has the following outstanding loans and stock certificates and a combined tax rate of 28.1%. Estimate the organization's MARR.

Short term loans $1,000,000 @ 5%	*Solution*: Total Liabilities and Equity = $20,000,000
Bonds $9,000,000 @ 7%	MARR = [(1/20) 5% + (9/20) 7%](1 − 0.281) +
Preferred stock $3,000,000 @ 9%	(3/20) 9% + (7/20) 14%
Common stock $7,000,000 @ 14%	MARR = 2.44% + 1.35% + 4.90% = 8.69%

Like the annual budgeting finance cycle, all products, processes, and systems progress through a **life cycle** and accrue costs at each stage.

Product life cycle—all the time from conception to death and disposal of a product, process, service, or system.

Live cycle costs—sum of all cost incurred during the product, process, service, or system life cycle.

Life cycle costing—designing with an understanding of and accounting for all the costs associated with a product, process, service, or system during its life cycle.

The typical product, process, service, or system life cycle is illustrated in Fig. 1.4. In general, the typical life cycle proceeds through six stages.

1. **Needs assessment and justification**—(a) Define the technological, economic, legal, or socio-human performance gap. (b) Identify and research alternatives and their requirements. (c) Identify technological, economic, legal, socio-human, and political feasibility of each alternative. (d) Develop an initial conceptual design for each feasible alternative.

Market Analysis	Conceptual Design	Detailed Design	Project Management	Production	Retire & Disposal
Competitive Deficiency / Opportunity	Function / Form	Resource Requirements	Project Initiation	Human Factors & Safety	Decline & Retirement
Design Options	Design Alternatives	Risk, Safety, Sustainability	Project Planning	Production Transforation	Phase Out
Feasibility Study	Design Proof	Stakeholder Analysis	Project Execution	Product Delivery	Retirement
Design Concept	Cost Analysis	Specifications	Project Monitoring	Maintenance & Support	Responsible Disposal
	Development, Testing, & Prototyping	Supplier Selection	Closure / Turnover	Customer Satisfaction	

Fig. 1.4 Typical product, process, service, or system life cycle

2. **Conceptual design**—(a) Impact analysis of each feasible alternative to measure and predict the change in life cycle costs attributable to the alternative. Impact analyses are based on validated and reliable cause-and-effect design models that control for factors other than the proposed design parameters that might account for design performance and life cycle costs. (b) Concept proof via CAD or simulation modeling of design performance and life cycle costs. Select the design alternative with the highest performance/cost ratio. (c) Prototype the design by building a small-scale tangible model or full-scale simulation to validate the design concept by presenting an early version of the design solution to real users. (d) Based on performance and failure feedback from prototyping, correct design deficiencies and optimize the performance/cost ratio. (e) Refine performance design and life cycle cost estimates into initial drawings, specifications, and total life cycle cost estimates.

3. **Detailed design**—(a) Identify economic, human, physical, energy, and information resources needed for the production/construction and operational phases. (b) Perform risk, safety, and sustainability analyses for the production/construction and operational phases. (c) Refine initial drawings, specifications, and total life cycle cost estimates into final drawings, specifications, and total life cycle cost estimates. (d) Select reliable suppliers of economic, human, physical, energy, and information resources needed for the production/construction and operational phases. (d) Develop the project management plan. A project management plan is a formal, approved document that defines how the production/construction phase project will be executed, monitored, and controlled. (e) Develop the transfer to operation plan. The transfer to operation plan is the final document of the project plan. It specifies project completion criteria, legal/regulatory closeout, complete product support documents, the transfer to operation documents and training.

4. **Production/construction** (project management/engineering)—(a) Build, add on, or upgrade production facilities for the specified products or services. (b) Build, add on, or upgrade supporting systems infrastructure to achieve productive efficiency and effectiveness. (c) Develop operational use plans with future management and process engineering.

5. **Operational**—Acquire materials, human, energy, and information inputs, produce products and services, manage human factors and safety, and maintain facilities and equipment until retirement.

6. **Decline and retirement**—(a) Develop a phaseout project management plan. (b) Phase out and retire products and services, equipment, and facilities. (c) Assure responsible disposal of non-hazardous and hazardous materials and securing production records and intellectual property.

There are three basic approaches to product life cycle design.

- *Bottom-up design* starts with the construction of components, bringing components together in subassemblies, and assembling the final product (CAD-centric).

- ***Top-down design*** starts with the high-level functional requirements and decomposes and allocates requirements to intermediate levels and finally to components at the physical implementation layer (concept products).
- ***Both-ends-against-the-middle design*** components are constructed within the context of intermediate levels and intermediate levels within the high-level functionality (emergent technologies).

The major issue in life cycle engineering is the cost impacts of design changes. Two key concepts in life cycle engineering are: (1) Design decisions made early in the life cycle "lock in" later benefits and costs. (2) The later in the life cycle that design changes are made, the higher the costs of the changes. Figure 1.5 illustrates the general relationship between committed total life cycle costs and life cycle costs spent and the life cycle phase in which the change is made.

Figure 1.6 illustrates the general relationship between the ease and cost of design changes and the life cycle phase in which the change is made.

From Figs. 1.5 and 1.6, it can be inferred that the needs assessment and conceptual design phases are the minimum life cycle cost impact times to make design changes before design costs are committed. In these early phases, engineers should consider the life cycle cost impacts of materials, testing and quality assurance, production, maintenance, warranty, and liability costs in the operational phase.

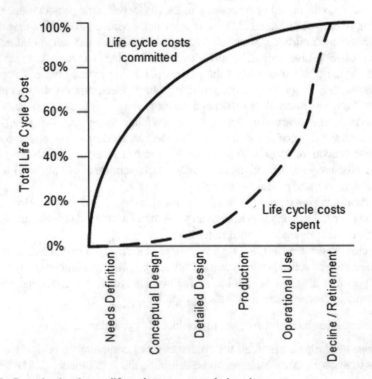

Fig. 1.5 Committed and spent life cycle costs versus design phase

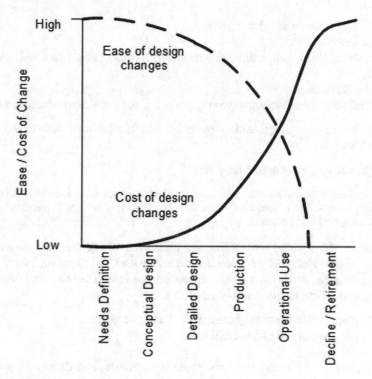

Fig. 1.6 Ease and cost of design changes versus design phase

1.7   Engineering Life Cycle Cost Management

Design inputs are required from multiple engineering disciplines to maximize the differential life cycle benefits relative to differential design costs. This section briefly summarizes the life cycle design contributions required by different engineering disciplines to maximize the life cycle benefit–cost ratio.

Research Engineering Management

> *Research Engineering*—the systematic, intensive study of natural phenomena to achieve fully scientific understanding of the phenomena.

- Basic Research—research devoted to attaining fuller knowledge of the natural phenomena under study, rather than practical application, although it may be of present or future interest in solving applied problems.
- Applied Research—research directed toward applying knowledge from basic research and other applied research to solve practical problems and has commercial potential for new products, processes, or systems.

Planning and organizing for research engineering activities includes:

- Staffing for the required scientific, engineering, and technical skills.

- Laboratory, shop, and office facilities.
- Raw materials.
- Computers; CAD, simulation, and modeling software; and data and knowledge bases.
- Technical libraries.
- Publication support including word processing and conference attendance.

Typically, research engineering precedes the "market analysis" phase of the product life cycle.

Development Engineering Management

Development Engineering—the systematic application of scientific knowledge directed toward the production of useful materials, devices, methods, or systems through prototype and scale-up to full processes.

Research and development, R&D, are often lumped together as a single function with dual objectives of scientific discovery and development of materials, devices, methods, or systems with commercial potential. Planning and organizing for development engineering activities includes:

- Staffing for the required engineering and technical skills.
- Laboratory, shop, and office facilities.
- Raw materials.
- Computers; CAD, simulation, and modeling software; and data and knowledge bases.
- Technical libraries.
- Copyright, patent, and trademark legal support.

Typically, development engineering precedes the "market analysis" phase of the product life cycle.

Design Engineering Management

Design Engineering—the recognition of a human or societal need, conception of an idea to meet that need, definition of the problem, coordination with development (or R&D) to produce a product or process or system to meet that need, and construction of a prototype to test physical, human perception, economic, and commercial feasibility.

Planning and organizing for design engineering activities includes:

- Staffing for the required engineering and technical skills.
- Prototype shop and office facilities.
- Materials and component input from development.
- Computers; CAD, simulation, and modeling software; and specification databases.
- Copyright, patent, trademark, and product safety and liability legal support.

Design engineering involves the first three phases (market analysis, conceptual design, and detailed design) of the product life cycle.

Product Engineering Management

Product Engineering—developing a device, assembly, or system such that it can be produced as an item for sale through some production manufacturing process.

Product engineering usually entails activities dealing with issues of cost, producibility, quality, performance, reliability, safety, serviceability, user features, and product disposal with an objective of making the resulting product attractive to its intended customers. Planning and organizing for design engineering activities includes:

- Staffing for the required engineering and technical skills.
- Prototype shop and office facilities.
- Test equipment and tools.
- Materials and component input from development.
- Computers; CAM software; spreadsheets; and ERP, quality/reliability, statistical modeling software, and databases.

Product engineering involves all phases of the product life cycle.

Process Engineering Management

Process Engineering—the design, operation, control, and optimization of chemical, physical, and biological processes through the integration of technology and humans with the aid of systematic computer-based methods.

There are three fundamental types of process engineering:

- Mechanical and assembly processes.
- Powders, liquids, and gas chemical processes.
- Biological processes.

Planning and organizing for process engineering activities includes:

- Staffing for the required engineering and technical skills.
- Small-scale process laboratories and office facilities.
- Test equipment and tools.
- Product design requirements.
- Computers; simulation software; spreadsheets; large-scale linear and nonlinear programming; and facility process control and feedback software.

Process engineering is initiated in the detailed design phase and implemented in the production/construction and operational phases.

Information Engineering Management

Information Engineering—an architectural approach to planning, analyzing, designing, and implementing computer hardware, network, and software applications within an enterprise.

In the twenty-first-century organization, information engineering supports all phases of the product life cycle with data and information technology. Planning and organizing for information engineering activities includes:

- Staffing for the required engineering and technical skills.
- Computer rooms, servers, computers, network equipment, operating systems, data/information/knowledge bases, systems monitoring software.
- Test equipment and tools.
- Data, information, and knowledge design requirements.

Information engineering supports all phases of the product life cycle.

Project Engineering Management

Project Engineering—It bridges the boundaries between engineering and project management, leading the technical workers who contribute to the building of structures or products.

In some cases, the project engineer is the same as a project manager but in most cases these two professionals have joint responsibility for leading a project. A project engineer's responsibilities include schedule preparation, pre-planning, and resource forecasting for engineering and other technical activities relating to the project. Planning and organizing for project engineering activities includes:

- Staffing for the required engineering management skills.
- Project management office.
- Software: project management, data/information/knowledge base management, spreadsheets, and configuration management.

Project engineering supports project management phase of the product life cycle, possibly early in the operational phase, and the decline and retirement phase.

Facilities Engineering Management

Facilities Engineering—utilities engineering, maintenance, environmental, health, safety, energy, controls/instrumentation, civil engineering, and HVAC needs.

In the twenty-first-century organization, facilities engineering is essential for designing and maintaining controlled facility environments demanded by personnel and processes for optimized performance and product manufacture. Planning and organizing for facilities engineering activities includes:

- Staffing for the required engineering and technical skills.
- Offices, tool rooms, facilities rooms, supplies, and equipment storage.
- Test equipment and tools.
- Process control data gathering software, facilities monitoring and control software, maintenance resource planning software, spreadsheets.

Facilities engineering supports all phases of the product life cycle.

Systems Engineering Management

Systems Engineering —an interdisciplinary field of engineering that focuses on how to design and manage complex engineering projects over their life cycles.

Fig. 1.7 Systems
engineering integration of
life cycle costs (*Source*
Systems Engineering
Fundamentals, Defense
Acquisition University Press,
2001)

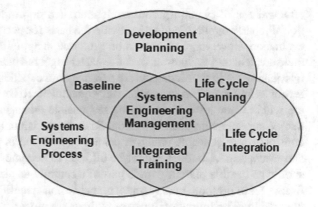

Systems engineering deals with work processes, optimization methods, and risk management tools in projects. It overlaps technical and human-centered disciplines such as control engineering, industrial engineering, organizational studies, and project management. Planning and organizing for systems engineering activities includes:

- Staffing for the required engineering management skills.
- Offices.
- Software: data/information/knowledge base management and graphical browsing; simulation; artificial intelligence, decision, and reasoning; neutral networks; and configuration management.

Figure 1.7 illustrates the concept that systems engineering integrates all phases of the product life cycle into a holistic and integrated function.

1.8 Budgeting for Engineering Operations and Projects

Budget—a quantitative expression of planned sales volumes, revenues, resource quantities, costs/expenses, assets, liabilities, and cash flows for a defined period of time.

A budget is summary of planned revenues and expenses for some future period, usually a fiscal year, quarter, or month. The first use of a budget is as a guide for organizational management toward attainment of financial and operational goals. Budgets predict fiscal outcomes given current strategic, marketing, and operational plans. Hence, budgets provide guidance on the amounts and timing of expected revenues and expenditures. The second use of a budget is for performance evaluation and control. Amount and timing variances from a planned budget provide signals of opportunities (positive variances) and under performance (negative variances). When properly applied, budgets should focus management attention on the effectiveness of planning, communication, and coordination within the organization. Properly applied, budgets provide the basis for performance measurement and evaluation. However, care should be exercised in applying budgets for performance measurement and evaluation, because each budget is a quantification of many subjective judgments in predicting future revenues and expenditures. Many future risks to revenues and expenses cannot be fully quantified. Risks that can be quantified and are within some level of control by responsible management are subject to performance measurement and evaluation. Risk, imprecision, and uncertainty outside the control be a responsible manager cannot be included in performance measurement and evaluation. A budget cannot be allowed to become an inflexible tool. Budgets should be flexible planning tools part of a continuous planning process during each fiscal year to promote management response non-quantifiable risk, imprecision, and uncertainty. The following types of budgets are most prevalent in use.

- *Master budget*—the projection of sales, revenues, and expenses for an organizational unit or product line.
- *Capital budget*—long-term planning of capital expenditures for facilities and equipment.
- *Departmental budget*—a cost center plan of expenditures for resources, materials, consumable assets, and services and the liability or equity financing required capital assets.
- *Flexible budget*—an overhead budget where overhead costs are established for a relevant range of activities rather than a single activity level. As a result, the overhead budget is adjustable to activity levels.
- *Special budgets*—prepared as needed to support investment and operating decisions outside of normal budgeting.
- *Governmental budgets*—expenditures made based on appropriations authorized by a legislative body. The revenue portion of the budget indicates funding sources for authorized expenditures. The expenditure portion establishes ceilings on the amount of expenditures by category.

Typically, ongoing engineering operations are controlled with departmental budgets, and engineering projects are controlled through capital budgets.

The organizational annual budgeting cycle is illustrated in Fig. 1.8. The budget cycle allows the organization to absorb and respond to new information arising from unquantifiable risk, imprecision, and uncertainty. The budget cycle consists of four phases: (1) preparation and submission, (2) approval, (3) implementation, and (4) audit and evaluation.

The responsibility for budget preparation varies with the type of budget. The responsible manager starts by evaluating variances from the prior period budget relative to the strategic or operations plan. She or he then estimates the revenues and expenditures needed to support the next period's long-term strategic plan and operations. She or he then adjusts the cash flows to account for variances encountered in the prior period's budget and finalizes the budget for the next operating period. All submitted operational budgets are summarized, submitted to organizational top management, and compared to the master budget for the next operating

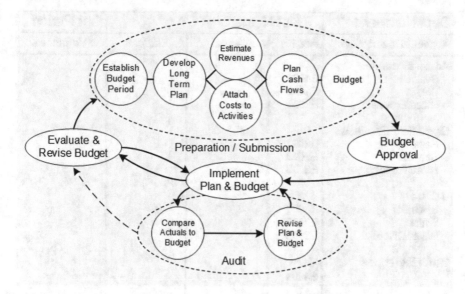

Fig. 1.8 Four-phase Budget Cycle

period. Variances between summarized operational budgets and the master budget are resolved, and the master budget and operational budgets are adjusted to reflect agreed resolutions. During the implementation phase, budgets are continually updated as revenues and expenses are incurred and compared to the approved operational and master budgets. Negative variances (shortfalls in revenues or overages in expenses) are addressed, and the master and operational budgets are revised. Positive variances (overages in revenues or shortfalls in expenses) are likewise noted but provide management flexibility in adjusting strategic plans. The final phase of the budgetary process is audit and evaluation. The main goal is to assure compliance to strategic and operational plans through the assessment of revenue and expenditure variances. The secondary goal is to provide feedback to responsible managers and improve their budget forecasting skills.

As noted, engineering operations are treated as cost centers, and only expenditure variances are controlled. A typical cost center budget template is shown in Fig. 1.9. Actual, budget, and variance in expenditures are accumulated by account codes. This allows engineering managers to further analyze the sources of positive and negative variances.

Likewise, as noted, engineering projects are controlled by a capital budget. There are two types of engineering projects. In the first type, the addition of productive capacity is transitioned to operations management with no scale-up operations. In the second type, scale-up operation of the new productive capacity is required under the project plan to resolve scale-up issues and optimize the productive capacity before turning the new productive capacity over to operations. In the second case, the engineering project will generate sales volumes and revenues in the early part of the operational phase of the product life cycle to scale up to full production for line

Dept Number	Dept Name			Opr Period
Account Description.	Acc No.	Actual	Budget	Variance
Admin Expenses				
Salaries	E1001			
Wages	E1002			
Office Expenses	E1003			
:	:			
Opr&Maint Expenses				
Payroll	E2001			
Supplies	E2002			
HVAC Repairs	E2003			
:	:			
Utilities				
Electricity	E3001			
Water	E3002			
Waste Disposal	E3003			
:	:			
Other Expenses				
	E4001			
	:			

Fig. 1.9 Engineering operations budget template

or facility testing and acceptance. Thus, product engineering and project engineering may operate as profit centers, and in addition to costs and expenses, their budgets must also account for:

- Sales revenues from products.
- Consulting and professional fee revenues.
- Other management fee revenues.

A typical product and project budget template is shown in Fig. 1.10. Actual, budget, and variance in revenues and expenditures are accumulated by account codes. This allows product and project engineering managers to further analyze the sources of positive and negative variances.

1.9 Summary

This chapter established the context of engineering managerial decision making within for-profit and not-for-profit organizations. Engineering managerial economic risk decisions are constrained by organizational business goals and objectives, the resources needed and available to produce its products and services, and the allocation of those resources.

To frame the engineering managerial economic risk decision-making context, the chapter set forth definitions of engineering management and engineering managerial economics.

Project Number	Project Name			Opr Period
Account Description.	Acc No.	Actual	Budget	Variance
Project Revenues				
Sales	P1001			
Management Fees	P1002			
Other Revenues	P1003			
:	:			
Operating Expenses				
Salaries	P2001			
Payroll	P2002			
Supplies	P2003			
:	:			
Travel Expenses				
Travel	P3001			
Lodging	P3002			
Per Diem	P3003			
:	:			
Professional Fees				
Legal	P4001			
Consulting	:			

Fig. 1.10 Product and project engineering budget template

Engineering management combines technical expertise with managerial knowledge to coordinate operational performance of engineering disciplines within an organizational setting or coordinate and control technical project outputs.

Engineering managerial economics is the body of knowledge and methods devoted to the systematic evaluation of the economic benefits resulting from ongoing engineering activities or proposed engineering projects in relation to the economic costs of those activities or projects.

The fundamental types of business for-profit organizations were specified.

A *proprietorship* is an unincorporated business owned by one individual.

A *partnership* is owned by two or more individuals, who jointly invest money, skills, and assets and who share in profits or losses in accordance with the terms set forth in their partnership agreement.

A *corporation* is a legal entity recognized under state or federal law.

The fundamental types of governmental not-for-profit agencies were identified as federal civilian and military, state, and local. This book addresses the economic decisions for the set of engineers and engineering managers employed in corporations, partnerships, state and local governmental agencies, and non-military federal governmental agencies.

In the design phase, engineers' primary task is to plan **capital expenditure** for equipment, facilities, and supporting infrastructure needed for the organization to produce its products or services.

Capital expenditure is the purchase of an asset with a life greater than one fiscal year.

An *asset* is land, physical facilities, equipment, money, monetary instruments, and intellectual property owned or controlled by a business organization for the purpose of producing products or services to provide a return on investment in the asset.

A *fiscal year* is a 12-month time period that a business organization uses for the accounting purpose of preparing financial statements. The fiscal year may or may not be equal to the calendar year.

In the implementation and maintenance phases, engineers' primary task is **cost control** for the efficient operation and maintenance of productive assets. In the retirement and disposal phase, engineers' primary task is defining all activities necessary for responsible disposal and minimizes the project costs associated with disposal.

Engineering decisions, and the corresponding cash flow implications, exist along three timescales. *Short term* decision for expenditures within the current fiscal year. *Intermediate-term* economic decisions are tied to the budget planning period, which itself is determined by the organization's product life cycle. *Complexlong range* economic decisions to develop future products and services for future budget planning periods.

The *product life cycle* is the expected life of a product or service from the time of its conception to its termination and disposal.

Engineering economic principles:

1. Money has time value.
2. All investments in productive assets must be economically justified.
3. All economic decisions must be based on differences in cash flows among the alternatives considered.
4. Use a common unit of measurement (metric) across alternative to make consequences commensurable.
5. Alternatives must be examined over a common planning horizon.
6. Only differences are relevant.
7. Only feasible alternatives can be considered.
8. Investment criteria must include the time value of money due to capital acquisition and rationing.
9. Separable decisions should be made separately.
10. The relative degrees of uncertainty associated with future cash flows must be incorporated into the economic decision process.
11. The effectiveness of capital budgeting procedures is a function of their implementation across organizational levels.
12. Post-decision audits are required to improve the quality of decisions over time.

Three fundamental exchanges must take place for an organization to produce a product.

1. Securing the capital (financing).
2. Purchasing assets (investing) by which to produce products.
3. Generating returns (producing) from the sales of its products.

The finance manager oversees the financing and investing activities.

Finance—The **acquisition of capital** from lenders and investors, **budgeting scarce resources effectively**, and **investing funds** in portfolios of assets and projects that achieve market and financial objectives **yielding the maximum return/risk trade-offs with the goal of maximizing the wealth or value of the organization** to its stockholders.

The **minimum attractive rate of return** (MARR) is the risk-adjusted, weighted interest rate on loans and rate of return on stocks that the portfolio of products, services, and processes must earn for the organization to remain viable.

$$\text{MARR} = \sum_i w_{Li} \times I_{Li} \times (1 - T_C) + \sum_j w_{Sj} \times \text{RoE}_{Si}$$

All products, processes, and systems progress through a **life cycle** and accrue costs at each stage.

Product life cycle—all the time from conception to death and disposal of a product, process, service, or system.

Live cycle costs—sum of all cost incurred during the product, process, service, or system life cycle.

Life cycle costing—designing with an understanding of and accounting for all the costs associated with a product, process, service, or system during its life cycle.

The typical life cycle proceeds through six stages

1. Needs assessment and justification.
2. Conceptual design.
3. Detailed design.
4. Production/construction.
5. Operation.
6. Decline and retirement.

Two key concepts in life cycle engineering are: (1) Design decisions made early in the life cycle "lock in" later benefits and costs. (2) The later in the life cycle that design changes are made, the higher the costs of the changes.

Design inputs are required from multiple engineering disciplines to maximize the differential life cycle benefits relative to differential design costs.

Research Engineering   the systematic, intensive study of natural phenomena to achieve fully scientific understanding of the phenomena.

Development Engineering—the systematic application of scientific knowledge directed toward the production of useful materials, devices, methods, or systems through prototype and scale-up to full processes.

Design Engineering—the recognition of a human or societal need, conception of an idea to meet that need, definition of the problem, coordination with development (or R&D) to produce a product or process or system to meet that need, and construction of a prototype to test physical, human perception, economic, and commercial feasibility.

Product Engineering—developing a device, assembly, or system such that it can be produced as an item for sale through some production manufacturing process.

Process Engineering—the design, operation, control, and optimization of chemical, physical, and biological processes through the integration of technology and humans with the aid of systematic computer-based methods.

Information Engineering—an architectural approach to planning, analyzing, designing, and implementing computer hardware, network, and software applications within an enterprise.

Project Engineering—bridges the boundaries between engineering and project management, leading the technical workers who contribute to the building of structures or products.

Facilities Engineering—utilities engineering, maintenance, environmental, health, safety, energy, controls/instrumentation, civil engineering, and HVAC needs.

Systems Engineering—an interdisciplinary field of engineering that focuses on how to design and manage complex engineering projects over their life cycles.

A budget is summary of planned revenues and expenses for some future period, usually a fiscal year, quarter, or month.

Budget—a quantitative expression of planned sales volumes, revenues, resource quantities, costs/expenses, assets, liabilities, and cash flows for a defined period of time.

The budget cycle consists of four phases: (1) preparation and submission, (2) approval, (3) implementation, and (4) audit and evaluation. Engineering operations are treated as cost centers. Engineering projects are controlled by a capital budget.

1.10 Key Terms

Asset
Budget and budget cycle
Capital expenditure
Corporation
Cost center
Engineering
Engineering Management
Engineering managerial economics
Fiscal year
Finance
Life cycle cost and costing
Management
Partnership
Product
Product life cycle
Production
Proprietorship
Service.

Problems

1. The term engineering economic decision refers to any investment related to an engineering project or investment. What facet of the decision is most important from the engineering point of view?
2. 2. List the two properties of professional engineers.
3. List the three fundamental types of business organizations.
4. Define the product life cycle.

5. Given that an intermediate economic decision involves costs in the $10,000s, are of intermediate risk, affect organizational and stakeholder impacts, and require formal analysis, define simple and complex decisions.
6. List the five (5) basic classes of engineering economic analyses.
7. From the 12 engineering economic principles, explain the statement, "money has time value" relative to the purchase and deployment of productive assets.
8. What are the three fundamental exchanges that must take place for an organization to produce a product?
9. An organization has $22,000,000 in outstanding common stock on which stockholders expect 15% return on equity and $5,000,000 in outstanding preferred stock with a dividend rate of 12%. It has $9,000,000 in short-term loans at 4% interest and $14,000,000 in bonds at an annual coupon rate of 8%. The organization has a combined tax rate of 30%. What is the organization's MARR?
10. A project manager is completing the sixth quarter of a two-year project to construct a new production facility. Per the project plan, in the sixth quarter she initiated scale-up production on the first production line. The project budget for the sixth quarter is shown below. Estimate the variances and note potential corrective actions needed in the seventh quarter to keep the project cash flows on schedule.

Proj: MF-20##	Name: New manufacturing facility			Period: 6th Qtr
Account desc.	Acct. no.	Actual $1K	Budget $1K	Variance $1K
Scale-up production				
Sales	SR1001			
Operating Expenses		$200	$500	
Materials	OE2001	$18	$40	
Wages/salaries	OE2002	$120	$280	
Utilities	OE2003	$6	$15	
Fixed costs	FC3001	$6	$15	
Overhead	OH4001	$30	$30	
Production variance				
Project expenses				
Construction	PC1001	$19,800	$20,000	
Equipment	PC1002	$5,600	$5,500	
Materials	PC1003	$900	$890	
Utilities	PC1004	$260	$250	
Wages/salaries	PC1005	$1200	$1200	
Services	PC1006	$500	$480	
Project variance				

Chapter 2
General Accounting and Finance Fundamentals

Abstract Revenues and expenses along with capital investment costs drive decision making in all organizations, for-profit and not-for-profit. The accounting function collects, summarizes, and reports revenue, expense, and cost transactions. Hence, accounting is the financial language of all organizations. The objective of this chapter is to provide current and future engineering managers and project engineering managers with the accounting knowledge necessary to communicate in an organization's financial language and apply accounting and financial reports in managing engineering operations and capital projects. First, this chapter sets forth the definition of accounting and the information it provides for organizational decision making. Next, the sources of operations financing are specified, and the minimum attractive rate of return is formally defined. The role of accounting and the accounting cycle are discussed. The three fundamental accounting reports are discussed with an example that illustrates how the three reports integrate within the accounting cycle. Finally, financial ratios that measure an organization's financial health are specified.

2.1 Introduction

Definition: *Accounting* is the discipline that systematically measures, records, interprets, and communicates the financial activities of an organization, providing insight into the trustworthiness, profitability, and solvency of an organization.

Accounting is concerned with the measurement, recording, interpretation, and communication of economic event, financial transactions, measurement of economic value, and determination of periodic changes in economic values. Accounting provides information on

- Allocation of scarce resources to organizational productive processes.
- Management and direction of resources within the organizations.
- Reporting on the custodianship of resources under management control.

As a discipline, accounting can be classified according to the four primary organizations that it serves: corporate, public, government, and forensic. Similarly, accounting can be classified according to the reporting purpose: (1) Financial accounting of organizational performance to lenders, stockholders, and governmental agencies. (2) Managerial accounting of organizational financial performance relative to strategic plans and budgets to internal management. (3) Cost accounting estimates costs, allocates overhead costs, and develops standard product costs. (4) Tax accounting assures that all relevant tax laws and regulations are followed by the organization. (5) Auditing provides independent analyses of organizational financial activity to ensure that records transactions in accordance with Generally Accepted Accounting Principles (GAAP) and applicable standards, laws, and regulations.

2.2 Sources of Operations Financing

From Chap. 1, the organization has three sources financing:

- Lenders—loans from banks and financial institutions.
- Investors—cash investment in the organization from the purchase of preferred and common stock.
- Retained earnings—the portion of net income which is retained and invested in productive assets by the organization rather than distributed to its owners as dividends.

Organizational sources of financing for new products and services or for the expansion of existing products and services are either lenders (banks) or investors (stockholders). From Fig. 1.3, banks and stockholders provide financing through the annual finance cycle.

- Begins with the present financial position.
- Examines financial and competitive trends and risks in the macro, business, and microeconomic environments.
- Evaluates competitive capacities of current and proposed products as a new portfolio of products and processes.
- Develops strategic financial plans.
- Acquires financing from lenders (banks) and investors which sets the MARR.
- Monitors financial results of operations.

Definition: *Minimum Attractive Rate of Return* (*MARR*)—the risk adjusted, weighted, minimum acceptable rate of return, or hurdle rate that must be earned on a project or investment, given its opportunity cost of foregoing other projects or investments.

From Chap. 1, MARR may be measured as

$$\text{MARR} = \sum_i w_{Li} \times I_{Li} \times (1 - T_C) + \sum_j w_{Sj} \times \text{RoE}_{Sj} \qquad (2.1)$$

where $w_{Li} = L_i/(\sum_i L_i + \sum_j S_j)$, $w_{Sj} = S_j/(\sum_i L_i + \sum_j S_j)$, I_{Li} = interest rate charged on loan i, T_C = the organization's combined tax rate, and RoE_{Sj} = return on equity interest rate for stock S_j.

2.3 The Role of Accounting in the Organization

The overriding purpose of financial accounting is to summarize all financial activity in the profit and loss income statement, balance sheet, retained earnings statement, and cash flow statement. Accounting practices are standardized to ensure compliance to all state and federal laws and to avoid fraud. Accounting principles are referred to as Generally Accepted Accounting Principles (GAAP). In the corporate organization, the four fundamental questions that accounting statements must answer are as follows:

1. What were the organization's revenues and expenses during the fiscal period and how much was retained as net profit and cash profit—answered by the income statement?
2. How did the organization reinvest its profit into new products or services—answered by the statement of retained earnings?
3. How much cash did the organization generate from sales and spend during the fiscal period—answered by the statement of cash position?
4. What is the organization's financial position in productive assets, liabilities (loans) against those assets, and stockholder ownership at the end of the fiscal period—answered by the balance sheet?

To obtain the answers to these questions:

Financial accounting looks at past financial data with the objective of determining an organization's value as a whole and at future investment opportunities with the goal of maximizing the wealth or value of the organization.

- Shareholders and investors will use financial information to decide if a public company is undervalued and worth investing in or overvalued and should be avoided.
- Creditors will use financial information to decide whether an organization is a good risk before lending money.
- Governmental agencies use financial information to levy taxes on for-profit organizations.

Managerial accounting is used internally for planning and for moving an organization forward in a financially sound manner.

- Accountants look at the historical financial data stream as well as the current economy and make assumptions about trends and what these trends mean for the organization's future.
- Managers predict an organization's financial future and make sound decisions based upon those expectations.

As part of financial transactions, accounting data tracks the revenues and costs of projects and products, which become the basis for estimating future revenues and costs in management planning and implementation and in the engineering economic valuation of future products, services, processes, and systems.

- **Accounting** measures and records financial transactions and provides the basis for assessing the economic viability of the organization.
- **Management** allocates available investment funds to projects, evaluates unit and firm performance, allocates resources, and selects and directs personnel.
- **Engineering** analyzes the economic impact of design and project alternatives over their present and future life cycles.

Figure 2.1 summarizes the interrelationship among accounting, management, and engineering.

Accounting	Management	Engineering
Recording	Planning & setting goals	Identify engineering needs
Measuring		
Interpreting	Capital budgeting	Collect/analyze data
Communicating	Operations budgeting	Feasibility of design alternatives
	Organizing & Staffing	Evaluating projects
		Benefit-costs tradeoffs
	Analyzing risks	Recommending designs and projects
	Assessing impacts	
	Controlling & Decision Making	Audit project results
Financial Activities	Implementation	Prediction

Fig. 2.1 Financial interrelationship among accounting, management, and engineering

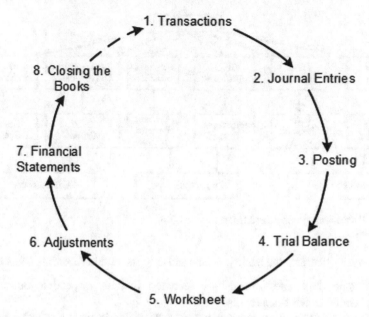

Fig. 2.2 Transactions accumulation in the accounting cycle

The central tenet of transaction accounting is objective costing (i.e., traceable to a transaction record). A transaction is an exchange of money or credit for a product or service documented by a receipt, invoice, bill of lading, etc. Accordingly,

- Accounting records financial transactions in nominal or stable actual dollars (A$)—reference the chapter on inflation.
- Assets are valued at their acquisition cost and adjusted for depreciation and improvement costs. If the market value is lower than the adjusted cost, the market value is used.

Transactions are accumulated and recorded for each accounting cycle. The accounting cycle is a set of rules within a specified methodology to ensure the accuracy and conformity of financial statements. Figure 2.2 illustrated the accounting cycle for accumulating and recording transactions.

1. Transaction—financial exchange.
2. Entries—listing in original journals.
3. Posting—entering into affected accounts in the general ledgers.
4. Trial balance—end of period initial balance.
5. Worksheet—corrections tracked in worksheets.
6. Adjustments—verified corrections posted to general ledger accounts.
7. Financial statements.
8. Close the books—reduce revenue and expense accounts to zero.

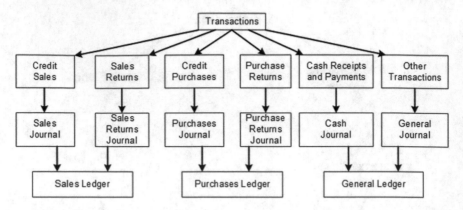

Fig. 2.3 Transactions journals and ledgers

Figure 2.3 illustrates the basic journals and ledgers used in recording transactions.

- Credit sales and sales returns are recorded in their respective journals and accumulated in the sales ledger.
- Credit purchases and purchase returns are recorded in their respective journals and accumulated in the purchases ledger.
- Cash receipts and payments for cash sales are recorded in the cash journal.
- Other transactions are recorded in the general journal.
- Cash receipts and payments and other transactions are accumulated in the general ledger.

Figure 2.4 shows that the input to the current accounting cycle is the four reports of financial performance and position and that the output is the same four reports updated for operational transactions during the current accounting cycle.

2.4 The Balance Sheet

The *balance sheet* summarizes the value of the organization's assets available to carry out its economic activities and the liabilities and ownership financing claims against those assets on each specific date of the end of an accounting cycle.

- Assets are the plant, equipment, supplies, etc., owned by the firm. Tangible assets (physical) include land, buildings, and equipment. Intangible assets (non-physical) include patents, copyrights, trademarks, and amounts owned by customers.
- Liabilities are claims against the firm from external sources of financing (notes, bonds, loans, etc.).
- Equity is internal financing provided by the firm's owners (stockholders).

Under double-entry accounting, the fundamental accounting equation is

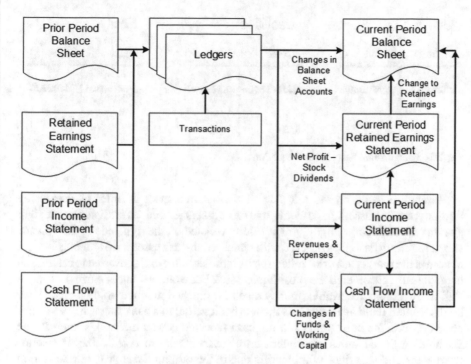

Fig. 2.4 Inputs and outputs of each accounting cycle

$$\text{Assets value} = \text{Liabilities value} + \text{Equity value} \qquad (2.2)$$

$$\text{Assets (debits, credits)} = \text{Liabilities (debits, credits)}$$
$$+ \text{Equity (debits, credits)}$$

On the assets side of the equation, debits increase assets value, and credits decrease assets value. On the liabilities + equity side, debits decrease the value of liabilities or equities, and credits increase their value. The fact that the firm's resources (assets) of revenue generation are balanced by its sources of financing (liabilities + equity) is the basis for the name "balance sheet." The fundamental accounting equation, assets = liabilities + equities, forms the basis of double-entry accounting (Fig. 2.5).

- Any financial transaction that increases (decreases) assets must also increase (decrease) liabilities and equity by the same amount.
- Every financial transaction must have an offsetting debit and credit.
- Debits increase assets, whereas credits increase liabilities and equity maintaining equality.
- Credits decrease assets, whereas debits decrease liabilities and equity maintaining equality.

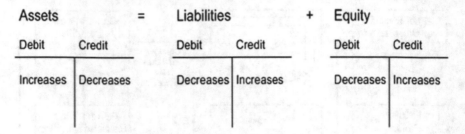

Fig. 2.5 Fundamental balance sheet equation

Assets are listed on the left side of the balance sheet in order of decreasing liquidity; current assets, fixed assets, and other assets. Liabilities are listed on the right side in order of decreasing liquidity. Equity is listed on the right side of the balance sheet. Liquidity describes the ease with which a non-cash asset can be quickly bought or sold in the market at a price reflecting its intrinsic value and converted to cash. Cash is universally considered the most liquid asset. For example, under current assets in Fig. 2.6, accounts receivable (money owed for finished products sold on credit) can be discounted (sold at a value less than its book value) to a bank for cash. The difference between the book value and the cash received covers the bank's interest rate for holding the accounts receivables until payments are received. Likewise, short-term notes and securities with payoffs due in the current fiscal year can be sold at discount early in the year to generate cash for operations. Conversely, inventory is not liquid. Raw materials inventories may be liquid if they are commodities that can easily be sold on the open market. Raw materials and components inventories that are specially made to meet the organization's specifications will not be liquid. Specialty raw materials and components must be integrated into finished products for sale to customers. Work in process (WIP) inventories are the least liquid, because they have been partially transformed into the organization's products. Due to intellectual property specification differences, partially transformed WIP cannot be integrated into competitor's products. WIP must be processed into finished products for sale to customers to be converted to cash. Likewise, on the liabilities side of the balance sheet, accounts payable and notes payable in the current fiscal year are more liquid than accrued expenses, which may or may not be due in the current accounting cycle. The balance sheet structure is illustrated in Fig. 2.6. An example balance sheet is illustrated in Fig. 2.7. Note that the fundamental balance sheet holds with total assets = \$629,699,100 and total liabilities plus total equity = \$629,699,100.

Balance Sheet Example Layout

Assets	**Liabilities**
Current Assets	Current liabilities
Cash	Accounts payable
Accounts Receivable	Notes payable
Securities	Acrued expenses
Inventories	Total current liabilities
Less: Bad debt account provision	
Total current assets	Long term liabilities
	Bank notes
Fixed assets	
Land	**Total liabilities**
Plant and equipment	
Less: Accumulated depreciation	Equity
Total fixed assets	Preferred stock
	Common stock
Other assets	Capital Surplus
Long term securities	Retained earnings
Prepays/deferred charges	
Intangibles	**Total equity**
Total other assets	
Total assets	**Total liabilities and equity**

Fig. 2.6   Balance sheet format

2.5   The Income Statement

The *income statement* summarizes the firm's revenues and expenses from ongoing operations over a stated period (usually month, quarter, or fiscal year). These are intervals between consecutive balance sheet statements. The fundamental outcome of the accrual income statement is the estimate of net profit.

The AAA Company
Balance Sheet as of December 31, 20##

Assets			Liabilities	
Current Assets			Current Liabilities	
Cash		$40,597,100	Accounts Payable	$90,000,000
Accounts Receivable		$76,259,000	Wages Payable	$20,000,000
Inventory		$38,753,000	Bond Dividend Payable	$1,000,000
Total Current Assets		$155,609,100	Total Current Liabilities	$111,000,000
Fixed Assets			Long Term Liabilities	
Equipment	$153,590,550		Bank Notes	$10,000,000
Less depreciation	($40,000,000)	$113,590,550		
Building	$345,499,450		Bonds	$40,000,000
Less depreciation	($20,000,000)	$325,499,450		
Land		$10,000,000	Total Long Term Liabilities	$50,000,000
Total Fixed Assets		$449,090,000		
			Owner's Equity	
Other Assets			Preferred Stock	$60,000,000
Patents		$15,000,000	Common Stock	
Copyrights		$10,000,000	Par Value	$40,000,000
Total Other Assets		$25,000,000	Capital Suplus	$360,000,000
			Retained Earnings	$8,699,100
			Total Equity	$468,699,100
Total Assets		$629,699,100	Total Liabilities and Equity	$629,699,100

Fig. 2.7 Example balance sheet

$$\text{Net Income}_n = \text{Revenue}_n - \text{Expenses}_n \tag{2.3}$$

$$\text{Net Income}_n = R_n - (O_n + M_n + D_n + I_n) = \text{EBT}_n - T_n$$

$$\text{Net Income}_n = R_n - (O_n + M_n + D_n + I_n) - [R_n - (O_n + M_n + D_n + I_n)]\text{TR}_n$$

where R_n = revenue in period n, O_n = operating expenses, M_n = maintenance expenses, D_n = depreciation expenses, I_n = interest expenses, T_n = tax expenses, and TR_n = combined tax rate.

In the accrual income statement in Fig. 2.8, the cost of goods sold includes labor, materials, and indirect costs directly related to production. Under the accrual basis of accounting (or accrual method of accounting), revenues and expenses are reported on the income statement in the accounting period in which they are earned or incurred. Revenues may be in both cash and credit sales. In the case of credit sales, the organization allows the customer to take possession of its products in the current accounting period but make cash payment(s) in subsequent accounting periods. The revenue is recognized in the current accounting period, and the balance due to the organization is carried in an accounts receivable journal under current assets until the balance due is paid off. Likewise, on the cost of goods sold side, the organization may order and take possession of materials and supplies needed for production in the current accounting period paying its suppliers with some combination of cash and credit. The amount owed is carried in an accounts payable journal under current liabilities until

Operating revenuse
>   Sales
>   Less: Returns and allowances
>   **Net revenues**

Operating Expenses   Eng. Managerial
>   Cost of goods sold   Operations/Project
>   Sales promotion   Income Statement
>   Depreciation
>   General and administrative expenses
>   Lease payments
>   **Total operating expenses**

Earnings before interest and taxes (EBIT)
Operating interest expenses
Earnings before taxes (EBT)
Taxes
>   **Total operating income (EAT)**
- -
Non-operating revenues
Less: Non-operating expenses
>   Non-operating interest expenses
>   **Total non-operating income**

EBIT + total non-operating income
Taxes
Net income

Fig. 2.8 Accrual income statement format

the balance owed is paid off. Cost of goods sold includes labor, materials, and indirect expenses. The economic design of production systems must include estimates of labor, materials, and indirect (supervision, maintenance, operating overhead, and other support) expenses. Also, of interest to engineering managerial economics is depreciation expenses, which will be discussed in its own module. Note in Fig. 2.8 that engineering managerial economic projects account for only through Total Operating Income or Earnings After Taxes (EAT). The organization itself may have additional non-operating revenues and expenses associated with non-operating investments in strategic suppliers or partners or temporary investment of excess cash in notes, loans, or stock. These accrual revenues and expenses are outside the scope of typical engineering managerial operations or projects and not included in their revenue and expenses estimates. Figure 2.9 illustrates an example accrual income statement.

The AAA Company
Income Statement for December 20##

Operating Revenues		
Sales		$47,800,000
Less: Returns and Allowances		($1,740,000)
Total Operating Revenues		$46,060,000
Operating Expenses		
Cost of Goods and Services Sold		
Labor	($10,280,000)	
Materials	($9,280,000)	
Indirect Costs	($4,560,000)	
Total CoGS		($24,120,000)
Selling and Promotion Expenses		($1,860,000)
Depreciation		($900,000)
General and Administrative		($4,320,000)
Lease Payments		($1,020,000)
Total Operating Expenses		($32,220,000)
Earnings Befort Interest and Taxes		$13,840,000
Operating interest		($2,000,000)
Earnings Before Taxes		$11,840,000
Less: State Taxes at	9.1%	($1,077,440)
Federal Taxable Income		$10,762,560
Federal Taxes at	21.0%	($2,260,138)
Net Profit		$8,502,422

Fig. 2.9 Example accrual income statement

2.6   The Retained Earnings Statement

When a corporation has positive net profit, it must decide on how to reinvest that profit. It has three options as follows:

- Pay dividends to stockholders if it does not have internal investments that have a rate of return greater than the MARR.
- Pay some dividends to stockholders and invest in a limited number of internal investments that have a rate of return greater than the MARR.
- Pay no dividends to stockholders, and invest all net profits into internal investments that have a rate of return greater than the MARR.

When an organization declares stock dividends, preferred stock dividends must be paid before common stock dividends. Much like bonds, preferred stocks have a stated dividend. The dividend is not a legal liability until the corporation's board of

The AAA Company
Retained Earnings Statement for December 20##

Beginning Balance	
Retained earnings Jan. 1, 20##	$711,000
Net income Dec 31, 20##	$8,502,422
Total	$9,213,422
Dividends declared and paid fiscal year 20##	
Preferred stock dividends	($205,700)
Common stock dividends	($308,622)
Total dividends deducted	($514,322)
Retained earnings ending balance Dec 31, 20##	$8,699,100

Fig. 2.10 Example retained earnings statement

directors declares it; however, many corporations view preferred stock dividends as an indicator to the stock market of corporate financial health and viability. Not paying dividends on preferred stock may be seen as an indicator of poor financial health and cause devaluation of outstanding common stock market value. Net profit remaining after preferred stock dividends is available for distribution as dividends to common stockholders are or may be reinvested in internal investments with rates of return greater than the MARR. As illustrated in Fig. 2.4, the *retained earnings statement* is the link between the income statement and the balance sheet. The equation for the addition of net profit to the balance sheet retained earnings is

$$\text{Retained Earnings (current)} = \text{Retained Earnings (prior)}$$
$$+ \text{New Profit} + \text{new stock} - \text{dividends} \quad (2.4)$$

where "current" = current accounting period balance sheet, "prior" = prior accounting period balance sheet, net profit from the current accounting period income statement, "new stock" = cash inflow from the sale of new shares of preferred or common stock—cash outflow for stock buybacks, and dividends = cash outflow for preferred stock and common stock dividends. Figure 2.10 illustrates the retained earnings statement for The AAA Company.

2.7 The Cash Flow Income Statement

The accrual income statement indicates net profit or net loss regardless of how products were sold (cash or credit) and cost of goods sold was incurred (cash or credit). In the extreme case in a given accounting period, it is possible for an organization to sell its products on credit and make all its purchases on credit with no change in cash position. In the worst-case scenario, an organization could sell all its product on credit and make all of its purchases with cash, that is, until the cash is depleted. Lack of cash to maintain ongoing operations is one of the major causes of business failure. Additionally, the accrual income statement ignores cash from financing and investments. Hence, it is essential that an organization maintains a cash flow income statement to track its cash position.

The *cash flow income statement* reports the difference between the sources and uses of cash during a fiscal accounting period. The fundamental outcome of the cash flow income statement is the estimate of cash net profit.

$$\text{Net profit (cash flow)} = \text{Revenues} - \text{Expenses}$$
$$+ \text{Financing} - \text{Investments}$$

$$\text{Net Profit}_n = R_n - (O_n + M_n + I_n) - T_n + R_F - C_I \qquad (2.5)$$

where R_F = revenues from financing, C_I = costs due to investments, and all transactions are stated in cash flows. Accrual revenue from accounts receivable, debts from accounts payable, and depreciation expenses are omitted from the cash flow income statement.

In preparing the cash flow income statement, accountants identify the sources and uses of cash according to the type of business activity.

- Operating activities—those cash flows from the production and sale of products and services.
- Financing activities—cash from borrowing, the sale of stock, or interest payments on notes and bonds the organization may own.
- Investing activities—cash for the purchase of new or used fixed assets (facilities and equipment), the sale of old equipment, the purchase of stocks in other organizations, or the purchase of bonds let by other organizations.

Summarizing the cash inflows and outflows from the three activities results in the change in cash flow position for the accounting period.

Figure 2.11 presents the cash flow income statement for The AAA Company for the period ending December 31, 20##. Note in this example, investment in capital expenditures for new equipment and facilities ($5,000,000) was financed by issuing bonds ($2,000,000) and common stock ($4,100,000–$1,100,000 repurchase preferred stock and declared dividend). The difference between the savings on future dividends not paid on preferred stock and current dividends must be less than or equal to the return on investment from future cash flows from the new equipment and facility as justified by engineering economic analysis.

The AAA Company
Cash Flow Income Statement for December 20##

Operating Revenues
    Cash Sales $43,566,400
    Less: Cash Returns and Allowances   ($1,545,120)
        Total Operating Revenues $42,021,280

Operating Expenses
    Cash Cost of Goods and Services Sold

Labor	($9,904,640)	
Materials	($6,885,760)	
Indirect Costs	($4,049,280)	
Total CoGS		($20,839,680)
Cash Selling and Promotion Expenses		($1,651,680)
Cash General and Administrative		($3,836,160)
Lease Payments		($1,020,000)
Total Operating Expenses		($27,347,520)

Financing

Issue Bonds	$2,000,000
Repurchase Preferred Stock	($1,000,000)
Issue Common Stock	$4,100,000
Declared Dividend	($100,000)
Net Financing	$5,000,000

Investments

Capital Expenditures	($2,000,000)
Purchase Facility	($4,000,000)
Net Investments	($6,000,000)

Cash Flow Before Taxes

Net Cash Flow	$13,673,760
Less: Taxes Paid	($2,260,138)
Net Increase(Decrease) Cash	$11,413,622
Cash at Beginning of Period	$29,183,478
Cash at End of Period	$40,597,100

Fig. 2.11 Example cash flow income statement

2.8   Financial Ratios Important to Engineering Managerial Economics

An engineering project to install new facilities and equipment or upgrade existing facilities and equipment must account for its effect on the organization's financial health. An organization's current financial health, or liquidity, is measured by working capital. Working capital is the difference between current assets and current liabilities and indicates the firm's liquidity or ability to pay its current debts.

$$\text{Working Capital } = \text{ Current Assets } - \text{ Current Liabilities} \qquad (2.6)$$

An alternate measure of an organization's current financial health, or liquidity, is its current ratio which provides insight into the organization's short-term solvency.

$$\text{Current Ratio } = \text{ Current Assets/Current Liabilities} \qquad (2.7)$$

Although working capital and the current ratio indicate the organization's ability to meet maturing debts in the current accounting period, they do not account for the structure of current assets in meeting those debts. Recall from discussion of the balance sheet that current assets are ordered by liquidity, and it was observed that inventory is not liquid. Hence, the greater proportion of current assets tied up in inventory the lower the organization's ability to make payments on current debts. The **acid-test ratio** (also termed the **quick ratio**) indicates the organization's ability to pay its currently due debts by subtracting the value of inventory from current assets.

$$\text{Acid-test ratio} = \text{ (Current Assets } - \text{ Inventories)/Current Liabilities} \qquad (2.8)$$

Thorough financial analyses must consider working capital, the current ratio, and the quick ratio, including comparison to industry or sector normal values for these indicators over time. Trends will indicate favorable or adverse changes in organizational financial health.

Example 2.1

From The AAA Company balance sheet:

The AAA Company
Balance Sheet as of December 31, 20##

Assets		Liabilities	
Current Assets		Current Liabilities	
Cash	$40,597,100	Accounts Payable	$90,000,000
Accounts Receivable	$76,259,000	Wages Payable	$20,000,000
Inventory	$38,753,000	Bond Dividend Payable	$1,000,000
Total Current Assets	$155,609,100	Total Current Liabilities	$111,000,000

$$\text{Working Capital} = \text{Current Assets}$$
$$- \text{Current Liabilities}$$
$$= \$155,609,000$$
$$- \$111,000,000$$
$$= \$44,609,000$$

$$\text{Current Ratio} = \text{Current Assets/Current Liabilities}$$
$$= \$155,609,000/\$111,000,000$$
$$= 1.40$$
$$\text{Quick Ratio} = (\text{Current Assets} - \text{Inventories})/\text{Current Liabilities}$$
$$(\$155,609,000 - \$38,753,000)/\$111,000,000$$
$$= 1.053$$

A third measure of an organization's current financial health is its profit margin from current operations. Profit margin is the ratio of net profit to net sales revenue and indicates the efficiency at which sales are converted into profits which can be reinvested into productive assets or paid as dividends.

$$\text{Profit margin on sales} = \text{Net profit/Net sales revenue} \qquad (2.9)$$

Example 2.2

From The AAA Company income statement:

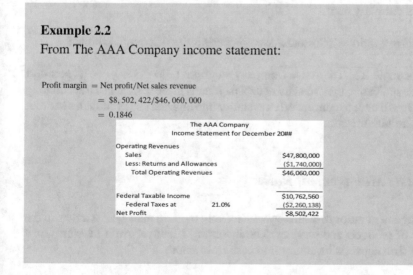

$$\text{Profit margin} = \text{Net profit/Net sales revenue}$$
$$= \$8,502,422/\$46,060,000$$
$$= 0.1846$$

The AAA Company
Income Statement for December 20##

Operating Revenues	
Sales	$47,800,000
Less: Returns and Allowances	($1,740,000)
Total Operating Revenues	$46,060,000
Federal Taxable Income	$10,762,560
Federal Taxes at 21.0%	($2,260,138)
Net Profit	$8,502,422

2.9 Financial Ratios—Measuring Organizational Health and Performance

The four primary financial statements report on an organization's financial position at a point in time and on its operations over some past period. The same financial statements can be applied to predict the organization's future earnings and financial health. Analysis of an organization's financial ratios provides predictive capability because financial ratios are designed to reveal relationships among the four primary financial statements. There are five categories of financial ratios.

- Liquidity ratios measure firm solvency or how able the firm is in meeting its maturing debt obligations.
- Asset management ratios measure how effectively the firm manages its assets.
- Debt management ratios measure the extent to which the firm uses debt financing.
- Profitability ratios measure the combined effects of liquidity, asset management, and debt management on operating results.
- Market value ratios relate the firm's stock price to its earnings and book value per share.

2.9.1 Liquidity Ratios

Liquidity ratios measure firm solvency or how able the firm is in meeting its maturing debt obligations. The **current ratio** (2.7) provides insight into the firm's short-term solvency.

$$\text{Current Ratio} = \text{Current Assets/Current Liabilities}$$

The **quick** or **acid-test ratio** (2.8) provides insight into an organization's ability to pay debts currently due.

$$\text{Acid-test ratio} = (\text{Current Assets} - \text{Inventories})/\text{Current Liabilities}$$

From Example 2.1, The AAA Company's current ratio is 1.40, but its acid-test ratio is 1.053 indicating that it can meet debt payments in the current fiscal accounting period, but it will have to convert WIP inventory into finished goods and sales in order to maintain minimum cash.

2.9.2 Asset Management Ratios

Asset management ratios measure how effectively the firm manages its assets in the production of products and services. The **inventory turnover ratio** measures how quickly the firm converts inventory into cash from sales.

$$\text{Inventory turnover ratio} = \frac{\text{Sales}}{\text{Inventory}} \qquad (2.10)$$

From its income statement and balance sheet, The AAA Company inventory turnover ratio is \$47,800,000/\$38,753,000 = 1.233, which indicates that it takes AAA about 3.5 weeks to convert inventory into sales. As with other ratios, AAA's inventory turnover ratio must be compared to industry or sector normal values and trends over time.

The **average collection period** measures how quickly accounts receivable is converted into cash from sales.

$$\text{Average collection period} = \frac{\text{Receivables}}{\text{Average sales/day}} \qquad (2.11)$$

Assuming a 30-day month, for the AAA Company, the average collection period was \$76,259,000/(\$47,800,000 / 30) = 47.86 days.

The **fixed assets turnover ratio** measures the utilization of plant and equipment in generating sales.

$$\text{Fixed assets turnover ratio} = \frac{\text{Sales}}{\text{Net fixed assets}} \qquad (2.12)$$

For the AAA Company, the fixed assets turnover ratio was \$47,800,000/\$449,090,000 = 0.106, which indicates that it takes AAA about 40.7 weeks production to convert plant and equipment value into sales.

The **total assets turnover ratio** measures the utilization of the organization's assets in generating sales.

$$\text{Total assets turnover ratio} = \frac{\text{Sales}}{\text{Total assets}} \qquad (2.13)$$

For the AAA Company, the total assets turnover ratio was \$47,800,000/\$629,699,100 = 0.0759, which indicates that it takes AAA about 57.1 weeks production to convert total asset value into sales.

2.9.3   Debt Management Ratios

Debt management ratios measure the extent to which the firm uses debt financing or financial leverage. The use of financial leverage has three implications.

1. The use of debt allows stockholders to maintain control of the organization with minimal investment.
2. Banks and financial bankers consider the proportion of owner-supplied financing as an indication of the risk of lending. The less stockholder supplied financing the greater are the risks bourne by its lenders.

3. The more debt is used in financing assets, the greater the return on investment for stockholders.

When examining an organizations financial statements, financial analysts consider two types of debt management ratios.

- Balance sheet ratios to determine how much debt has been used to finance assets.
- Income statement ratios to determine the number of turns that fixed charges are covered by operating profits.

The total debt ratio measures the total financing provided by creditors. Total debt = current liabilities + long-term debt.

$$\text{Debt ratio} = \frac{\text{Total debt}}{\text{Total assets}} \qquad (2.14)$$

The **interest coverage ratio** indicates how much revenue must decrease to affect the firm's ability to finance its debt from ongoing operations.

$$\text{Interest coverage ratio} = \frac{\text{EBIT}}{\text{Interest payments}} \qquad (2.15)$$

From its income statement, The AAA Company's interest coverage ratio = $13,840,000/$2,000,000 = 6.92.

The **fixed charge coverage ratio** indicates how much revenue must decrease to affect the firm's ability to finance its debt plus leases from ongoing operations.

$$\text{Fixed charge coverage ratio} = \frac{\text{EBIT} + \text{Lease payments}}{\text{Interest payments} + \text{Lease payments}} \qquad (2.16)$$

The AAA Company's fixed charge coverage ratio = ($13,840,000 + $1,020,000)/($2,000,000 + $1,020,000) = 4.92.

2.9.4 Profitability Ratios

Profitability ratios measure the combined effects of liquidity, asset management, and debt management on operating results. The **profit margin on sales** measures firm efficiency in generating profits from sales.

$$\text{Profit margin on sales} = \frac{\text{Net Income}}{\text{Sales}} \qquad (2.17)$$

The AAA Company's profit margin on sales = $8,502,422/$46,060,000 = 0.1846.

The **basic earning power ratio** measures firm efficiency in generating income from assets.

$$\text{Basic earning power} = \frac{\text{EBIT}}{\text{Total assets}} \tag{2.18}$$

The AAA Company's basic earning power = \$13,840,000/\$629,699,100 = 0.022.

The **return on total assets** measures firm efficiency in generating profits from assets.

$$\text{Return on total assets} = \frac{\text{Net Income}}{\text{Total assets}} \tag{2.19}$$

The AAA Company's return on total assets = \$8,502,422/\$629,699,100 = 0.0135.

The **return on common equity** measures actual rate of return to common stockholders from investment in assets.

$$\text{Return on common equity} = \frac{\text{Net Income}}{\text{Common stock equity}} \tag{2.20}$$

The AAA Company's return on common equity = \$8,502,422/\$400,000,000 = 0.021.

2.9.5 Market Value Ratios

Market value ratios relate the firm's stock price to its earnings and book value per share. The **price/earnings ratio** indicates how much investors are willing to pay per dollar of reported profits.

$$\text{Price/Earnings Ratio} = \frac{\text{Price per share}}{\text{Net income/Number of shares}} \tag{2.21}$$

The **market/book ratio** indicates how investors regard the firm's future potential as in investment.

$$\text{Market/Book Ratio} = \frac{\text{Price per share}}{\text{Book value per share}} \tag{2.22}$$

2.10 Integrated Ratio Analysis

The chart in Fig. 2.11 shows how the financial ratios integrate into a unified analysis of organizational financial health. The left side of the chart develops the profit margin on sales analysis. On the right side, the chart lists asset categories, totals them, and divides sales by total asset to estimate the number of times an organization turns over its assets. The profit margin multiplied by the total assets turnover yields the rate of

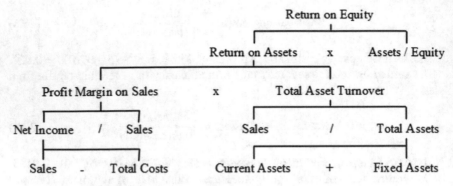

Fig. 2.12 Integrated ratio analysis yielding ROA

return on assets (ROA), and ROA multiplied by the ratio of total assets to equity yields return on equity (ROE) (Fig. 2.12).

$$\text{ROE} = \frac{\text{Net Income}}{\text{Total Assets}} \times \frac{\text{Total Assets}}{\text{Common Equity}}$$

$$\text{ROE} = \frac{\text{Net Income}}{\text{Sales}} \times \frac{\text{Sales}}{\text{Total Assets}} \times \frac{\text{Total Assets}}{\text{Common Equity}}$$

$$\text{ROE} = \text{Profit Margin} \times \text{Total Asset Turnover} \times \frac{\text{Total Assets}}{\text{Common Equity}}$$

2.11 Summary

Definition: *Accounting* is the discipline that systematically measures, records, interprets, and communicates the financial activities of an organization, providing insight into the trustworthiness, profitability, and solvency of an organization.

Accounting is concerned with the measurement, recording, interpretation, and communication of economic event, financial transactions, measurement of economic value, and determination of periodic changes in economic values. The four fundamental types of accounting are (1) financial accounting, (2) managerial accounting, (3) cost accounting, and (4) tax accounting.

Sources of operations financing include (1) lenders, (2) investors, and (3) retained earnings.

Definition: *Minimum Attractive Rate of Return* (*MARR*) —the risk adjusted, weighted, minimum acceptable rate of return, or hurdle rate that must be earned on a project or investment, given its opportunity cost of foregoing other projects or investments.

$$\text{MARR} = \sum_i w_{\text{Li}} \times I_{\text{Li}} \times (1 - T_C) + \sum_j w_{\text{Sj}} \times \text{RoE}_{\text{Sj}}$$

The overriding purpose of financial accounting is to summarize all financial activity in the profit and loss income statement, balance sheet, retained earnings statement, and cash flow statement.

- **Accounting** measures and records financial transactions and provides the basis for assessing the economic viability of the organization.
- **Management** allocates available investment funds to projects, evaluates unit and firm performance, allocates resources, and selects and directs personnel.
- **Engineering** analyzes the economic impact of design and project alternatives over their present and future life cycles.

The central tenet of transaction accounting is objective costing (i.e., traceable to a transaction record). A transaction is an exchange of money or credit for a product or service documented by a receipt, invoice, bill of lading, etc.

The **balance sheet** summarizes the value of the organization's assets available to carry out its economic activities and the liabilities and ownership financing claims against those assets on each specific date of the end of an accounting cycle. The fundamental accounting equation is

$$\text{Assets (value)} = \text{Liabilities (value)} + \text{Equity (value)}$$

Assets		=	Liabilities		+	Equity	
Debit	Credit		Debit	Credit		Debit	Credit
Increases	Decreases		Decreases	Increases		Decreases	Increases

The **income statement** summarizes the firm's revenues and expenses from ongoing operations over a stated period (usually month, quarter, or fiscal year). These are intervals between consecutive balance sheet statements. The fundamental outcome of the accrual income statement is the estimate of net profit.

$$\text{Net profit (accrual)} = \text{Revenues} - \text{Expenses}$$

Under the accrual basis of accounting (or accrual method of accounting), revenues and expenses are reported on the income statement in the accounting period in which they are earned or incurred.

The *retained earnings statement* is the link between the income statement and the balance sheet. The equation for the addition of net profit to the balance sheet retained earnings is

$$\text{Retained Earnings (current)} = \text{Retained Earnings (prior)}$$
$$+ \text{New Profit} + \text{new stock} - \text{dividends}$$

The *cash flow income statement* reports the difference between the sources and uses of cash during a fiscal accounting period. The fundamental outcome of the cash flow income statement is the estimate of cash net profit.

$$\text{Net profit (cash flow)} = \text{Revenues} - \text{Expenses} + \text{Financing}$$
$$- \text{Investments}$$

An organization's current financial health, or liquidity, is measured by working capital. Working capital is the difference between current assets and current liabilities and indicates the firm's liquidity or ability to pay its current debts.

$$\text{Working Capital} = \text{Current Assets} - \text{Current Liabilities}$$

An alternate measure of an organization's current financial health, or liquidity, is its **current ratio** which provides insight into the organization's short-term solvency.

$$\text{Current Ratio} = \text{Current Assets/Current Liabilities}$$

The **acid-test ratio** (also termed **quick ratio**) indicates the organization's ability to pay its currently due debts by subtracting the value of inventory from current assets.

$$\text{Acid} - \text{test Ratio} = (\text{Current Assets} - \text{Inventories})/\text{Current Liabilities}$$

Profit margin is the ratio of net profit to net sales revenue and indicates the efficiency at which sales are converted into profits which can be reinvested into productive assets or paid as dividends.

$$\text{Profit margin on sales} = \text{Net profit/Net sales revenue}$$

2.12 Key Terms

Accounting
Accrual
Assets

Balance Sheet
Cash Flow
Cash Flow Income Statement
Current Ratio
Equity
Finance
Financial Ratios
Income Statement
Investors
Lenders
Liabilities
Minimum Attractive Rate of Return
Profit Margin
Quick Ratio
Retained Earnings
Retained Earnings Statement
Stock
Stockholders

Problems

1. The following data is taken from Yetep Factory Works balance sheet. Determine the working capital, current ratio, and quick ratio.

Cash	$100,000
Net accounts receivable	$285,000
Inventories	$220,000
Prepaid expenses (long-term)	$6,000
Accounts payable	$210,000
Notes payable (short term)	$110,000
Accrued wage expenses	$150,000
Accrued salary expenses	$47,000

2. Account data ($\times$$1,000) for Telsa Engineering Corporation for the fourth month of its fiscal year is given below. Construct Telsa's balance sheet, and estimate its working capital, current ratio, quick ratio, and the proportion of debt provided by lenders.

    * Prepaid expenses cover a three-year fixed minimum purchase contract for specialty consulting to exclusively for Telsa. The original contact was signed at the start of the current fiscal year for $36,000,000 with minimum $1,000,000 services delivered each month.

Accounts receivable	$160,000	Accum depreciation building	$84,000
Accrued wage expenses	$80,000	Accum depreciation equip.	$126,000
Accrued salary expenses	$9000	Building	$200,000
Cash	$110,000	Equipment	$310,000
Inventories	$250,000	Capital surplus	$145,000
Prepaid expenses (short term)*	$1000	Land	$25,000
Prepaid expenses (long term)*	$32,000	Long-term liabilities	$100,000
Securities (short term)	$40,000	Common stock par value	$15,000
Accounts payable	$120,000	Retained earnings	Not given

3. The income statement data ($1,000) for Telsa Engineering Corporation for the fourth month of its fiscal year is given below. Construct Telsa's income statement, and estimate its profit margin, basic earning power, inventory turnover ratio, fixed assets turnover, total assets turnover, interest coverage ratio, return on total assets, and return on common equity.

Sales	$480,700	Depreciation building	$5200
Returns allowances	$10,400	Depreciation equipment	$54,000
Labor expense	$160,000	Operating interest	$400
Salary expense	$18,000	State income tax rate	6.0%
Materials	$8900	Federal income tax rate	21.0%
Indirect costs	$4200		
Selling and Adm expenses	$6400		

4. The cash flow income statement data ($1,000) for Telsa Engineering Corporation for the fourth month of its fiscal year is given below. Construct Telsa's cash flow income statement.

Sales	$360,500	Depreciation building	$5,200
Returns allowances	$7800	Depreciation equipment	$54,000
Labor expense	$120,000	Operating interest	$400
Salary expense	$13,500	State income tax rate	6.0%
Materials	$6,700	Federal income tax rate	21.0%
Indirect costs	$3200		
Selling and Adm expenses	$4800		

Chapter 3
Cost Accounting Fundamentals

Abstract Whereas managerial accounting tracks financial performance relative to strategic plans and budgets, cost accounting estimates costs, allocates overhead costs, and develops standard product costs. This chapter will provide an overview of cost accounting fundamentals. First, the chapter will define the different types of costs and cost purposes. Next, cash flow diagram conventions and uses are discussed in terms of breakeven, profit, and loss. Finally, the fundamentals of cost accounting for materials and components, labor, and overhead allocation are presented.

3.1 Introduction

Definition: *Cost accounting* is the discipline that captures an organization's total cost of production by assessing the variable costs of each production step, the associated fixed costs needed to support production, and the allocation of those fixed costs to each production step to determine standard production costs and product costs.

- Unlike general accounting, which provides information to external financial statement users, cost accounting is not required to adhere to set standards and can be flexible to meet the needs of management.
- Cost accounting considers all input costs associated with production, including both variable and fixed costs.
- Cost accounting is used internally by management to make fully informed business decisions.
- Types of cost accounting include standard costing, activity-based costing, lean accounting, and marginal costing.

Cost accounting was first developed during the Industrial Revolution. The economics of industrial supply and demand forced manufacturing organizations to track their fixed and variable costs to understand cost structure, determine the breakeven point their production processes, and decide on ways to increase profitability. Cost accounting allowed manufacturing organizations to control costs, become more efficient, and provide return on investment to stockholder owners. By the beginning of the 20th century, cost accounting had become accounting discipline. The general types of cost accounting include

- *Activity-based costing* (ABC) identifies the support and production activities needed to produce a unit of product or service in an organization and assigns the production cost of each activity to all products and services according to the actual consumption. This model assigns more indirect costs (overhead) into direct costs compared to conventional costing.
- *Lean accounting*, an extension of the philosophy of lean manufacturing, has the goals of improving financial management practices within an organization, minimizing waste, and maximizing productivity.
- *Marginal costing* (also termed cost–volume–profit analysis) assesses the impact on the cost of a product by adding one additional unit into production. Marginal costing assists management in identifying the impact of varying levels of costs versus varying levels of production volume on operating profit.
- *Standard costing* develops "standard" process and product costs used to determine the value of inventory and the cost of goods sold (COGS). Standard costs are based on an efficient use of labor and materials to produce the product or service under standard operating conditions. Assessment of the difference between the standard cost and actual cost incurred is termed "variance analysis."

3.2 Cost Terms and Purposes

Definition: *Manufacturing expense* is the monetary consumption of labor, materials, tools, equipment, and facilities to increase the value of resultant products and services.

In theory, private firms pursue two economic objectives: (1) At minimum, economic survival. (2) Optimally, maximize wealth creation value to its shareholder owners.

Definition: *Service or governmental cost* is the monetary value of resources that have been, or must be, consumed to yield a particular service.

In theory, service organizations exist to fill intangible client or public needs such as entertainment, food, transportation, communications, or health care. Democratic governmental and regulated organizations exist to promote the general welfare of its citizens by maximizing benefits to those who receive them while minimizing the costs to those who pay for them. A service organization or government's cost objectives are a function of its service objectives to its clients or citizens. Cost is measured in monetary units (exchange price) attached to the resources (raw materials, goods, and services) consumed by those activities necessary to attain a specific performance objective.

Types of costs include

- **Cash costs**: movement of money from one owner to another—also known as a cash flow (engineering economy).

  - Payment this month on a short-term loan.
  - Purchase of supplies paid in cash.

- **Accrued costs**: expenses which are incurred but for which no payment is made during an accounting period. They are shown in the balance sheet as a current (short term) liability.

  - Wages payable.
  - Salaries payable.
  - Utility charges.

- **Book cost**: Cost of a past transaction that is recorded in an accounting journal.

  - Asset initial cost.
  - Asset depreciation.
  - Stock par value.

- **Operating expenses**. associated with the productive operations of an organization. Operating costs may be either fixed or variable.
- **Direct costs**: non-operating costs directly related to producing a product or service.
- **Indirect costs**: cannot be directly linked to a product or service but are required for productive activities to exist.
- **Controllable costs**: can be influenced or regulated by the manager responsible for it (i.e., budgeted costs for labor, materials, and processes, controlled by the manager).
- **Non-controllable costs**: those costs that a manager cannot affect or change (i.e., space rental cost, equipment lease costs, and insurance costs).
- **Period costs**: are not directly related to the productive activities. Overhead sales, general, and administrative (SG&A) costs are considered period costs.

A general framework for cost terminology is illustrated in Fig. 3.1. Cost information is required for various purposes and costing objects. Cost purposes include

- *Planning*: determines the economic feasibility of a new or upgraded productive facility or line, process, or product or service.
- *Control*: the management budgeting process of identifying and reducing business expenses with a focus on increasing profit.

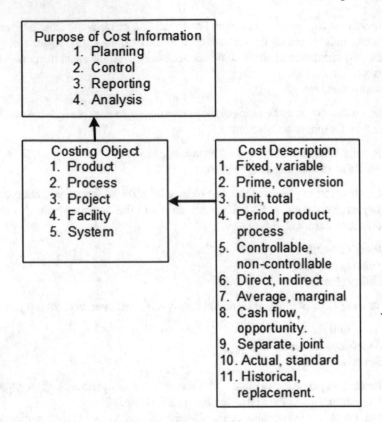

Fig. 3.1 Cost terminology framework

- **_Report_**: a process used to inform management about the magnitude of a process's or product's budgeted versus predicted cost or a client about the magnitude of a project's budgeted versus actual cost.
- **_Analysis_**: the process of decomposing summarized costs into their components for the purpose if identifying cost drivers.

Cost objects include

- **Product**: the per-unit cost incurred to produce a product or service. Generally, product costs include labor, raw materials, indirect consumable manufacturing supplies, and general overhead.
- **Process**: the allocation of the total costs of production to homogenous units produced via a continuous process that usually involves multiple steps or departments.
- **Project**: all estimated and actual costs or monetary obligations incurred or estimated to be incurred to complete a project that has a specified beginning and end time, scope of resources and objective.

- **Facility**: building maintenance, property taxes, insurance, cleaning services, net costs of cafeteria services, depreciation, utilities, security, parking, and all other non-capital expenditure costs relating to a facility as a whole or to the common areas of a facility.
- **System**: all costs incurred in connection with the financing, development, construction, care, custody, control, and retirement of a system.

Only manufacturing and retail merchandising organizations must consider product and period costs for direct materials, direct labor, and manufacturing overhead (supporting supervisory, engineering, maintenance, and logistics support personnel necessary for process functioning). Manufacturing overhead includes additional costs for indirect materials, indirect labor, depreciation, extraordinary maintenance and repairs, utilities, taxes, and insurance.

Conversely, service organizations and governments need not consider product and period costs. Due to the non-repetitive nature of services, the costs which should be considered in service and governmental management budgeting are not past recorded costs but rather predicted future costs that will differ among the possible alternative courses of action. Data for budgeting costs should be found in well-constructed, full-cost databases of past transactions. The full-cost information on past transactions must then be adjusted for cost trends, anticipated general inflation, and changes in service offerings or governmental programs.

Fixed, Variable, and Overhead Cost

The engineering economic analyst's most important task is the analysis differences in cost behaviors among feasible alternatives. This analysis reveals how costs respond to changes in activity levels within a relevant range. The relevant range is the bounded (minimum, maximum) of activity level over which the costs (dependent variable) and activity level (independent variable) relationships are expected to hold. Outside of that relevant range, the cost versus activity-level relationship will likely differ. Expected costs include

Fixed—constant (relatively) over a time period or range of activity level.

- Investment in a manufacturing facility.
- Rent for additional warehouse space.

Variable—depends on activity level.

- Raw materials—greater production levels require more materials.
- Direct labor—greater production levels require more direct labor.

Overhead—all operating costs that are not raw materials or direct labor.

The summation of fixed, variable, and overhead costs yields total costs of production.

$$C_T = C_F + \sum_U C_{V/U} \times U + \{C_{OH}\} \tag{3.1}$$

where C_T = total cost for the activity level, C_F = fixed cost for the activity level, $C_{V/U}$ = variable cost per unit, U = number of units produced, and C_{OH} = allocated overhead cost for the activity level. C_{OH} is shown in brackets because its estimate may not be applicable to all engineering projects.

Figure 3.2 illustrates four possible relationships between fixed cost and variable cost per unit. In the upper left corner, both the fixed cost level and variable cost per unit remain constant over the relevant range. In the upper right corner, the fixed cost increases by steps (semifixed cost) from relevant range to relevant range, but the variable cost per unit remains constant across the relevant ranges. In the lower left corner, the fixed cost remains constant across the ranges, but the variable cost per unit decreases across the relevant ranges. This is termed economies of scale and occurs in manufacturing operations as a result of cost savings gained by increased production activity. Economies of scale allow suppliers to offer "price breaks" on differing purchase volumes of materials or components. Price breaks may also occur when suppliers total cost is planned to cover a fixed cost plus the variable cost per unit up to an upper bound activity level, and beyond that upper bound, the fixed cost is covered, and only additional cost per unit is incurred. In the lower right corner, both the fixed cost and variable cost per unit vary across the relevant ranges.

Prime, Conversion, Total, and Unit Costs

In manufacturing, cost components are also classified as prime costs, conversion costs, total cost, and unit cost.

> **Prime**—cost of direct material and direct labor to transform raw materials and components into finished products.
> **Conversion**—direct labor and manufacturing overhead costs.

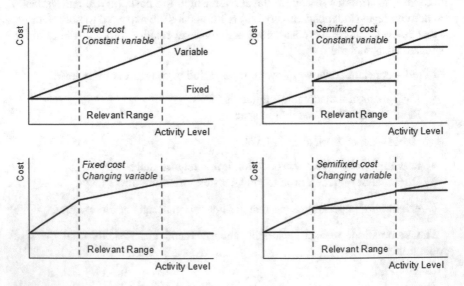

Fig. 3.2 Cost behaviors over a relevant range

Total—indirect product costs + direct product costs + fixed product costs
Unit—total product cost/number of units
Direct cost—direct material and direct labor (prime costs) traceable directly to the manufacture of a particular product.
Indirect cost—general administrative, marketing and sales, factory management and staff, indirect materials, and other fixed costs not traceable directly to a particular product.

Product costs are accumulated on either:

Full absorption costing—direct material, direct labor, fixed overhead, and factory overhead.
Direct variable costing—direct material, direct labor, and factory overhead.

The relationships among these costs and their accumulation into total product costs and cost per unit are illustrated in Fig. 3.3 and described in Table 3.1.

Average, Marginal, and Incremental Costs

Engineering economic analysts use average cost as an estimate of expected costs. The average cost = total product cost/number of units produced. **Incremental cost** is the cost incurred to add another unit of activity or the difference in total cost between two alternatives. **Marginal cost** is the change in the total cost that arises when the quantity produced changes by one unit.

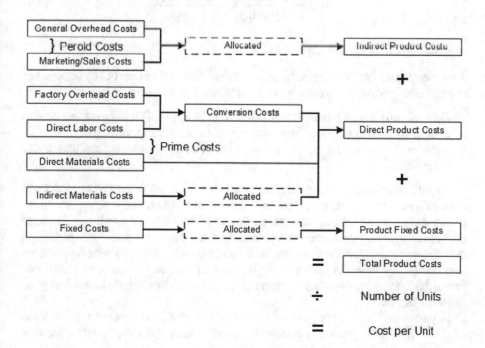

Fig. 3.3 Relationship among prime, conversion, total, and unit cost

Table 3.1 Description of prime, conversion, total, and unit cost

Category	Type	Source	Measurement	Cost
General management	Fixed	Gen Mgt/staff	Allocated	Indirect production
Marketing/sales	Fixed	Sales/staff	Allocated	
Factory overhead	Fixed	Mgt/super/staff/Eng	Allocated	
Direct Labor	Variable	Labor	Measured	
Direct materials	Variable	Materials	Measured	
Indirect Materials	Variable–Fixed	Supplies	Allocated	Direct Production
Fixed costs	Fixed	Gen facilities	Allocated	Product fixed

$$\frac{\mathrm{d}C_T}{\mathrm{d}Q} = \frac{\mathrm{d}\big(C_F + C_{V/U} \times U + \{C_{\mathrm{OH}}\}\big)}{\mathrm{d}Q} = \frac{\mathrm{d}C_{V/U}}{\mathrm{d}Q} \tag{3.2}$$

Since $U = 1$-unit change, CF is the fixed cost, and C_{OH} is a fixed overhead allocation charge, marginal cost may be approximated in the production quantity range as

$$\mathrm{MC} = \frac{\mathrm{TotalCost_2} - \mathrm{TotalCost_1}}{\mathrm{Quantity_2} - \mathrm{Quantity_1}} \tag{3.3}$$

Opportunity and Sunk Costs

Two costs for which there are no journal or ledger accounts but are of key importance to engineering economic analysis are opportunity cost and sunk cost.

Opportunity cost—a benefit that is foregone by engaging a resource in a chosen activity instead of engaging that same resource in the foregone activity.
Sunk cost—money spent due to a past decision. We cannot do anything about these costs.

Organizations apply resources to accomplish tasks. Each resource has associated costs of maintenance and use. However, the cost of owning, maintaining, and using resources is not just the monetary value. Resource costs also include the opportunity cost of not applying a resource to a potentially more profitable use. An organization that uses a resource for one task chooses to forego benefits from applying the resource to a potentially more profitable use. The potential benefit cash flow or return that would be realized by applying the resource to the alternate task is the **foregone opportunity cost**.

A **sunk cost** is monetary value of a past decision for the acquisition or application of a resource. In engineering economic analysis, sunk costs are ignored, because engineering decisions are current improvements or future projects to expand organizational capacity or capability. There is one exception to the rule to ignore sunk cost in engineering economic analyses. When an asset resource is disposed of or sold, U.S. Internal Revenue Service tax code requires that the original sunk cost monetary value paid for the asset be accounted for in determining the organization's income taxes for the accounting period in which the asset is disposed or sold.

Additional Cost Terms

Six additional cost terms are defined by the way in which they are incurred or recognized in the accounting system.

Separate—costs incurred that can be exclusively identified with the production of an individual product.

Joint—costs of products of relatively significant sales values that are simultaneously manufactured by a process or series of processes.

Actual—actual direct labor, direct material, and overhead costs directly traceable and charged to a product or process.

Standard—a predetermined cost derived from cost analysis of a product or process.

Historical—the original monetary value of an economic item based on the stable measuring unit assumption.

Replacement—or replacement value refers to the amount that an entity would have to pay to replace an asset at the present time, according to its equivalent current worth.

Cash Flow Diagram Conventions

The cash flow revenues and costs resulting from engineering projects occur over time and may be summarized on a cash flow diagram. A cash flow diagram graphically illustrates the magnitude, sign, and timing of individual cash flows. A cash flow diagram is created by drawing a time-incremented horizontal line. Time increments may be days, weeks, months, quarters, or years. Cash flows are accumulated over each time increment and represented at the end of each time increment by an arrow drawn to scale whose length indicates the magnitude and direction indicated the direction (negative or positive). The cash flows resulting from engineering projects may be categorized typically by one of the following categories.

First cost—time 0 capital expense to build or to buy and install an asset.

Operations costs—annual expenses for the manufacture of a product or service.

Maintenance costs—annual maintenance expenses in support of manufacturing.

Overhaul—major capital expenditure that occurs during an assets life possibly to extend its life beyond the original project termination.

Revenues—annual receipts due to sale of products or services from the project.

Market value—the most likely trading price that may be generated for an asset under market circumstances that exist at the time of exchange during the asset's useful life.

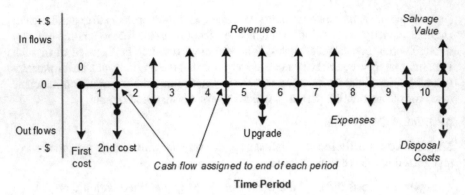

Fig. 3.4 Typical cash flow diagram

Salvage value—the estimated value that will be realized upon the sale or other disposition of an asset at the end of its useful life.

Figure 3.4 illustrates a typical cash flow diagram.

Related Cost Terminology—Profit, Loss, and Breakeven

Equation (3.1) defined total cost as the summation of fixed, variable, and overhead costs. When total cost is subtracted from total revenue (either on an accrual or cash basis), the organization's profit or loss regions and breakeven point may be determined for any accounting period.

- **Breakeven**: total revenue = total costs. Zero profit.
- **Profit region**: total revenue > total costs. Positive profit.
- **Loss region**: total revenue < total costs. Negative profit or increasing debt.

Example 3.1

Halley, Inc., produces one product with a fixed cost $160,000 and variable cost of $3.00 per unit. In its current facility, Halley can produce a maximum of 100,000 units per month. The price is $6.20 per unit. Halley has the potential to move into a new market at the same price and increase demand by 50,000 units per month; however, Halley will have to add production capacity. The new production capacity will increase Halley's fixed cost to $200,000 but will reduce the variable cost to $2.25 per unit. Should Halley undertake this expansion. Estimate Halley's breakeven number of units under both capacities. Estimate Halley's total cost, total revenue, and total profit at 100,000, 110,000, 120,000, 130,000, 140,000, and 150,000 units

Breakeven: Revenue = Total Cost(1) $6.20 U = $160,000 + $3.00 U ($6.20 – $3.00) U = $160,000 $3.20 U = $160,000 U = 50,000	Revenue = Total Cost(2) $6.20 U = $200,000 + $2,25 U ($6.20 – $2.25) U = $200,000 $3.95 U = $200,000 U = 50,633		
Units	Total cost	Revenue	Profit
100,000	$425,000	$620,000	$195,000
110,000	$447,500	$682,000	$234,500
120,000	$470,000	$744,000	$274,000
130,000	$492,500	$806,000	$313,500
140,000	$515,000	$868,000	$353,000
150,000	$537,500	$930,000	$392,500

3.3 Cost Accounting in the Organization

Definition: *Cost accounting* is the systematic recording and analysis of direct material and direct labor costs and allocation of indirect overhead costs to the production of an organization's products or services.

Cost accounting objectives include

- Develop product costs.
- Determine the mix of labor, materials, and overhead costs for an organization.
- Evaluate opportunities for outsourcing or subcontracting.

Process Transformations and Cost Flows

Costs are incurred to produce and distribute products and services. When the final product is sold or service delivered, the sum of product costs becomes the expense that is subtracted from revenues by the products or services to estimate the operations profitability. Figure 3.5 summarizes the cost flows necessary to support product transformation processes.

Direct Material and Components—All materials and components that are directly integrated into the finished product or are necessary as part of a service represent direct materials. Usually, only the more significant items are costed as direct materials. Minor items (screws, nuts, fasteners, tape, glue, thread, labels, etc.) that are common across products or services are too costly to trace and are accounted for as factory overhead.

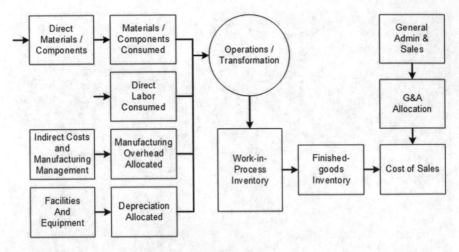

Fig. 3.5 Cost flows into transformation processes

Direct Labor—Labor used in the transformation process that directly affects or controls the conversion of materials, components, energy, and information into the finished product is considered direct labor. From transaction accounting, direct labor must be physically traceable to the finished product.

Manufacturing Overhead—All other fixed and variable costs necessary to directly support the transformation process are considered as manufacturing overhead. These costs typically include fixed and variable costs for indirect materials, indirect labor, factory equipment depreciation, maintenance, repairs, and process engineering.

Facilities and Equipment Depreciation—As will be discussed in Chap. 7, depreciation is an annual allocation for the initial expense of a capital asset (facility or equipment) as set forth by governmental tax code. Depreciation must be considered in estimating after-tax cash flow net income for each accounting fiscal period, because the depreciation amount reduces income taxes that must be paid.

Work-in-Process Inventory—As materials, components, energy, and information flow into each process transformation step, their value is accumulated in the partially finished product or service exiting each step. This accumulated value is termed "work-in-process" inventory and re-valued as partially finished product passes each transformation step.

Finished goods Inventory—As completed products or services exit the final production step, their work-in-process value is transferred into finished goods inventory value. This transfer recognizes that work-in-process inventory value is only a book value, because work-in-process inventory does not have a market. Conversely, finished goods inventory is available for sale to customers and clients and can generate revenue.

General Administration and Sales Overhead—Corporations are required to have top management who, as part of their responsibility, legally represent the organization. A proprietor assumes legal liability as the owner, and partnerships assign legal liability as part of the partnership agreement. Sales personnel provide place and time value-added to finished products and services by making the customer or client aware of the benefits of the organization's products and services and negotiating sales contracts that specify the price, time, and place of delivery. Since general administrative and sales costs cannot be directly traced through transactions to each individual product, they must be allocated to the finished goods cost of each product or service, which is the subject of this chapter.

Cost of Goods or Services Sold—The total cost = Sum(indirect product costs + direct product costs + product fixed costs) = Sum(G&A + sales allocation + Sum(direct materials + direct labor + manufacturing overhead + depreciation) as shown in Fig. 3.3.

3.4 Cost Accounting—Material and Component Analysis

Material and components cost analyses are based on engineering bill of materials and bill of components which specify materials or components descriptions for standard and customized products. Manufacturing organizations may have bill of materials and bill of components for thousands, if not millions, of material and component types. A vertically integrated processing organization must identify the raw and intermediate materials cost values for partially and completely processed products. A horizontally integrated organization must identify the work-in-process product value for transfer between process steps or between production facilities. Fluctuation in raw materials and commodity prices due to supply and demand and inflation must be continually tracked and incorporated into costs. Cost determination includes historical organizational records and current market information.

Material/Component Cost—quantities of materials or components that can be specifically identified with a product, contract, project, organizational subdivision or function for which a unit price is to be determined.

Direct Material—an input substance or information or service element to be altered into a component of the finished product or service.

Direct Component—a designed grouping of partially altered raw materials, information, services, or other components which will be further altered into a component of the finished product or service.

The scope of what constitutes a direct material or component depends on the product to be manufactured. Raw materials or components are those that have been purchased from external supplier and not manufactured by the organization. Design documents are the engineering bill of materials or bill of components and design engineering drawings and specifications.

- *Engineering bill of materials/components*—itemized list with part numbers, descriptions, and units.
- *Specifications*—drawings, physical, and performance technical requirements.
- Purchasing *units* and *quantities* per unit or lot of finished product.

Direct materials or components may be further classified into raw materials, standard commercial-off-the-shelf (COTS) materials or components, subcontracted materials or components, and interdivisional transfer items. Raw materials are fabricated, intermediate, or processed material in a form that will be further converted by the organization's transformation processes. Raw components are components of the organization's design and specification that are manufactured on contract by other organizations and may require further conversion. Raw materials or components include partially converted or fabricated intermediate items that require final conversion into the finished product by the organization. Standard commercial materials and components are other organizations' finished product materials or components that will be incorporated into the finished product with no further conversion (i.e., brakes or tires assembled onto a finished automobile, earphones added to the packaging of smartphones, etc.). Standard purchased materials and components may incorporate a lower purchasing overhead rate than the general overhead rate of the organization's manufacture products. Standard purchased materials and components may be costed by the purchasing department rather than design engineering.

Subcontracted materials and components are items that are manufactured by a supplier to the organization's purchasing and engineering requirements. Interdivisional transferred materials or components are items transferred at cost (no mark-up profit) between divisions, subsidiaries, or affiliates wholly owned and controlled by the organization. *Commodities* are raw materials that are traded on commodities markets or exchanges (metals, timber, minerals, food, etc.). Accordingly, commodity prices are volatile and costed at spot prices. Semi-engineered materials (steel, copper, plastic pellets, etc.) pricing may behave as either like a standard engineering item or a commodity. *Normative materials* or components costs are fixed by governmental regulation or cartel price controls.

Indirect materials are those materials that are necessary for the transformation of direct materials into finished products and are not directly traceable to any single product (cleaning chemicals, fluxes, thinners, catalysts, gasses, lubricants, and perishable fixtures). Operating supplies (pens, pencils, brooms, rags, etc.) are too inexpensive to be costed in engineering drawings and specifications. Operating supply costs are usually allocated as part of product fixed costs.

The problem of estimating direct materials costs is decomposed into three parts: (1) shape measurement, (2) cost per-unit shape, and (3) salvage value of recyclable materials (blanking or machining webs, chemical byproducts, etc.). The cost of direct materials is estimated as

$$\text{Cost direct material} = \text{unitshape} \times \frac{\text{cost}}{\text{unitshape}} - \text{salvage value} \qquad (3.4)$$

Unit shape implies engineering dimensional units such as mass, area, length, or another physical or service units or count. Salvage value is a recovered material having a debit or credit applied against the original direct material cost.

3.5 Cost Accounting—Labor Analysis

Depending on the level of automation, labor can comprise one of the most important cost components of an organization. To estimate labor cost, we must identify an objective unit of labor time. Time study, man-hour reports, and work sampling are the methods applied to identify the units of labor time consumed to manufacture a unit of finished product or service. Once the unit of labor time is determined, it is multiplied by the wage and fringe cost per unit of labor time.

Labor can be in classified as (1) direct or indirect, (2) non-recurring or recurring, (3) non-designated or designated, non-exempt or exempt, line or supervisory or management, or non-union or union. Wages and salaries may be based on time attendance, performance, or services. This discussion of labor cost will be restricted to direct and indirect labor.

Labor cost is the wage or payroll per person-hour or fraction thereof that can be specifically and consistently assigned to or associated with the manufacture of a product, a particular work order, or provision of a service.

Direct labor is the people who directly contribute to manufacture a specific product, complete a work order, or provide a service.

Indirect labor is the people who do not directly produce products or services, but who make their production possible or more efficient. Indirect labor costs are not readily identifiable with a specific task, work order, or service. They are termed indirect costs and are charged to overhead accounts.

The simple formula for direct, non-exempt labor cost is

$$\text{Labor cost} = \text{number time units worked} \times \text{wage/timeunit} \qquad (3.5)$$

Inputs for direct labor cost analysis include

- Measured time—predetermined time standards, time study analysis, work sampling, standard data, man-hour reports. May be stated in units of hours, quarter hours, pieces, bags, bundles, or standard number of units.
- Wage and benefit rates—drawings, physical, and performance technical requirements.
- Efficiency improvement—learning curve estimates as adjustments to measured time.

Labor Analysis of Measured Time

Time standards development is the analysis of an operation to eliminate unnecessary elements and to determine the most cost-effective method of performing the operation. Time standards development is accomplished in the following sequence.

1. Identify operation input and output boundaries.
2. Standardize the operation's methods, equipment, and conditions.
3. Decompose each operation into its motion elements, and identify motion boundaries.
4. Determine by measurement the standard time for each element and the number of standard hours required for an average worker to complete the operation.
5. File motion study and record standard hours into the labor time database.
6. Layout the operation in the correct and most efficient motion sequence and the correct presentation of tools.
7. Redesign tooling to support quick changeover, and contact mistake proofing.
8. Redesign materials input system to automatically present materials ergonomically and in the correct sequence.
9. Verify that the process and product meet quality requirements and standards.

There are four basic techniques used on time standards development.

- Predetermined time standards system.
- Time study analysis.
- Work sampling.
- Man-hour reports.

Predetermined time standards are most useful in the conceptual design phase of new products when there is only limited information on future production. Predetermined time standards are based on standard work elements encoded in measurements time methods (MTM) work factors. MTM was developed by Maynard, Stegemarten, and Schwab in 1948 and is still the most widely used predetermined time system.

- MTM-1 has ten elements. Each element is assigned a number of measured time unit (TMUs), which are in 0.00001 h. Requires 350 cycle times to analyze and operation. MTM-1 accuracy is ±7.0%.
- MTM-2 uses the same ten elements and TMUs but requires only 150 cycle times. MTM-2 accuracy is ±15.6%.
- MTM-3 uses the same ten elements and TMUs but requires only 50 cycle times. MTM-3 accuracy is ±20.0%.

The ten MTM time motion elements are as follows:

Reach—move a hand or finger to a destination.
Move—transport an object to a destination.
Turn—hand is turned or rotated about the long axis of the forearm.
Apply Pressure—apply a force to an object.
Grasp—secure sufficient control of one or more objects with the fingers or the hand.

PREDETERMINED TIME STANDARDS ANALYSIS

Operation: 11		PN: ASY101		Description			
Date: 04/09/x3		Time:		Assemble & bolt 2			
By: TST				supports to frame			

LHand	Freq	LH	Time	RHand	Freq	RH	Time
Reach	1	R30	38	Move	1	M30	41
Grasp	1	G1	8	Release	1	RL	6
Move	1	M30	22	Reach	1	R30	15
				Grasp	1	G2	8
Press	1	AP1	5	Press	1	AP1	5
			73				75
Reach	1	R12	9	Reach	1	R12	9

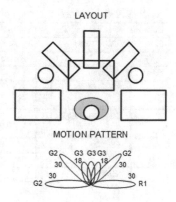

LAYOUT

MOTION PATTERN

Fig. 3.6 Example predetermined time standard analysis sheet

Position—motion is employed to align, orient, and/or engage one object with another.

Release—relinquish control of an object.

Disengage—break contact between two objects.

Eye Times—eyes direct hand or body motions.

Body Motions—motions made by the entire body, not just the hands, fingers, or arms.

Figure 3.6 shows an example of a predetermined time standard analysis sheet for a workstation.

Time study analysis is defined as an engineering study to determine the time required by a skilled, properly trained operator working at a normal pace to complete a specific task. The general time study procedure is

Preparation:

1. Analyze the operation's methods, equipment, and conditions; improve where necessary.
2. Define the operation's start and end boundaries and list the sequence of significant transformation points.
3. Decompose each transformation point into motion elements for a sample of *n* operations.

   Acquire time study information.
4. Time study—record the time consumed by each motion element for a sample of *n* operations.
5. Rate the pace of each motion element.
6. Determine process and personal allowances.
7. Convert each rated motion element into normal times with allowances.
8. Express the standard time as the normal time per common unit(s) production.

Time study equipment is just a clipboard, standardized time study worksheet, and a stopwatch of a stated resolution. There are two basic time study procedures: (1)

TIME STUDY WORKSHEET

Operation: 11	Description	Assemble & bolt 2 supports to frame															
Date: 04/09/x3	Time:		PN: ASY101		Drawing: ASY101-03			Machine: PRESS			Machine No: A03						
By: TST			Opr: J. B P.		On Job: 38 months			Dept: Assembly			Tool No: M61						
Element	Description		1	2	3	4	5	6	7	8	9	10	Avg	%	Nrm	UN
	ASSEMBLY	R	19	51	81	1.17	1.48	1.87	2.18	2.55	2.94	3.25	0.14	90	0.12	0.12
		E	.19	.13	.12	.13	.11	.11	.13	.17	.18	.09				
	DRIVE SCREW	R	25	56	89	1.23	1.53	1.92	2.24	2.63	3.04	3.31	0.06	110	0.07	0.07
		E	.06	.05	.08	.06	.05	.05	.06	.08	(10)	0.6				
	PRESS	R	38	69	1.04	1.37	1.76	2.05	2.38	2.76	3.16	3.45	0.13	100	0.13	0.13
		E	.13	.13	.15	.14	(23)	.13	.14	.13	.12	.14				
	INSPECT	R														
		E														

Fig. 3.7 Completed continuous time study worksheet

continuous time study used for short-duration tasks, and (2) long-cycle time study used for processing that requires long-duration tasks. Figure 3.7 shows an example of a completed continuous time study worksheet.

Work sampling gathers task information from large samples of a work force population. Work sampling involves taking observations of specific activities within an operation at random intervals.

1. Each observation is classified into predefined work categories such as "transport," "setup," "work," "inspection," "clean up," "idle," or "absent."
2. The number of observations within each category are divided by the total number of observations to yield the proportion of time spent in each category.
3. The binomial distribution is applied to estimate expected proportions and standard deviations.

$$p_i = \frac{N_o}{N} \tag{3.6}$$

$$\sigma_{pi} = \sqrt{\frac{p_i(1 - p_i)}{N}} \tag{3.7}$$

4. Given very large sample sizes, the standard normal distribution approximation of the binomial distribution is applied to estimate the sample size N_i necessary to achieve a desired sampling interval I at a stated confidence.

$$I = 2Z \left(\frac{p_i(1 - p_i)}{N_i} \right)^{1/2} \tag{3.8}$$

$$N_i = \frac{4Z^2 p_i(1 - p_i)}{I^2} \tag{3.9}$$

Work sampling continues until the maximum N_i is reached. Z values for various areas under the standard normal distribution curve are as follows:

Area(%)	-Z to +Z	Area Outside Limits(%)
68	±1.000	32
90	±1.645	10
95	±1.960	5
99	±2.576	1

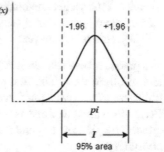

5. The standard hours per category are estimated as

$$H_S = \frac{(N_i/N)\mathrm{HR}(1 + \mathrm{PF\&D})}{U} \qquad (3.10)$$

where H_s = standard man-hours per unit for each job element, H = total man-hours worked during the study, R = rating factor, PF&D = personal fatigue and delay allowance decimal, and U = total number of units produced during the study. The following table presents the estimated work time in each category from preliminary work sampling and the required N_i for a 95% confidence interval $I = 0.10$ for a ±0.05 or ±5.0% error allowance.

Element	Observations	p_i
Setup	138	0.10
Transport 1	138	0.10
Machining	384	0.50
Inspection	138	0.10
Transport 2	196	0.15
Clean up	73	0.05

The following table presents work sampling results for the machining maximum $N_i = 384$ observations and actual/estimates (3.8) for the observations.

Element	Observations	p_i	I
Setup	59	0.075	0.134
Transport 1	90	0.113	0.131

(continued)

(continued)

Element	Observations	p_i	I
Machining	384	0.485	0.100
Inspection	71	0.090	0.133
Transport 2	147	0.186	0.126
Clean up	41	0.051	0.135

A total of $H = 293$ man-hours were worked producing $U = 510$ units during the study yielding the following H_s standard man-hour per-unit estimates (3.10).

Element	Observations	R	PF&D	H_S
Setup	59	1.00	0.00	0.0428
Transport 1	90	1.00	0.10	0.0718
Machining	384	1.20	0.20	0.4011
Inspection	71	0.90	0.20	0.0556
Transport 2	147	1.00	0.10	0.1173
Clean up	41	0.85	0.05	0.0265

Man-hour reports are used to gather work data and estimate time standards for non-repetitive work (construction, long-cycle production or processing, job shops, maintenance, or professional). The basic metrics for man-hour reports are as follows:

Man-hour—one worker working for 1 hour.

Man-month—based on a standard 40-h work week and 4.33 work weeks per month. This yields one man-month = $40 \times 4.33 = 173.33$ man-hours per month.

Man-year—based on a standard 40-h work week and 52 work weeks per year. This yields one man-year = $40 \times 52 = 2080$ man-hours per year.

In practice, quantification reflects either (1) actual clock hours or (2) effort adjustment to allow time. The first quantification captures the complete time including productive effort, wait time, and idle time. In the second quantification, a standard PF&D allowance decimal is added to productive time. Man-hour reports may be replaced by the term "person-hour" reports to reflect modern concepts of equal opportunity.

Man-hour report information is compiled from job tickets. A job ticket is a printed form that accompanies a work order. A job ticket provides instructions for recording time spent on the work tasks, traveling to and from inventory stores, extraordinary time spent on unplanned tasks, and identifying materials and components used. An example job ticket is illustrated in Fig. 3.8.

Job tickets are collected for similar work tasks, and the data is entered into a database or spreadsheet for analysis. Average time and 95% or 99% prediction intervals are estimated for each activity that comprise the common task. These times then

JOB TICKET

Employee Name: Joe Brown   Employee ID: 12122   Ticket Date: 04 / 23 / 15

Work Order: RP 070   Machine Name: Federal Punch Press 60 ~~Ton~~

Department: Frame Fabrication   Part Name: Clutch

Location/Zone: FF-B12   Asset Code: 447-053

Work Code: PM-17

Location/Zone: Preventive Maintenance – Replace worn clutch at 22,500 hours ~~operation.~~

Transit Time Start: 8:45 AM   Setup Time Start: 9:05 AM   Work Time Start: 9:30 AM

Transit Time End: 9:05 AM   Setup Time End: 9:30 AM   Work Time End: 11:35 AM

Job Notes: Extra time – 30 minutes to break rusted bolts on housing.

Work Order: IS 051   Machine Name: KUKA Arc Welding Robot

Fig. 3.8 Example job ticket

Work Code	Work Order	Date	Transit Start	Transit End	Transit Time	Setup Start	Setup End	Setup Time	Work Start	Work End	Work Time	Extra Time
PM-17	RP-068	19-Jan-15	3:17 AM	3:36 AM	19.00	3:36 AM	4:00 AM	24.00	4:00 AM	5:34 AM	94.00	0.00
	RP-069	4-Mar-15	3:26 PM	3:38 PM	12.00	3:30 PM	4:02 PM	32.00	4:02 PM	5:27 PM	85.00	0.00
	RP-070	23-Apr-15	8:45 AM	9:05 AM	20.00	9:05 AM	9:30 AM	25.00	9:30 AM	11:35 AM	95.00	30.00
	RP-071	9-Jun-15	7:14 PM	7:28 PM	14.00	7:28 PM	7:58 PM	30.00	7:58 PM	9:37 PM	99.00	0.00
	RP-072	9-Jul-15	6:32 AM	6:58 AM	26.00	6:58 AM	7:19 AM	21.00	7:19 AM	8:53 AM	94.00	0.00
	RP-073	18-Aug-15	4:58 PM	5:26 PM	28.00	5:26 PM	6:42 PM	16.00	6:42 PM	8:20 PM	88.00	10.00
	RP-074	24-Sep-15	1:48 PM	5:06 PM	18.00	5:06 PM	5:33 PM	27.00	5:33 PM	7:28 PM	85.00	20.00
	RP-075	1-Nov-15	3:26 PM	3:49 PM	23.00	3:49 PM	4:14 PM	25.00	4:14 PM	5:52 AM	88.00	0.00
	RP-076	14-Dec-15	8:45 AM	9:01 AM	16.00	9:01 AM	9:18 AM	17.00	9:18 AM	10:51 AM	93.00	0.00
Average					19.56			24.11			91.22	6.67
Std Dev					5.34			5.40			4.89 NA	
<= 97.5% PI					30.02			34.69			100.81 NA	
>= 2.5% PI					9.09			13.54			81.63 NA	

Fig. 3.9 Example man-hour report analysis

become guidelines for planning future similar non-repetitive tasks and estimating the number of personnel required to support non-repetitive tasks. For man-hour analysis to be of value, provision must be made for the original job tickets, databases, or spreadsheet to be retained as permanent backups. This permits engineering managers to identify future unusual task times and re-estimate average times and prediction intervals as needed to reflect non-repetitive task conditions. An example man-hour report analysis with 95% prediction intervals is illustrated in Fig. 3.9.

3.6 Cost Accounting—Overhead Cost Allocation

An organization's cost accounting function collects, analyzes, and summarizes raw cost data into performance cost data (overhead costs allocation, product unit cost, utilization rated, etc.). Cost accounting performance data is then applied to allocate overhead and establish product unit cost, determine the mix of materials, labor, and indirect costs for future production, and evaluate the outsourcing and subcontracting possibilities. There are two primary methods of expensing allocated fixed overhead costs: (1) direct or contribution expensing, and (2) absorption or functional expensing. Under the direct expensing method, overhead costs are said "to expire immediately" and are accounted for in the fiscal period in which they are incurred and are charged against product sold or services delivered, regardless of when the products were placed in finished goods inventory or the services completed. Under the absorption expensing method, overhead costs are charged to products or services and become expenses when the product is sold or the service complete. Direct versus absorption expensing allocation is illustrated in Fig. 3.10. Regardless of the expensing allocation method, overhead allocation accounting methods are the same.

The choice between direct and absorption expensing of allocated fixed overhead costs depends on cost accounting purpose. Direct expensing evaluates only the costs directly associated with production and makes it easier for an organization to compare the potential profitability of manufacturing one product over another (cost–volume– profit analysis). However, direct expensing makes it more difficult for an organization to determine profit maximizing pricing for its goods and services, since it does not directly consider all the costs that must be covered to be profitable. On the other hand, absorption expensing provides a more accurate accounting of net profitability, especially when an organization does not sell all its products in the same accounting

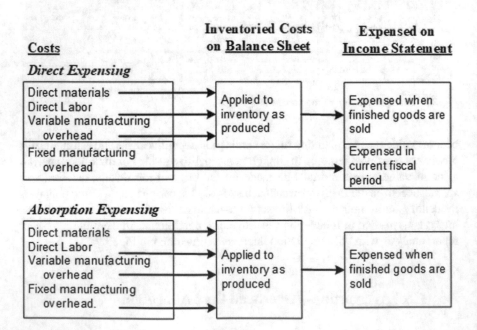

Fig. 3.10 Expensing allocated fixed overhead costs

period in which they are manufactured. However, absorption expensing is not as helpful as direct expensing for comparing the profitability of different product lines. The major advantage of absorption expensing is that it is required for an organization to comply with Generally Accepted Accounting Principles and U.S. tax law.

Since it is required legally, this overhead cost allocation discussion will present only the absorption expensing method. As previously defined, the costs incurred to produce a product or service are either direct or indirect.

- Direct costs for labor and materials are activity-based and can be linked directly to specific products or projects.
- Indirect costs for management, support functions, sales, and administrative expenses cannot be linked directly to products or projects.

Generally, the proportion of direct labor hours, direct labor costs, direct material costs, or total direct (prime) costs is used as the burden vehicle. The burden vehicle is just the denominator by which total indirect overhead cost is divided to yield the burden rate that is then applied to absorb the indirect overhead cost. The selection of the burden vehicle is based on its dominance in determining total product cost.

- Direct labor hours—large quantity of direct labor hours at a low hourly rate for manual labor.
- Direct labor costs—small quantity of direct labor hours at a high hourly rate for highly skilled or technical labor.
- Material costs—total material cost is much greater than direct labor cost.
- Prime costs (direct labor and material costs)—total direct labor cost approximately equals total material costs.

Overhead Cost Allocation Based on Direct Labor Hours

1. Determine base period total overhead cost and direct labor hours for the manufacturing unit.
2. Calculate the rate per direct labor hour.

$$\text{Rate} = \frac{\text{Total overhead cost}}{\text{Total direct labor hours}} \qquad (3.11)$$

3. Estimate the direct labor hours for a product.
4. Calculate the overhead cost for the job as

$$\text{Overhead cost} = \text{Rate estimated direct labor hours} \qquad (3.12)$$

Overhead Cost Allocation Based on Direct Labor Cost

1. Determine base period total overhead cost and total direct labor cost for the manufacturing unit.
2. Calculate the rate per direct labor cost.

$$\text{Rate} = \frac{\text{Total overhead cost}}{\text{Total direct labor cost}} \qquad (3.13)$$

3. Estimate the direct labor cost for a product.
4. Calculate the overhead cost for the job as

$$\text{Overhead cost} = \text{Rate estimated direct labor cost} \qquad (3.14)$$

Overhead Cost Allocation Based on Direct Material Cost

1. Determine base period total overhead cost and total direct material cost for the manufacturing unit.
2. Calculate the rate per direct material cost.

$$\text{Rate} = \frac{\text{Total overhead cost}}{\text{Total direct material cost}} \qquad (3.13)$$

3. Estimate the direct material cost for a product.
4. Calculate the overhead cost for the job as

$$\text{Overhead cost} = \text{Rate estimated direct material cost} \qquad (3.14)$$

Overhead Cost Allocation Based on Prime Cost

1. Determine base period total overhead cost, direct labor cost, and direct material cost for the manufacturing unit.
2. Calculate the rate per direct labor cost plus direct material cost.

$$\text{Rate} = \frac{\text{Total overhead cost}}{\text{Total direct material cost} + \text{Total Direct Material Cost}} \qquad (3.15)$$

3. Estimate the total direct labor and total direct material cost for a product.
4. Calculate the overhead cost for the job as

$$\text{Overhead cost} = \text{Rate estimated(direct labore} + \text{direct material)cost} \quad (3.16)$$

Example 3.2

Robotics Control, Inc. (RCI), manufactures computer control systems for industrial robots. The individual robot controller (IRC) computer is a generic personal computer with software customized to accommodate all robotic control languages and industrial robot types. The IRC computer can work as a stand-alone controller for an individual robot application or as a front-tend controller integrated into RCI's systems robot controller (SRC) computer system. SRC is a server-based system that runs all robotic control languages and can control multiple industrial robots directly or can act as a central controller for a computer controller distributed system. RCI's units produced, total direct labor cost, and total direct components cost for each controller are given in the following table. RCI's fixed overhead cost for the last fiscal year was $4,250,000. Estimate the per-unit cost of each controller allocating the fixed overhead cost by direct labor cost, direct components cost, and prime cost as the burden vehicles. Given the per-unit cost differentials, which fixed overhead allocation method is applicable to each controller type?

Units/cost	IRC	SRC
Units produced	7500	200
Total direct labor cost each	$450	$1100
Total direct components cost each	$500	$3900

Allocation by Direct Labor Cost:

	IRC	SRC	Total
Units produced	7500	200	
Total direct labor cost each	$450	$1100	
Total direct labor cost	$3,375,000	$220,000	$3,595,000

$$\text{Rate} = \frac{\$4,250,000}{\$3,595,000} = 1.1822$$

Overhead allocation for the base production period is

$$\text{Overhead(IRC)} = \$3,375,000 \cdot 1.1822 = \$3,989,917$$

$$\text{Overhead(SRC)} = \$220,000 \cdot 1.1822 = \$260,083$$

Cost/Unit Estimates:

Cost	IRC	SRC
Total direct labor cost	$3,375,000	$220,000
Total direct components cost	$3,750,000	$780,000
Allocated overhead	$3,989,917	$260,083
Total cost	$11,114,917	$1,260,083
Units	7500	200
Cost/Unit	$1481.99	$6300.42

Allocation by direct components cost:

	IRC	SRC	Total
Units produced	7500	200	
Total direct components cost each	$500	$3900	
Total direct components cost	$3,750,000	$780,000	$4,530,000

$$Rate = \frac{\$4,250,000}{\$4,530,000} = 0.93819$$

Overhead allocation for the base production period is

$$Overhead(IRC) = \$3,750,0000.93819 = \$3,518,212$$

$$Overhead(SRC) = \$780,0000.93819 = \$731,788$$

Cost/Unit Estimates:

Cost	IRC	SRC
Total direct labor cost	$3,375,000	$220,000
Total direct components cost	$3,750,000	$780,000
Allocated overhead	$3,518,212	$731,788
Total cost	$10,642,212	$1,731,788
Units	7500	200
Cost/Unit	$1,418.96	$8,658.94

Allocation by Prime Costs:

	IRC	SRC	Total
Units produced	7500	200	
Total direct labor cost each	$450	$1100	
Total direct components cost each	$500	$3900	
Total labor cost	$3,375,000	$220,000	
Total components cost	$3,750,000	$780,000	
Prime (total) cost	$7,125,000	$1,000,000	$8,125,000

$$Rate = \frac{\$4,250,000}{\$8,125,000} = 0.52308$$

Overhead allocation for the base production period is

$$\text{Overhead(IRC)} = \$7,125,0000.52308 = \$3,726,923$$

$$\text{Overhead(SRC)} = \$1,000,0000.52308 = \$523,077$$

Cost/Unit Estimates:

Cost	IRC	SRC
Total direct labor cost	$3,375,000	$220,000
Total direct components cost	$3,750,000	$780,000
Allocated overhead	$3,726,923	$523,077
Total cost	$10,851,923	$1,532,077
Units	7500	200
Cost/Unit	$1,446.92	$7615.39

Comparison of Cost/Unit by Allocation Methods:

Burden vehicle	IRC cost/Unit	SRC cost/Unit
Direct labor cost	$1481.99	$6300.42
Direct components cost	$1418.96	$8658.94
Prime cost	$1446.92	$7615.39

Given that the labor cost per unit ($450) approximately equals the components cost per unit ($500) for the IRC computer, allocating overhead using prime cost per unit provides a near average cost per unit that reflects both direct cost drivers. For the SRC server, the components cost per unit is the cost driver and should be used to allocate fixed overhead; however, prime cost per unit as a burden vehicle biases the SRC cost/unit upward also reflecting the components cost per unit. If the IRC computer and SRC server are manufactured on the same production line, prefer prime cost as the burden vehicle. If they are manufactured on separate production lines, prefer direct labor cost per unit as the burden vehicle for the IRC computer, and prefer direct components cost per unit as the burden vehicle for the SRC server.

3.7 Summary

Definition: *Cost accounting* is the discipline that captures an organization's total cost of production by assessing the variable costs of each production step, the associated fixed costs needed to support production, and the allocation of those fixed costs to each production step to determine standard production costs and product costs.

Definition: *Cost accounting* is the systematic recording and analysis of direct material and direct labor costs and allocation of indirect overhead costs to the production of an organization's products or services.

Definition: *Manufacturing cost* is the monetary consumption of labor, materials, tools, equipment, and facilities to increase the value of resultant products and services.

Definition: *Service* or *governmental cost* is the monetary value of resources that have been, or must be, consumed to yield a particular service.

Types of costs include

- **Cash costs**: movement of money from one owner to another—also known as a cash flow (engineering economy).
- **Accrued costs**: expenses which are incurred but for which no payment is made during an accounting period. They are shown in the balance sheet as a current (short term) liability.
- **Book cost**: cost of a past transaction that is recorded in an accounting journal.
- **Operating costs**: associated with the productive operations of an organization. Operating costs may be either fixed or variable.
- **Direct costs**: non-operating costs directly related to producing a product or service.
- **Indirect costs**: cannot be directly linked to a product or service but are required for productive activities to exist.
- **Controllable costs**: can be influenced or regulated by the manager responsible for it (i.e., budgeted costs for labor, materials, and processes, controlled by the manager).
- **Non-controllable costs**: those costs that a manager cannot affect or change (i.e., space rental cost, equipment lease costs, and insurance costs).
- **Period costs**: are not directly related to the productive activities. Overhead sales, general, and administrative (SG&A) costs are considered period costs.

A general framework for cost terminology:

- *Planning*: determines the economic feasibility of a new or upgraded productive facility or line, process, or product or service.
- *Control*: the management budgeting process of identifying and reducing business expenses with a focus on increasing profit.
- *Report*: a process used to inform management about the magnitude of a process's or product's budgeted versus predicted cost or a client about the magnitude of a project's budgeted versus actual cost.
- *Analysis*: the process of decomposing summarized costs into their components for the purpose if identifying cost drivers.

Cost objects include

- **Product**: the per-unit cost incurred to produce a product or service. Generally, product costs include labor, raw materials, indirect consumable manufacturing supplies, and general overhead.
- **Process**: the allocation of the total costs of production to homogenous units produced via a continuous process that usually involves multiple steps or departments.
- **Project**: all estimated and actual costs or monetary obligations incurred or estimated to be incurred to complete a project that has a specified beginning and end time, scope of resources, and objective.
- **Facility**: building maintenance, property taxes, insurance, cleaning services, net costs of cafeteria services, depreciation, utilities, security, parking, and all other non-capital expenditure costs relating to a facility as a whole or to the common areas of a facility.
- **System**: all costs incurred in connection with the financing, development, construction, care, custody, control, and retirement of a system.

Expected costs include

- **Fixed**—constant (relatively) over a time period or range of activity level.
- **Variable**—depends on activity level.
- **Overhead**—all operating costs that are not raw materials or direct labor.

$$C_T = C_F + \sum_U C_{V/U} \times U + \{C_{OH}\}$$

In manufacturing, cost components are also classified as prime costs, conversion costs, total cost, and unit cost.

- **Prime**—cost of direct material and direct labor to transform raw materials and components into finished products.
- **Conversion**—direct labor and manufacturing overhead costs.
- **Total**—indirect product costs + direct product costs + fixed product costs.
- **Unit**—total product cost/number of units.
- **Direct cost**—direct material and direct labor (prime costs) traceable directly to the manufacture of a particular product.
- **Indirect cost**—general administrative, marketing and sales, factory management and staff, indirect materials, and other fixed costs not traceable directly to a particular product.

Product costs are accumulated on either:

- **Full absorption costing**—direct material, direct labor, fixed overhead, and factory overhead.
- **Direct variable costing**—direct material, direct labor, and factory overhead.

The **average cost** = total product cost/number of units produced.

Incremental cost is the cost incurred to add another unit of activity or the difference in total cost between two alternatives.

Marginal cost is the change in the total cost that arises when the quantity produced changes by one unit.

$$\frac{dC_T}{dQ} = \frac{d\left(C_F + C_{\frac{V}{U}} \times U + \{C_{OH}\}\right)}{dQ} = \frac{dC_{\frac{V}{U}}}{dQ}$$

$$MC = \frac{TotalCost_2 - TotalCost_1}{Quantity_2 - Quantity_1}$$

Opportunity cost—a benefit that is foregone by engaging a resource in a chosen activity instead of engaging that same resource in the foregone activity.

Sunk cost—money spent due to a past decision. We cannot do anything about these costs.

A *cash flow diagram* graphically illustrates the magnitude, sign, and timing of individual cash flows over the lifetime of a project.

- **Breakeven**: total revenue = total costs. Zero profit.
- **Profit region**: total revenue > total costs. Positive profit.
- **Loss region**: total revenue < total costs. Negative profit or increasing debt.

Material and Component Analysis

- **Material/Component Cost**—quantities of materials or components that can be specifically identified with a product, contract, project, organizational subdivision, or function for which a unit price is to be determined.
- **Direct Material**—an input substance or information or service element to be altered into a component of the finished product or service.
- **Direct Component**—a designed grouping of partially altered raw materials, information, services, or other components which will be further altered into a component of the finished product or service.

$$\text{Cost direct material} = \text{unitshape} \times \frac{\text{cost}}{\text{unitshape}} - \text{salvage value}$$

Labor Analysis

- **Labor cost** is the wage or payroll per person-hour or fraction thereof that can be specifically and consistently assigned to or associated with the manufacture of a product, a particular work order, or provision of a service.
- **Direct labor** is the people who directly contribute to manufacture a specific product, complete a work order, or provide a service.
- **Indirect Labor** is the people who do not directly produce products or services, but who make their production possible or more efficient. Indirect labor costs are not readily identifiable with a specific task, work order, or service. They are termed indirect costs and are charged to overhead accounts.

$$\text{Labor cost} = \text{number time units worked} \times \frac{\text{wage}}{\text{timeunit}}$$

Overhead Cost Allocation

1. Determine base period total overhead cost and burden vehicle hours or costs for the manufacturing unit.
2. Calculate the rate per burden vehicle hours our costs.

$$\text{Rate} = \frac{\text{Total overhead cost}}{\text{Total burden vehicle hours or costs}}$$

3. Estimate the direct labor hours for a product.
4. Calculate the overhead cost for the job as

$$\text{Overhead cost} = \text{Rate estimated burden vehicle hours or cost}$$

3.8 Key Terms

Accrued cost
Actual cost
Average cost
Book cost
Breakeven
Burden vehicle
Cash cost
Controllable cost
Conversion cost
Direct component cost
Direct cost
Direct labor
Direct material cost
Direct variable cost
First cost
Fixed cost
Full absorption costing
Government cost
Historical cost
Indirect cost
Indirect labor
Joint costs
Loss region
Maintenance cost
Manufacturing cost

Marginal cost
Market value
Non-controllable cost
Operating cost
Opportunity cost
Overhead cost
Predetermined time
Period cost
Prime cost
Profit region
Replacement cost
Salvage value
Standard costs
Service cost
Sunk cost
Time standard
Total cost
Variable cost
Work sampling.

Problems

1. Chapman Automotive Remanufacturers refurbish 23,000 brake pads per month using only one daytime shift. CAR's fixed cost per month to support brake refurbishing is $200,000, and the total labor cost is $910,900. CAR is considering doubling its brake refurbishing capacity to 46,000 pads per month by adding a second shift. Second shift labor will require a 10% premium, and the fixed cost will increase to $240,000. (a) Estimate the current manufacturing cost and the cost per brake pad for the daytime shift operation. (b) Will adding the second shift increase or decrease the brake pad cost per unit?

2. Company A1 has total indirect fixed expenses of $150,000 per year, and each product unit has $2.00 per-unit variable cost. Company A2 has total indirect fixed expenses of $50,000 per year, and each product unit has $5.00 per-unit variable cost. What is the breakeven number of units for the two companies in comparison with each other?

3. Kitchen Gadget's assembly line produces 160 blenders per hour at a cost of $9,000 per hour on straight time (the first 8-h of work). Operators are guaranteed a minimum of 6 hours of work each day. Overtime is paid at 150% of straight time for each hour worked beyond straight time. Industrial engineering time study indicates that productivity drops by 2% for all hours worked after the first 6 hours and by 5% for all hours worked after 8 hours. Estimate the average and marginal cost per unit for the 6, 8, 9, and 10-h workdays.

4. Collate Commodities is considering adding a hand sanitizer production line. Industrial engineering has identified three alternative line configurations with the following fixed and variable cost over the relevant range of 1–50,000 unit/day

production. Determine the ranges of production (units/day) over which each alternative would yield the minimum total cost.

Alternative	Fixed cost	Variable cost/unit
A	$150,000	$ 7.00
B	$ 70,000	$14.00
C	$250,000	$ 4.00

5. A medium-size manufacturing company produces a product with fixed cost of $32,400 and variable cost of $15 per unit. The Marketing Department has determined that the quantity-demanded versus price relationship is $P = \$425 - \$0.35$/unit, where P = unit price. Using the following financial relationships,
    Total cost = Fixed cost + Variable cost/Unit Units
    Revenue = Price Units sold
    Profit = Revenue—Total cost
    Estimate the (a) total cost equation and total revenue equation, (b) breakeven quantities, (c) the number of units sold that maximizes total revenue and the maximum total revenue, and (d) the number of units sold that maximizes net profit and the maximum net profit.

6. An air handling unit has failed and must be replaced for a cleanroom production area that itself will be replaced in 5 years. An equivalent air handling unit, with an expected life of 5 years, can be purchased and installed for $12,000. However, the maintenance shop has a refurbished unit of greater air handling capacity in stock. The refurbished unit cost $23,000 new, but the accounting department indicates that its current value is $14,000 today. The maintenance manager indicates that it will cost $1000 to reconfigure the refurbished unit for the cleanroom application. He also says that the refurbished unit can be sold for $10,000 market value. (a) What is the book cost of the refurbished unit? (b) What is the opportunity cost of the refurbished unit? (c) What is the cost differential to install the refurbished unit over purchasing the equivalent unit?

7. The following work standard table sets for the standard times (minutes) for a machining operation. The operation has an 11% fatigue allowance for an 8-h shift. Estimate the standard minutes, absolute minimum case minutes, absolute maximum case standard minutes, and the 95% prediction interval for a completed machined piece. What is the expected number of units in for an 8-hour shift with two 10-min personal breaks?

Seq. No.		Element frequency	Unit normal time	Range	Standard deviation
1	WSP	1	0.048	0.02	0.02256
2	C15C	1	0.159	0.04	0.03546

(continued)

(continued)

Seq. No.		Element frequency	Unit normal time	Range	Standard deviation
3	W10P	2	0.093	0.04	0.03546
4	C10C	1	0.115	0.04	0.03546
5	W15P	3	0.140	0.04	0.03546
6	CSC	1	0.070	0.02	0.02256

8. The following table summarizes the man-hour reports for the last calendar year. Estimate the average, standard deviation, and 95% prediction interval for transit time, setup time, and work time.

Work order	Date	Transit start	Transit end	Setup start	Setup end	Work start	Work end
WO-059	3-Feb	9:44 AM	9:55 AM	10:00 AM	11:44 AM	11:44 AM	2:20 PM
WO-060	4-Mar	3:39 PM	3:45 PM	3:45 PM	5:42 PM	5:54 PM	7:04 PM
WO-061	18-Mar	6:19 PM	6:24 PM	6:26 PM	7:17 PM	7:20 PM	9:51 PM
WO-062	3-Apr	10:59 AM	11:14 AM	11:20 AM	12:12 PM	12:12 PM	1:28 PM
WO-063	10-May	9:12 PM	9:19 PM	9:19 PM	10:56 PM	11:10 PM	12:23 AM
WO-064	20-May	10:31 PM	10:41 PM	10:44 PM	11:29 PM	11:42 PM	12:13 AM
WO-065	10-Jun	8:41 PM	8:51 PM	8:55 PM	10:03 PM	10:03 PM	12:42 AM
WO-066	8-Jul	10:22 AM	10:30 AM	10:31 PM	10:51 PM	11:07 PM	1:47 AM
WO-067	9-Jul	8:55 PM	8:57 PM	9:03 PM	10:12 PM	10:12 PM	11:36 PM
WO-068	25-Aug	5:13 PM	5:23 PM	5:23 PM	6:45 PM	6:54 PM	9:56 PM
WO-069	12-Sep	10:56 AM	11:03 AM	11:05 PM	11:18 PM	11:18 PM	11:39 PM
WO-070	15-Dec	9:37 AM	9:49 AM	9:49 AM	10:18 AM	10:24 AM	12:33 PM

9. Best Aluminum, Inc., manufactures aluminum tubing. Total direct labor hours, direct labor cost, direct material cost are given in the following table. Use direct labor hours, direct labor cost, direct material cost, and prime cost as the burden

vehicle to allocate overhead of $18,592,000 for the past fiscal year, and estimate the cost for unit of each aluminum tube product.

Item	Product A	Product A+	Product A++
Direct labor hours	128,000	40,000	64,000
Direct labor cost	$1,200,000	$760,000	$820,000
Direct material cost	$7,600,000	$3,060,000	$4,210,000
Units/year	200,000	100,000	64,500

Chapter 4
Cost Estimating Fundamentals

Abstract Cost engineering arises from need for engineering managers to act as stewards of organizational and project resources. Professional Certified Cost Engineers are often required for organizations to bid on US government or military contracts. Cost engineering is generally recognized as the application of engineering principles, techniques, judgment, and experience to cost estimation, engineering function or project planning and scheduling, and cost control for the purposes of contributing to organizational profitability or the measurement and management of project costs throughout project life cycles. This chapter provides an introduction to engineering cost estimating fundamentals. First, the chapter will set forth a definition of cost estimating and the reasons for cost estimates. Next, the chapter will provide an overview of cost models in the design maturity process. The cost estimating process will be described as a general level of detail. The fundamentals of product and operations costing will be presented. Finally, widely used cost estimating models will be defined with example calculations.

4.1 Introduction

As established in Chap. 2, introduction to accounting fundamentals, all production and service organizations are driven by the need to generate a positive return on investment (ROI) to cover the cost of borrowing money from lenders (interest rate) and obtaining investment money from stockholders (dividend rate). Governmental organizations are driven by the need to promote the general welfare of the citizens they serve (at least theoretically) through the efficient allocation of taxes and fees collected to public projects and services. Even nonprofit organizations must maintain a positive cash flow balance between short-term income and long-term debt. To remain economically viable, each of these organizational types must estimate and predict costs in budgeting their operations.

Definition: *Cost estimating* is the approximation of the probable future cost of a product, program, or project, computed on the basis of currently available information.

Cost estimates are made for a variety of reasons. The most common include:

- *Investment decisions*—accurate cost estimates are essential to determine the potential return on investment.
- *Comparing alternative investments*—selecting the investment or set of investments that yield a return on investment greater than the minimum attractive rate of return (MARR) given an available spending budget.
- *Deciding on which products should be produced or services delivered*—detailed cost estimates assist management in making maximizing return on investment development decisions in the selection of new products and in setting the mix of current products to produce.
- *Determining selling price*—the selling price must cover the indirect, direct, and fixed costs necessary to produce a product or service plus sufficient profit to allow reinvestment in maintenance and growth of the organization and cover debt costs and stock dividends.
- *Validating suppliers and contractors price quotes*—cost quotations must be validated to assure that all partied understand the scope of work and accurately price for materials, components, or services.
- *Make-or-buy decisions*—accurate cost estimates are essential to assist management in deciding whether to make components and modules internally or to purchase them from vendors who specialize in their production and can make them at a low per unit.
- *Estimating temporary work standards*—temporary work standards are usually necessary in the early production of new products until time–motion studies can establish permanent standards.
- *Cost control*—to fulfill its responsibility to be a good steward of stockholders' investments, management must be able to assess actual versus budgeted costs, interpret the effects (positive or negative) of variances, and adjust strategy and project or production plans to maintain the required return on investment. Accurate project and operational budgets require accurate cost estimates.
- *Performance evaluation and benchmarking*—during project or operations execution, management must also benchmark its performance against that of competitors or industry standards. Positive or negative variances indicate project or organizational health and viability.

As illustrated in chapter one, engineering cost estimation arises in development and design activities and continues through the product life cycle. Table 4.1 gives the fundamental classification of cost estimates.

Table 4.1 Cost estimates classification

Category	Fundamental characteristic	Measure
Product	Cost/unit and price/unit	Price—[(liabilities + capital)/unit] that achieves a stated level of performance
Operation	Worker, tool, equipment, space costs	Cost—consumption of labor, materials, and tools to increase the value of some object
Project	Project deliverable cost	Bid—$[C_{DL} + C_{DM} + C_{Cap} + C_{Eng} + C_{Oh} + C_{Int} + C_{Profit}]$ to win a project
System	Configuration cost	Effectiveness = Max(Benefits/Costs)

The cost estimate of an operation is a forecast of labor and material or components necessary to produce a component, module, or product at a transformation point. Labor may be further classified as unskilled, skilled, craftsman, apprentice, journeyman, or professional requiring different pay or wage grades. Depending on traceability to the product, operations labor may be direct or indirect. Cost estimating for a product accumulates the operations costs necessary to produce one unit of finished product. The purpose of product cost estimation is to set its unit price such that its value per-unit price is positioned against competitors' products yielding sufficient profit to allow reinvestment in maintenance and growth of the organization and cover debt costs and stock dividends. Project cost estimation predicts the quantity and cost of the resources required by the scope of a project. The purpose of project cost estimation is to produce a bid low enough to win the project contract but still cover all project costs. Systems cost estimation is the configuration of design, project, operations, and product flow costs to maximize systemic effectiveness as measured by the benefit–cost ratio. The fundamental element of a system is the configuration of its material, labor, product, and information flows within its interacting operations. Any system may have combinatorically multiple configurations; hence, systemic configurations benefit–cost ratios provide a normalized comparison of systemic effectiveness.

4.2 Cost Estimating Accuracy

As shown in Fig. 2.1, the focus of economic analyses of engineering projects is on predicting future cash flows. Prediction of future data always involves inherent uncertainty in the estimates. In the case of engineering projects, variance is induced in revenue and cost estimates, interest rates, stock dividend rates resulting from political, macroeconomic, and business economic uncertainty. Prediction uncertainty in estimated future cash flows is only as accurate and precise as the organization's existing financial data and records and external estimates of macroeconomic and business economic forecasts.

Definition: *Risk* is observed in those situations in which the potential outcomes can be described by well-known probability distributions.

Definition: *Imprecision* is observed in those situations in which the potential outcomes cannot be described by well-known probability distributions but can be estimated by subjective probabilities.

Definition: *Uncertainty* is observed in those situations in which the potential outcomes cannot be described by well-known probability distributions and cannot be estimated by subjective probabilities.

Definition: *Accuracy* is the difference between an average or median estimated cash flow at any life cycle design phase and its actual realized cash flow.

Definition: *Precision* is (1) the best-fit variance distribution under risk, (2) fuzzy membership or rough set membership under imprecision, or (3) fuzzy membership, rough set membership, or allowable greyness under uncertainty.

Thus, it is incumbent on the engineering or project manager to progressively improve the accuracy and precision of cash flow estimates as design moves through the first three phases of the product, process, service, or system life cycle in Fig. 1.4. The following accuracy of cash flow estimates is reported as observed in associated life cycle phases.

Needs Assessment and Justification—Rough estimates are composed of educated guesses from a high-level macro-feasibility perspective.

Conceptual Design—Refined estimates based on historical accounting and financial performance records and on the latest macroeconomic and business economic data. These estimates are reasonably accurate for initial budgeting purposes.

Detailed Design—Further refined estimates based on current revenue and cost cash flows of detailed design specifications. These estimates are used for contract bids.

Typically, the accuracy and precision of investment cash flow estimates must proceed from very rough to very refined as design proceeds through the first three life cycle phases.

Figure 4.1 shows the information and cost refinement as designs mature through the first three life cycle phases. The initial recognition of the problem results in a vague statement of a need to satisfy some gap in performance. At this stage, only rough technical, non-technical, and cost data bounds are necessary to answer the question of whether it is feasible to seek a solution to the problem. If the answer to the issue of feasibility is affirmative, technical and non-technical design concepts and their associated costs are refined into semi-detailed estimates. The semi-detailed estimates need to be only sufficient to filter out non-feasible designs. Retained designs may then be modeled and prototype tested based on ranked feasibility strength. The physical and technological constraints are traded off against economic and time constraints through design simulations and prototype performance testing. Ranking based on feasibility strength permits the organization to invest scarce development money efficiently in designs with the highest probability of success. Cost estimates must now

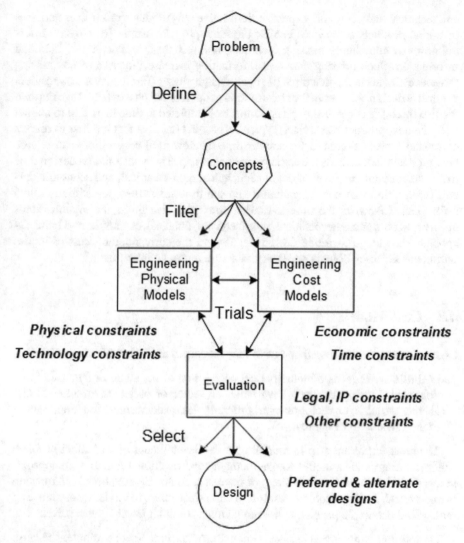

Fig. 4.1 Cost estimates design maturity

be bounded with sufficient accuracy to permit return on investment estimates. Tightly bounded detailed design estimates are necessary to minimize the risk of not attaining the required return on investment. Ultimately, the final design must be evaluated for legal, regulatory, intellectual property, and other non-technical constraints. The final cost estimate accuracy boundaries must be refined to the accuracy range necessary to seek external debt and stock financing and commit those funds to the construction of assets to produce the design.

Cost estimating in the first three life cycle design phases is further complicated by innovative new products or services, one-of-a-kind designs, the time-to-market, and technical and economic expertise. Innovative new products or services and one-of-a-kind products or services exhibit the same problem of novelty. Novelty infers unknown or completely missing technical, non-technical, and economic data and information. Such missing data and information increase the risks of committing large investments in the first three life cycle design phases (Fig. 1.5) that may result in expensive design changes in the construction or operational phases (Fig. 1.6). Further, the first three life cycle design phases must be completed in time to be first to market with the new product or service. Typically, product releases that are first to market command a profit premium for a short time window until competitors can release new products that address the same performance gap. This constraint on design time limits the amount and level of detail of technical, non-technical, and economic data and information that can be gathered and the time for review, re-estimation, and refinement. Likewise, the time-to-market constraint also limits the organization's ability to access or create technical, non-technical, financial, economic, and industry-specific knowledge to refine design details and the corresponding cost estimates within the accuracy bounds necessary to assure return on investment.

4.3   Cost Models

Cost estimates are expressed and cash flow models of cost components over time.

> **Definition**: *Model* is a mathematical description of the static or dynamic functional relationship between a variable, or vector of variables, of interest and another variable, or vector or matrix of variables that determine the functionality of the variable(s) of interest.

The most important step in modeling is the development of a succinct problem statement that yields a model whose mathematical functional relationship approximates the behavior of the variable(s) of interest such that the functional relationship is understood, and the problem resolution will be effective. In modeling engineering managerial economic problems, the model must admit all feasible alternatives:

- Do nothing; do not invest in any alternative product, process, or equipment because each return on investment is less than the MARR.
- Patch and delay the current product, process, or equipment. An alternative's return on investment justifies replacement, but external or internal economic constraints prohibit the investment.
- Continue with the current product, process, or equipment and re-evaluate the investment decision at a later time.
- Upgrade the current product, process, or equipment
- Purchase a replacement for the current process or equipment.
- Innovate a new product, process, or equipment.

Once the model's functional relationship is sufficiently described, the next problem is to gather relevant data. For engineering managerial economic analyses, relevant data includes:

- Hard-to-acquire as well as easily assessable data.
- Accounting, financial, macroeconomic, industry, engineering, sociotechnical, regulatory, and political data. The later six contribute primarily to risk assessment. This book is concerned primarily with the first two types of data.
- Market data.
- Intangible consequences data.

The final model must integrate the relevant data into the mathematical relationship such that:

- The functional relationship approximates physical and economic reality;
- All physical, economic, sociotechnical, regulatory, and political constraints are observed;
- The relevant data span the modeling space;
- The model admits all feasible alternative outcomes; and
- Selection criterion assists the decision-maker in choosing the "best" alternative.

Model validity is the degree that a model achieves these objectives. Finally, in engineering managerial economics, models must observe engineering principles, physical laws, and economic and financial theory.

By cost estimating model, we seek to establish *Cost estimating Relationships* (CERs) or *Time estimating Relationships* (TERs).

Definition: *Cost estimating relationship* is a model that estimates a cost or price by using an established relationship with one or more independent variables.

Definition: *Time estimating relationship* is a model that estimates activity duration by connecting an established relationship with one or more independent variables to the duration time of the activity.

In engineering managerial economics, we seek to identify the best-fit parameters that accurately describe the CER or TER {e.g., in $C_T = C_F + U \times C_V/U$, we seek to identify the values of the fixed cost and variable cost per-unit parameters that best estimates the total cost}.

4.4 Cost Estimating Process

Cost estimating is a managerial engineering process. Accordingly, the cost estimating process must observe managerial and engineering theory, principles, and best practices. Based on a survey of industry and governmental cost estimating processes, this work recommends the following cost estimation process:

1. **Define the Cost Estimate Scope**—determines the cost elements and structure of the model.
2. **Identify Assumptions and Constraints**—assumptions limit the scope of the model, and constraints are limiting boundaries.
3. **Develop Cost Element Structure**—all revenue and costs must be included, and none duplicated.
4. **Collect and Normalize Cash Flow Data**—apply relevant monetary exchange rates and adjust for inflation to the project base zero year. Normalization ensures that revenue and cost data are comparable across alternatives and time.
5. **Develop Cost Estimating Relationships**—equations that fully describe the cost model.
6. **Select the Discounted Cash Flow Criterion**—present worth, equivalent uniform annual worth, future worth, internal rate of return, external rate of return, benefit–cost analyses, or cost-effectiveness. The decision criterion must reflect the decision-maker or customer's economic drivers.
7. **Compute the Cost Estimate**—compare the criterion value against the decision-maker or customer's economic requirement.
8. **Document and Present the Cost Estimate**—only the decision-maker or customer can make the final decision to invest or not invest in the alternative.
9. **Audit Actual Life Cycle Cash Flows Versus Model Cash Flow Estimates**—explain all variances and use those explanations to improve future cost estimation models.
10. **Develop a Cost Estimation Database**—engineering and project managers use a cost database to store cost and model data in structured way which is easy to manage, retrieve, and apply to the formulation of future cost estimates.

4.5 Product Costing

Definition: *Product costing* is the estimation of the cost of a unit of product (or service delivered) through the determination or allocation of all expenses related to the creation of the product (or service).

Information required for product cost estimates include:

- Engineering bill of materials, specifications, and designs.
- Due date for completion of the estimate.
- Quantity, rate of production, and schedule for the product.
- Special test, inspection, and quality control requirements.
- Packaging, warehousing, and shipping instructions.
- Marketing information.

A *bill of materials* (BOM) is a complete list of all the materials, subcomponents, components, subassemblies, and assemblies and the exact quantities of each necessary to produce one unit of a finished product. There are two fundamental types of bill of materials. An *engineering bill of materials* (EBOM) defines one unit of finished product as it was originally designed. An EBOM may be generated by design or product engineers depending on the organizational size and complexity. An EBOM ensures that purchasing agents acquire the correct direct materials, subcomponents, components, and subassemblies from suppliers and have them in stock at the time of manufacture. A *manufacturing bill of materials* (MBOM) specifies all the direct and indirect materials, supplies, subcomponents, components, subassemblies, assemblies, and packaging materials required to build a unit of shippable product. Typically, and MBOM will have a valid date range to reflect a stable revision of the product. A correctly designed, hierarchical BOM will contain the following essential descriptions.

- **BOM level**—a number or ranking that codes where a unit of material or a part fits in the BOM hierarchy.
- **Part number**—a code that uniquely identifies each material or part and allows manufacturing personnel to reference and identify materials and parts throughout the production cycle.
- **Part name**—a detailed, unique name that allows manufacturing personnel to identify the part easily without having to reference the part number or engineering specifications.
- **Phase**—the life cycle stage of each material or part in the BOM; development, prototype, unreleased, released, production, or obsolete.
- **Description**—a narration of each material or part to aid in identification and distinguish between similar materials and parts.
- **Quantity**—the number or units of each material or part used in each manufacturing step.
- **Unit of measure**—quantification unit of material or part (e.g., each, cubic liter, gram, etc.). Quantity information assures that correct quantities are purchased, stocked, and delivered to manufacturing.
- **Procurement type**—source such as make or buy, commodity, commercial-off-the-shelf (COTS), modified-off-the-shelf (MOTS), or purchased according to engineering specifications.
- **BOM notes**—all additional information necessary for those who will use the BOM.

The last five elements required for product cost estimates are coordinated in a *materials requirements planning* (MRP) system. An MRP system is a software application that assists management in purchasing materials and parts, acquiring and staging resources, controlling inventory, and planning and tracking of production activities. MPR systems first input is independent demand by customers for finished products. The MRP system then uses BOM information to decompose the independent demand into internal dependent demand for packaging materials, finished product, assemblies, subassemblies, components, subcomponents, supplies, and direct and indirect materials working backwards through the manufacturing process to suppliers.

In product costing, only the materials, supplies, and resources that are required to produce a unit of finished product are included. How product costs are summarized costing method selected. The three primary product costing techniques are:

- Full absorption costing—tracing all direct and indirect manufacturing-related costs and allocating all indirect overhead costs for the product and absorbing them into the cost per finished product unit.
- Variable costing—variable costs are counted in the cost of a product unit. Fixed manufacturing costs are treated as period costs and expensed in the period they are incurred. The accumulated variable costs are divided by the number of product units produced to estimate the cost per finished product unit.
- Activity-based costing—accumulates the indirect, direct, and fixed costs of each activity in the production of on unit of finished product. The sum of activity costs divided by the number of finished units to estimate the cost per finished product unit.

Since it is required by GAAP and US tax law, only full absorption costing will be considered in this text.

The finished product is decomposed into individual components in a bill of material (BOM) tree. Starting at the lowest level cost estimates are made for each material or component, individual material and component costs are multiplied by respective quantities required, and the estimates are summed as the material cost for the next higher level in the product hierarchy.

$$M_{hi} = \sum_i Q_i(M_i + H_i \times \text{PHC}_i) \tag{4.1}$$

where M_{hi} = material cost for next higher assembly, Q_i = quantity of component i, M_i = unit material/component cost, H_i = hours per unit, and PHC_i = unit productive hour cost (standard operation cost). The unit productive hour cost is the sum of standard direct labor cost, standard direct equipment cost, and allocated cost per unit of space. PCH_i is estimated as:

$$\text{PCH}_i = \sum_j C_{(sdl)ij} + \sum_k C_{(sdm)ik} + C_{spi} \tag{4.2}$$

where $C_{(sdl)ij}$ = standard unit cost direct labor j for product i, $C_{(sdm)ik}$ = standard unit cost direct material or component k for product i, and C_{spi} = allocated direct unit cost of the manufacturing space for product i as established by industrial engineering studies.

Example 4.1

Below is the bill of material tree for the X-512 small, unmanned aircraft. Using Eq. (4.1), hierarchically compute the per-unit cost of each X-512 unmanned aircraft.

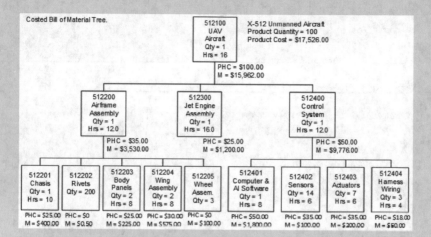

The costed BOM for X-512 small, unmanned aircraft:

Level 1,2,3	PN, RV	Description	UM BOM	Unit Qty	O P	Unit Matl Cost	Hr/Unit	PHC	Unit Labor Cost	Total Unit Cost	Assem. Matl Cost
1	512100	X-512 UAV	Ea	1	M	$15926	16.0	$100	$1600	$17526	—
2	512200	Airframe Assembly	Ea	1	M	$3530	12.0	$35	$420	$3950	$3950
3	512201	Chassis	Ea	1	M	$400	10.0	$25	$250	$650	$650
3	512202	Rivets	Ea	200	B	$0.5	0.0	$0	$0	$0.5	$100
3	512203	Body Panels	Ea	2	M	$225	8.0	$25	$200	$425	$850
3	512204	Wing Assembly	Ea	2	M	$575	8.0	$30	$240	$815	$1630
3	512205	Wheel Assembly	Ea	3	B	$100	0.0	$0	$0	$100	$300
2	512300	Jet Engine	Ea	1	M	$1200	16.0	$25	$400	$1600	$1600
2	512400	Control System	Ea	1	M	$9776	12.0	$50	$600	$10376	$10376
3	512401	Computer/Software	Ea	1	M	$1800	8.0	$50	$400	$2200	$2200
3	512402	Sensors	Ea	14	M	$100	6.0	$35	$210	$310	$4340
3	512403	Actuators	Ea	7	M	$200	6.0	$35	$210	$410	$2870
3	512404	Harness Wiring	Ea	3	M	$50	4.0	$18	$72	$122	$366

Harness wiring assembled cost = $3(\$50 + 4.0 \times \$18) = 3(\$50 + \$72) = 3(\$122) = \$366$.

Control system material cost = $\$366 + \$2870 + \$4340 + \$2200 = \$9776$.

Control system assembled cost = $1(\$9776 + 12.0 \times \$50) = 1(\$9776 + \$600) = \$10,376$.

Level 1,2,3	PN. RV	Description	UM BOM	Unit Qty	O P	Unit Matl Cost	Hr/Unit	PHC	Unit Labor Cost	Total Unit Cost	Assem. Matl Cost
2	512400	Control System	Ea	1	M	$9776	12.0	$50	$600	$10376	$10376
3	512401	Computer/Software	Ea	1	M	$1800	8.0	$50	$400	$2200	$2200
3	512402	Sensors	Ea	14	M	$100	6.0	$35	$210	$310	$4340
3	512403	Actuators	Ea	7	M	$200	6.0	$35	$210	$410	$2870
3	512404	Harness Wiring	Ea	3	M	$50	4.0	$18	$72	$122	$366

Jet engine assembled cost $= 1(\$1200 + 16.0 \times \$25) = 1(\$1200 + \$400) = 1(\$1600) = \$1600$.

Level 1,2,3	PN. RV	Description	UM BOM	Unit Qty	O P	Unit Matl Cost	Hr/Unit	PHC	Unit Labor Cost	Total Unit Cost	Assem. Matl Cost
2	512300	Jet Engine	Ea	1	M	$1200	16.0	$25	$400	$1600	$1600

Wing assembly assembled cost $= 2(\$575 + 8.0 \times \$30) = 2(\$575 + \$240) = 2(\$815) = \$1630$.

Airframe assembly material cost $= \$300 + \$1630 + \$850 + \$100 + \$650 = \$3530$.

Airframe assembly cost $= 1(\$3530 + 12.0 \times \$35) = 1(\$3530 + \$420) = \$3950$.

Level 1,2,3	PN. RV	Description	UM BOM	Unit Qty	O P	Unit Matl Cost	Hr/Unit	PHC	Unit Labor Cost	Total Unit Cost	Assem. Matl Cost
2	512200	Airframe Assembly	Ea	1	M	$3530	12.0	$35	$420	$3950	$3950
3	512201	Chassis	Ea	1	M	$400	10.0	$25	$250	$650	$650
3	512202	Rivets	Ea	200	B	$0.5	0.0	$0	$0	$0.5	$100
3	512203	Body Panels	Ea	2	M	$225	8.0	$25	$200	$425	$850
3	512204	Wing Assembly	Ea	2	M	$575	8.0	$30	$240	$815	$1630
3	512205	Wheel Assembly	Ea	3	B	$100	0.0	$0	$0	$100	$300

X-512 UAV material cost $= \$10,376 + \$1600 + \$3950 = \$15,926$.

X-512 UAV assembled cost $= 1(\$15,926 + 16.0 \times \$100) = 1(\$15,926 + \$1600) = \$17,526$.

Level 1,2,3	PN. RV	Description	UM BOM	Unit Qty	O P	Unit Matl Cost	Hr/Unit	PHC	Unit Labor Cost	Total Unit Cost	Assem. Matl Cost
1	512100	X-512 UAV	Ea	1	M	$15926	16.0	$100	$1600	$17526	
2	512200	Airframe Assembly	Ea	1	M	$3530	12.0	$35	$420	$3950	$3950
2	512300	Jet Engine	Ea	1	M	$1200	16.0	$25	$400	$1600	$1600
2	512400	Control System	Ea	1	M	$9776	12.0	$50	$600	$10376	$10376

To make a product decision, the product cost estimate (from the BOM), quantity, rate of production, and schedule are summarized into:

1. A cash flow statement.
2. Rate of return analysis.
3. Income statement.

This text focuses on developing the product cash flow statement. The income statement was discussed in lecture two. Rate of return analysis will be discussed in lecture six. The product cash flow statement summarizes the value of money flowing into and out of an organization from the production of the given product. The fundamental product cash flow equation is:

$$F_p = (R_p - C_p - D_p)(1 - \text{TR}) + D_p \qquad (4.3)$$

where F_p = total source or use of money (after-tax cash flow) per year, R_p = revenues from the sale of the product, C_p = cost to manufacture the product, D_p − depreciation cost due to the investment of capital related to the product, and TR = applicable tax rate. As will be discussed in Chap. 7, depreciation is a non-cash expense that reduces tax expense at the tax rate, hence the addition of the depreciation amount D_p to after-tax cash flow.

4.6 Operation Costing

Definition: *Operation costing* is the estimation of the total hourly cost of labor plus the operation hourly rate for indirect materials, tools, equipment/space, utilities, and factory overhead to operate a transformation point in value-increasing processes over a stated time frame.

Example 4.2

Assume that for the X-512 small, unmanned aircraft the firm needs to realize a profit sufficient to cover its 16% MARR (lecture two). The price per unit was estimated by trial and error in a spreadsheet to be $22,198 to achieve the 16% MARR. Also assume straight-line depreciation with a 5-year life and $0 salvage value (to be discussed in Chap. 7) and that the firm has a combined federal and state tax rate of 30% (to be discussed in lecture 8). The cash flow statement from the spreadsheet analysis for a five-year product life is shown below.

X-512 Small Unmanned Aircraft Product Costing						
	MARR	16.0%		Unit Price	$22,198	(Trial)
	Tax Rate	30.0%		Unit Cost	$17,526	
Cash Flow Statement						
	Year 0	Year 1	Year 2	Year 3	Year 4	Year 5
Initial assets cost	($700,000)					
Quantity		25	50	75	75	50
Revenue		$554,950	$1,109,900	$1,664,850	$1,664,850	$1,109,900
Cost (x $17,526)		$438,150	$876,300	$1,314,450	$1,314,450	$876,300
Depreciation		$140,000	$140,000	$140,000	$140,000	$140,000
Taxable cash flow		($23,200)	$93,600	$210,400	$210,400	$93,600
After tax income		($16,240)	$65,520	$147,280	$147,280	$65,520
Add depreciation		$140,000	$140,000	$140,000	$140,000	$140,000
After tax cash flow	($700,000)	$123,760	$205,520	$287,280	$287,280	$205,520
IRR	(16.00%)					

Operation costing is a microeconomic analysis that subdivides a facility into transformation points, activities needed to complete each transformation, and the motion tasks required to complete each activity. Transformations are a change in product value through the designed integration of labor, materials or components, energy, and information at each activity. The measure of the change of product economic value through the combination of inputs at each activity is measured as cost. Operation costing is limited as to their valid time horizon because they are sensitive to:

- *Time frame*—general and commodity-specific inflation affects operations costs.
- *Product mix*—differing mixes will demand differing combinations of materials, labor, and factory overhead.
- *Technology mix*—the increasing application of technology may increase or decrease operations costs depending on changes in efficiency and yields.

Figure 4.2 shows how the product mix/technology mix over time drives changes in PCH. To remain efficient and competitive, organizations must innovate new products and obsolete old products. The demand for new products drives the demand for new technologies, and new technologies force changes in PHC components (overhead, utilities, equipment/space, materials, and labor) to support each new technology mix. Conversely, constraints in PHC components limit technology mix, innovation, product mix, and competitiveness.

The general process for determining operation costs is as follows.

1. Determine the future time frame over which the PHC of the operation will be applied (determines the actual dollar value given inflation and the technology mix).
2. Determine standard categories of product mix.
3. Determine the operation input and output boundaries and process flow.
4. For each product mix category, estimate the productive hour cost PHC.

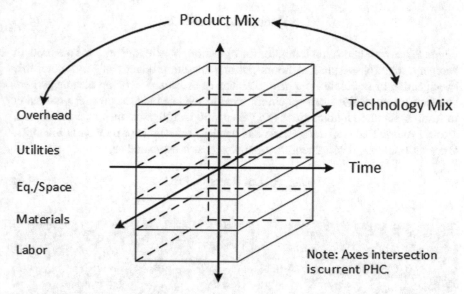

Fig. 4.2   Time horizon effects of operations costs and PHC

$$\text{PHC} = \text{SUM}(\text{Labor, Id. Materials, }[\text{Eq./Space}]\text{, Utilities, Factory OH})$$
$$/\text{HrPHC} = \text{SUM}(\text{Labor, Id. Materials, }[\text{Eq./Space}],$$
$$\text{Utilities, Factory OH})/\text{Hr} \qquad (4.4)$$

Figure 4.3 shows the typical flow of inputs into, through, and out of a value-added transformation process and the relationship of PHC components (overhead, utilities, equipment/space, materials, and labor) to needed to support the interacting activities in the process flow.

Once the productive hour cost standards have been determined, the operations cost for a given product category can be estimated. First, lot hours are found using the relationship,

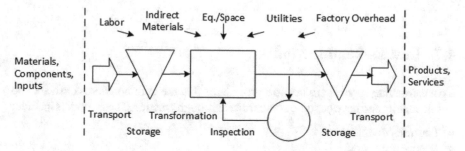

Fig. 4.3   PHC components in support of a typical transformation process

$$\text{Lot Hours} = \text{SU} + N \times H_s \qquad (4.5)$$

where SU = standard setup hours for the operation, N = lot quantity for the product category, and H_S = standard hours per unit for the product category. Note: this specification of Lot Hours is a generalization of the initial job shop meaning where standard lot quantities were specified. Even in the chemical or powder processing industries, standard lot quantities (55 gallon, 500 L, 100 cubic meters, 1 cubic yard, 1 ton, 1 metric ton, etc.) are specified as standard sales units or transportation units. Once lot hours are determined, the operation cost is estimated as,

$$C_{\text{opr}} = \text{Lot Hours} \times \text{PHC} \qquad (4.6)$$

Example 4.3
A numerically controlled, vertical end mill is used to automatically mill pump bases. The standard lot size is 200 bases. The operation has a standard setup time of 1.8 h, standard production hours per 100 units of 5.25 h, and PHC = $27.85. The unit material cost is $7.38. Estimate the lot hours, operation cost, and cost per unit.

$$
\begin{aligned}
\text{Lot Hours} &= \text{SU} + N H_S \\
&= 1.8 + 25.25 \\
&= 12.3 \text{ h } C_{\text{opr}} \\
&= \text{Lot HoursPHC} \\
&= 12.3\$27.85 \\
&= \$342.56 \text{ Cost per unit} \\
&= \text{Mat'l Cost} + \text{Copr/Units} \\
&= \$7.38 + \$342.56/200 \text{ units} = \$9.10
\end{aligned}
$$

4.7   Cost Estimating Models

Cost estimating models are costing procedures that are common across private and public sectors and have become standardized through practice. These models include:

- Per-unit Model
- Segmenting Model
- Cost Index Model
- Power-sizing Model

- Triangulation
- Learning Curve Model.

Per-unit Cost Model

Per-unit model costing uses a "per-unit" factor such as $/sq ft, cost/customer, benefits/employee, vehicle cost / mile, etc. Per-unit cost is based on a strong relationship between the unit cost and the total cost.

$$C_U = \frac{\sum C_i}{\sum n_i}$$

$$\text{Estimated cost} = C_U \times \text{Units} \tag{4.7}$$

Example 4.4

Agro Foods, Inc., needs to add a 10,000 square foot flash freezer warehouse. A pre-engineered steel enclosure will cost $13.60/sqft, whereas an insulated concrete form enclosure will cost $14.20/sqft. Spray foam insulation for the steel building will cost $1.50 per linear foot. With a 12 ft roof, there will be 4800 linear feet of walls to cover with spray foam insulation. Additional insulation will not be required for the insulated concrete form enclosure. Costs for the concrete slab and roofing insulation are the same for either enclosure type. What is the minimum total cost alternative?

$$\text{Steel enclosure} = \$13.60/\text{sqft } 10,000 \text{ sqft} + \$1.50/\text{ft } 4800 \text{ ft}$$
$$= \$143,200$$

$$\text{Insulated concrete enclosure} = \$14.20 \; 10,000 \text{ sqft} = \$142,000$$

Segmenting Cost Model

The *segmenting cost model* decomposes a new product into its individual components and assemblies, obtains or estimates the cost of each component (which typically can be obtained or estimated by suppliers), and sums the component costs into subassembly costs and subassembly costs into product cost per unit.

Example 4.5

A yard equipment manufacturer is planning to introduce a new general-use mini tractor with which attachments can be purchased for mowing lawns, brush hogging, backhoeing, and grading. The Accounting department has requested material cost estimates from the project engineer. The material cost estimate will be combined with labor and overhead cost estimates to evaluate the potential profitability of the proposed mower. To make the product cost estimate, the project engineer segments the component costs and estimates the total components cost as shown below.

Cost Item		Unit Mat'l Estimate	Cost Item		Unit Mat'l Estimate
A. Chassis			**B. Power train**		
A.1 Deck		$74.00	B.1 Engine		$385.00
A.2 Front steering suspension		$210.00	B.2 Starter assembly		$59.00
A.3 Rear axle/differential		$185.00	B.3 Transmission		$145.00
A.4 Front wheels ($20 ea)		$20.00	B.4 Drive disc assembly		$100.00
A.5 Rear wheels ($24 ea)		$48.00	B.5 Clutch linkage		$51.50
A.6 Engine housing		$120.00			$740.50
A.7 Rear fenders		$80.00			
		$737.00			

Cost Item		Unit Mat'l Estimate
C. Miscellaneous		
C.1 Seat assembly		$38.50
C.2 Instruments assembly		$85.50
C.3 Brake system		$74.00
C.4 Speed control		$21.50
C.5 Drive control assembly		$67.00
C. 6 Lighting system		$74.00
		$360.50
	Total cost	$1,838.00

Cost Index Model

Cost indexes are dimensionless multipliers that reflect relative price change from some base year t_0 to some later year t_n. Cost indexes can reflect relative price changes in specific commodities (labor, material, electricity, water, etc.) or in bundles of commodities (consumer price index or producer price index, combined utilities, etc.). A cost index is computed as

$$\frac{C_{t(n)}}{C_{t(0)}} = \frac{I_{t(n)}}{I_{t(0)}}$$

or as

$$C_{t(n)} = C_{t(0)} \frac{I_{t(n)}}{I_{t(0)}} \qquad (4.8)$$

where the initial cost $C_{t(0)}$ is historically documented and indexes $I_{t(n)}$ and $I_{t(0)}$ are obtained from a source such as the US Department of Commerce or the US Department of Labor.

Example 4.6

An industrial engineer needs to estimate the startup labor and material costs for a new production line with a capacity of 1,000,000 units/year. The following data was obtained from prior production estimates and the US Department of Commerce and US Department of Labor.

Index	8 years prior	Today	Cost	8 years prior	Today	
Labor		124	293		$ 455,500	?
Material		460	715		$2,575,000	?

$$C_{t(n)} = C_{t(-8)} \frac{I_{t(n)}}{I_{t(-8)}} = \$455,500 \frac{293}{124} = \$1,076,302$$

$$C_{t(n)} = C_{t(-8)} \frac{I_{t(n)}}{I_{t(-8)}} = \$2,575,000 \frac{715}{460} = \$4,002,446$$

Power-Sizing Model

The *power-sizing model* is used to estimate the costs of industrial plants and equipment by "scaling up" or "scaling down" known costs to account for economies of scale. The power-sizing model uses the exponent (x), called the power-sizing exponent, to reflect economies of scale.

$$C_A / C_B = (Size_A / Size_B)^x$$

or as

$$C_A = C_B (Size_A / Size_B)^x \qquad (4.9)$$

where C_B is the historically documented base cost. Costs C_A and C_B are estimated at the same point in time (same monetary basis), and the size is in the same physical units for A and B.

An exponent $x = 1.0$ indicates a linear cost-versus-size. An exponent >1.0 indicates diseconomies of scale, and an exponent <1.0 indicates economies of scale. Exponent values may be found in *Perry's Chemical Engineer's Handbook, Plant Design and Economics for Chemical Engineers*, or *Preliminary Chemical Engineering Plant Design*.

Power-sizing models provide scaling for only the same point in time. If time scaling must also be factored into the estimate, use the power-sizing model to scale up or down in the prior period then use the scaled estimate in the prior period in a cost index model.

Example 4.7

The industrial engineer in Example 4.6 has been asked to re-estimate the cost for the new production line with a capacity of 2,500,000 units/year. She acquired the following data.

- The prior production facility cost $50,000,000 eight years ago.
- Technological efficiency improvements indicate that the power-sizing exponent $x = 0.66$.
- Production facility $I_{t(-8)} = 1200$ and $I_{t(n)} = 1490$.

$$C_{(2,500,000)} = \$50,000,000 \left(\frac{2,500,000}{1,000,000} \right)^{0.66} = \$91,539,985$$

$$C_{t(n)} = C_{t(-8)} \frac{I_{t(n)}}{I_{t(-8)}} = \$91,539,985 \frac{1490}{1200} = \$113,662,148$$

Triangulation Model

Triangulation in cost estimating uses three or more different sources of data or different quantitative models to estimate. Triangulation is used for cost estimating innovative new products that have no predecessors. The triangulation technique involves acquiring cost data on three or more prior similar innovative products; comparing similarities and differences in design parameters, parameter performance; contrasting performance differences; and adjusting expected costs positive and negative differences to arrive at three cost estimates. The three cost estimates then provide a minimum, middle, and maximum cost range.

Triangulation is used in the Needs Assessment and Justification Phase or early in the Conceptual Design Phase of the product life cycle when insufficient information exists to use more formal cost estimation techniques. Triangulation has multiple benefits. Without prior cost data for the innovative product, triangulation provides a means of gaining an early indication of potential cash flows and return on investment before committing funds to a risky investment. Second, large discrepancies in the initial cash flow estimates indicate lack of confidence in the estimates. Examination of the causes of discrepancies can provide additional adjustments toward better estimates. Triangulation provides an initial structuring of cash flows, which can be further refined as more information is gained later in the Conceptual Design Phase. This allows a sequence of decision points to be incorporated into the risky investment decision process.

Example 4.8

A software development firm is bidding on its first military contract (i.e., it has no prior experience developing software for the military). But the firm has written code for the following similar commercial applications (next slide), and it knows that military contracts should be more expensive due to additional requirements and documentation.

Application	Purpose	Language	Size KLOC	Cost/KLOC
Graphics	CAD	C+	2500	$125.00
MS Office Ap	Office tools	Visual Basic	2000	$214.29 (max)
Acc Ledger	Business	C	1500	$107.00 (min)
Excel® Addin	Office tools	C++	2500	$150.00 (mean)
		Averages	2125	$149.07

The firm could consider bidding ~$150/KLOC expecting to write about 2125 KLOC. To account for additional expenses, the bid might be ~$200/KLOC.

Learning Curve Model

It has long been observed that the time and effort to perform a task decreases with repetition. The improved performance is termed "learning" and is generally modeled by a learning curve. The learning curve is based on the following observations.

1. Human performance usually improves when a task is repeated.
2. This happens by a fixed percent each time the production doubles.
3. Percentage is called the learning rate and is modeled by a power law.

For any repetitive task, the underlying hypothesis is that the direct labor person-hours necessary to complete a unit of product will decrease by a constant percentage each time the production volume doubles. For example, an improvement of 10%

between doubled quantities establishes a 90% learning curve. The time required to build the second unit will be 0.90 times that required for the first unit. The fourth unit will require 0.90 times that required for the second unit. The eighth unit will require 0.90 time that required for the fourth unit, and so on. That is, standard learning curve ratios are defined in terms of the time required to double output.

$$\Phi = \frac{T_n}{T_1} = n^r$$

$$r = \frac{\log \Phi}{\log 2}$$

$$T_n = T_1 n^r \tag{4.10}$$

$$\text{Total Time} = T_1 \sum_n n^r \tag{4.11}$$

Take the log() of Eq. (4.10) allows the fitting of a simple linear regression to estimate the learning rate r as the slope coefficient.

$$\log T_n = \log T_1 + r \log n$$

A table of decimal learning ratios to Φ is given as follows.

Table 4.2 Decimal learning ratios to Φ

Φ	r
1.0 (no learning)	0
0.95	−0.074
0.90	−0.152
0.85	−0.234
0.80	−0.322
0.75	−0.415
0.70	−0.515
0.65	−0.621
0.60	−0.737
0.55	−0.861
0.50	−1.000

Example 4.9

If we have no prior learning curve data, we evaluate the % learning by the ratio of the production hours for the unit doubling rate.

Unit No.	Hours	Dbl Rate	Total Hrs
1	141.3		141.3
2	120.11	0.850083	261.41
3	109.21		370.62
4	102.09	0.849971	472.71
5	96.89		569.60
6	92.83		662.43
7	89.54		751.97
8	86.78	0.850034	838.75
9	84.41		923.16
10	82.35		1005.51
11	80.53		1086.04
12	78.91		1164.95
13	77.44		1242.39
14	76.11		1318.50
15	74.89		1393.39
16	73.76	0.849965	1467.15

The ratio of unit 2 to unit 1:

$$\%\text{Learning} = \frac{120.11}{141.3} = 0.85004$$

The ratio of unit 4 to unit 2:

$$\%\text{Learning} = \frac{102.09}{120.11} = 0.84997$$

Using the estimated 85% learning rate, we would estimate r and complete the table for the project direct labor hours.

Learning curve rare		0.85	
	r =	−0.23446	
Unit No.	Hours	Dbl Rate	Total Hrs
17	72.72		1539.87
18	71.75		1611.62
19	70.85		1682.47
20	70.00		1752.47
21	69.20		1821.67
22	68.45		1890.12
23	67.74		1957.87
24	67.07		2024.94
25	66.43		2091.37

Using the 85% learning rate, the estimate of the time to complete the 25th unit is

$$T_{25} = T_1 \times n^r = 141.3 \times 25^{-0.23446} = 66.43$$

4.8   Summary

Definition: Cost estimating is the approximation of the probable future cost of a product, program, or project, computed on the basis of currently available information.

Definition: *Risk* is observed in those situations in which the potential outcomes can be described by well-known probability distributions.

Definition: *Imprecision* is observed in those situations in which the potential outcomes cannot be described by well-known probability distributions but can be estimated by subjective probabilities.

Definition: *Uncertainty* is observed in those situations in which the potential outcomes cannot be described by well-known probability distributions and cannot be estimated by subjective probabilities.

Definition: *Accuracy* is the difference between an average or median estimated cash flow at any life cycle design phase and its actual realized cash flow.

Definition: *Precision* is (1) the best-fit variance distribution under risk, (2) fuzzy membership or rough set membership under imprecision, or (3) fuzzy membership, rough set membership, or allowable greyness under uncertainty.

Definition: *Model* is a mathematical description of the static or dynamic functional relationship between a variable, or vector of variables, of interest and another variable, or vector or matrix of variables that determine the functionality of the variable(s) of interest.

Definition: *Cost estimating relationship* is a model that estimates a cost or price by using an established relationship with one or more independent variables.

Definition: *Time estimating relationship* is a model that estimates activity duration by connecting an established relationship with one or more independent variables to the duration time of the activity.

Cost Estimating Process.

1. Define the Cost Estimate Scope.
2. Identify Assumptions and Constraints.
3. Develop the Cost Estimate.
4. Collect and Normalize Cash Flow Data.
5. Develop the Cost Estimating Relationships.
6. Select the Discount Cash Flow Criteria.
7. Compute the Cost Estimate.
8. Document and Present the Cost Estimate.
9. Audit Actual Life Cycle Cash Flows Versus Model Cash Flow Estimates.
10. Develop a Cost Estimation Database.

Definition: *Product costing* is the estimation of the cost of a unit of product (or service delivered) through the determination or allocation of all expenses related to the creation of the product (or service).

- A *bill of materials* (BOM) is a complete list of all the materials, subcomponents, components, subassemblies, and assemblies and the exact quantities of each necessary for the production of one unit of a finished product.
- An *engineering bill of materials* (EBOM) defines one unit of finished product as it was originally designed.
- A *manufacturing bill of materials* (MBOM) specifies all the direct and indirect materials, supplies, subcomponents, components, subassemblies, assemblies, and packaging materials required to build a unit of shippable product.

Material cost for the next higher level in the product hierarchy.

$$M_{hi} = \sum_i Q_i (M_i + H_i \times \text{PHC}_i)$$

PHC_i = unit productive hour cost (standard operation cost). The unit productive hour cost is the sum of standard direct labor cost, standard direct equipment cost, and allocated cost per unit of space.

$$\text{PCH}_i = \sum_j C_{(\text{sdl})ij} + \sum_k C_{(\text{sdm})ik} + C_{\text{spi}}$$

The **product cash flow statement** summarizes the value of money flowing into and out of an organization from the production of the given product.

$$F_p = (R_p - C_p - D_p)(1 - \text{TR}) + D_p$$

Definition: **Operation costing** is the estimation of the total hourly cost of labor plus the operation hourly rate for indirect materials, tools, equipment/space, utilities, and factory overhead to operate a transformation point in value-increasing processes over a stated time frame.

The general process for determining operation costs is as follows.

1.   Determine the future time frame over which the PHC of the operation will be applied (determines the actual dollar value given inflation and the technology mix).
2.   Determine standard categories of product mix.
3.   Determine the operation input and output boundaries and process flow.
4.   For each product mix category, estimate the productive hour cost PHC.

$$\text{PHC} = \text{SUM}(\text{Labor, Id. Materials}, [\text{Eq./Space}], \text{Utilities, Factory OH})/\text{HrPHC}$$
$$= \text{SUM}(\text{Labor, Id. Materials}, [\text{Eq./Space}], \text{Utilities, Factory OH})/\text{Hr}$$

Lot hours are found using the relationship,

$$\text{Lot Hours} = \text{SU} + N \times H_s$$

Once lot hours are determined, the operation cost is estimated as,

$$C_{\text{opr}} = \text{Lot Hours} \times \text{PHC}$$

Cost estimating models are costing procedures that are common across private and public sectors and have become standardized through practice.

Per-unit model costing uses a "per-unit" factor such as \$/sq ft, cost/customer, benefits/employee, vehicle cost/mile, etc. Per-unit cost is based on a strong relationship between the unit cost and the total cost.

$$C_U = \sum C_i / \sum n_i$$
$$\text{Estimated cost} = C_U \times \text{Units}$$

The **segmenting cost model** decomposes a new product into its individual components and assemblies, obtains or estimates the cost of each component (which typically can be obtained or estimated by suppliers), and sums the component costs into subassembly costs and subassembly costs into product cost per unit.

Cost indexes are dimensionless multipliers that reflect relative price change from some base year t0 to some later year t_n.

$$C_{t(n)} = C_{t(0)} \frac{I_{t(n)}}{I_{t(0)}}$$

The **power-sizing model** is used to estimate the costs of industrial plants and equipment by "scaling up" or "scaling down" known costs to account for economies of scale. The power-sizing model uses the exponent (x), called the power-sizing exponent, to reflect economies of scale.

$$C_A = C_B (\text{Size}_A/\text{Size}_B)^x$$

Triangulation in cost estimating uses three or more different sources of data or different quantitative models to estimate.

Learning curve model—it has long been observed that the time and effort to perform a task decreases with repetition. The improved performance is termed "learning" and is generally modeled by a learning curve.

$$\frac{T_n}{T_1} = n^r$$

$$r = \frac{\log}{\log 2}$$

$$T_n = T_1 n^r$$

4.9 Key Terms

- Bill of materials
- Cost estimating
- Cost estimating relationship
- Cost index model
- Detailed estimate
- Lot hours
- Learning curve model
- Operating costing
- Power-sizing model
- Product costing
- Productive hour cost
- Rough estimate
- Segmenting model
- Semi-detailed estimate
- System estimating
- Triangulation.

Problems

1. Product 8718 has the following operational times.

Operation	Setup Hrs	Hours per 100 units	PHC
Punch holes	0.4	0.20	65.00
Deburr	0.1	0.015	35.60
Punch shape	0.4	0.20	65.00
End mill	0.8	0.50	34.75
Degrease & Clean	–	0.10	22.10

Material cost per unit = $0.125. Parts are shipped in quantities of 1000 per order.

a. Find the lot hours and cost per operation.
b. Find the total and cost per unit of manufacturing.
c. If the markup is 30%, find the price per unit.

2. The warehouse floor is worn with incomplete required OSHA markings due to forklift traffic. The forklift traffic floor area is 14,300 sq ft. A local contractor has bid $4.10 per square foot to resurface the forklift traffic area, $1.40 per square foot area to apply the required OSHA markings, and $1.75 per square foot to apply a surface finish coat. Two coats of the surface finish will be required for the 5-year warranty. Estimate the total cost to repair the worn warehouse floor.

3. A cable television company is installing a new 2,025 square foot substation to distribute programming to a new subdivision. The costs are as shown below.

Cost item	Cost
Construction permits, legal and title fees	$30,000
Driveway	$56,250
Foundation	$48,750
Flooring	$22,500
Framing, sheathing, wallboard	$45,000
Roofing	$22,500
Utilities	$18,750
Communications	$18,750
Painting, finishing	$63,750

a. What is the total cost of construction and the cost per square foot?
b. The subdivision contractor has gained approval from the city building department to increase the size of the subdivision by 60%, and the cable company will now need to increase the substation size proportionately. What are the new item cost, total cost, and cost per square foot of the substation?

4. Five years ago, the relevant cost index was 180 for an automated titration system, and its cost was $43,000.00. The current titration system had a capacity of processing 2250 samples per 24-h day. Today, the laboratory needs a capacity of 6750 samples per 24-h day, and the cost index for a new automated titration system is 450. Assuming a power-sizing exponent of 0.80 to reflect advances in titration technology, use the power-sizing model to determine the approximate cost (in today's dollars) of the new automated titration system.

5. Violet is an industrial engineer at the Alest electric automobile manufacturing facility. She has collected data on assembly of EV charging systems in a new startup facility. She had determined that the 10th system required 175 person-hours to assemble, and the 20th system required 140 person-hours to assemble. Estimate the % learning rate, the r exponent, and the system number at which the assembly will attain the 100 person-hours per system breakeven point.

Part II
Economic Analysis of Engineering Activities and Projects

Chapter 5
Time Value of Money

Abstract Investing in productive assets requires the availability of money. From the finance cycle in lecture two, organizations obtain financing for investment in assets from lenders (banks and other financial institutions) or investors (stockholders). Both sources of funds require a return on their investments that accounts for the risk of loaning money, banks in the form of contractual interest, and investors in the form of dividends or growth in stock value, for the time of the loan or the outstanding stock.

5.1 Interest Equivalence

From the finance cycle in chapter two, the risk-adjusted, weighted, minimum acceptable rate of return, or hurdle rate, that must be earned on a project or investment is termed the minimum attractive rate of return. In chapter one, the MARR was specified as

$$MARR = \sum_i w_{Li} \times I_{Li} \times (1 - T_C) + \sum_j w_{Sj} \times RoE_{Sj} \qquad (5.1)$$

where $w_{Li} = L_i/(\sum_i L_i + \sum_j S_j)$, $w_{Sj} = S_j/(\sum_i L_i + \sum_j S_j)$, I_{Li} = interest rate charged on loan i, T_C = the organization's combined tax rate, and RoE_{Si} = return on equity interest rate for stock S_j. The interest rate charged for a loan and the dividend rate on stocks each incorporates a risk premium that accounts for the riskiness of the investment (probably of loss). The risk premium is the interest rate in excess of the risk-free rate of return an investment is expected to yield.

$$\text{Risk Premium} = \text{Interest(Investment)} - \text{Interest(Riskfree)} \qquad (5.2)$$

© The Author(s), under exclusive license to Springer Nature Switzerland AG 2022
T. S. Cotter, *Engineering Managerial Economic Decision and Risk Analysis*,
Topics in Safety, Risk, Reliability and Quality 39,
https://doi.org/10.1007/978-3-030-87767-5_5

The risk-free rate of return is the theoretical rate of return of an investment with zero risk over a specified period. The current risk-free rate can be calculated by subtracting the current inflation rate from the current interest rate of the US Treasury bond matching an investment duration.

Risk   Uncertainty about the future causes a decline in the value of money. Risk increases into future times. As risk increases, the value of money decreases.

Inflation   The general price increase in an economy. As prices increase, the value of a monetary unit decreases.

Liquidity   Investors have a *preference for liquidity*, i.e., they prefer money today to the promise of future money. Investments in monetary returns generating assets are the same as a promise of future money. Investment must be made in the assets today for future returns.

The risk premium arises from the fundamental relationship between risk and return—the return on an investment should be proportional to the risk involved that the rate of return, RoR, will be less than the required MARR. The fundamental risk–return relationship is illustrated in Fig. 5.1.

In engineering economics, risk is the probability of the project's realized return which differs from the project's original return estimate. Risk is a measure of volatility. There at two components to risk volatility:

Structural difference between long run average realized return and the expected return.

Random   random error difference between realized return and realized long run average return.

Fig. 5.1 Risk–return relationship

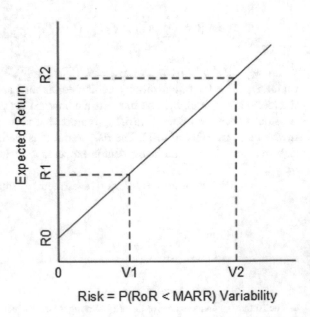

When we compare the cash flows from two or more projects, we must compare them on the same basis. By definition, any set of two or more cash flows are equivalent if they have the same economic value. More precisely, any set of two or more cash flows are equivalent at interest rate i if we can convert one cash flow into the equivalent value of the other(s) by using the proper interest factors.

5.2 Cash Flow Transactions with Interest

Interest elements common to all cash flow transactions include:

- **Principal**—the initial amount money for a loan, bond, stock offering, or asset purchase.
- **Interest rate**—the proportion cost of borrowed money per period of time.
- **Interest period**—the time over which interest is paid.
- **Number of interest periods**—duration of the loan, bond, stock offering, or asset purchase.
- **Receipts or disbursements schedule**—the cash flow pattern with interest over the number of interest periods.
- **Future amount of money**—cumulative effect of principal and interest over the number of interest periods.
- **Present amount of money**—the current value of a set of future cash flows discounted for interest costs.

These interest elements are represented by the following variables:

A_n = a discrete payment or receipt of money occurring at the end of an interest period.

i = interest rate per period.

N = number of interest periods.

P = present value or worth of a set of future cash flows discounted for interest costs.

F = single amount that is equivalent to the cumulative effect of principal and interest over the number of interest periods.

A = an end of period payment or receipt of money in a uniform series that continue for n periods; $A_1 = A_2 = \cdots = A_n$.

5.3 Cash Flow Income Statement

Chapter two introduced the cash flow income statement. Differences from an accrual income statement include:

- Credit sales and allowances are not included.
- Accounts payable amounts are not included in CoGS, selling, and G&A expenses.
- Depreciation is not included but accounted for in total taxes.

Cash flow diagram conventions were introduced in chapter three. Cash flow diagrams represent time by a horizontal line marked with an equal number of cash flow or interest periods. Cash receipts are represented by arrows pointed upward in the positive direction, and cash disbursements are represented by arrows pointed downward in the negative direction. The combination of positive and negative cash flows at each cash flow or interest period represents the **net cash flows**.

$$\text{Net Cash Flow}_n = R_n - O_n - M_n - I_n - T_n \tag{5.3}$$

where R_n = cash flow revenue, O_n = operations cash flow cost, M_n = maintenance cash flow cost, I_n = interest cash flow payment, and T_n = tax cash flow payment. Each cash flow is assumed to occur at the end of each interest period to capture a full interest payment. This is known as the **end of period convention**. The end of period convention also conforms with monthly, quarterly, semi-annual, and annual accounting statements.

Example 5.1

Given a set of cash flows, timing, and amounts, what interest rate equates the two cash flows?

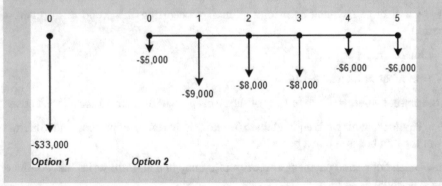

Example 5.2

Given an interest rate of 6.0% and an initial loan of $100, what are the future two payments required to repay principal equally plus interest?

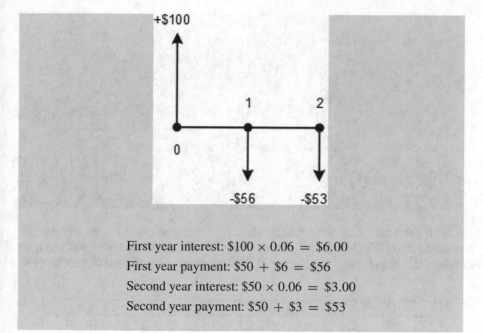

First year interest: $\$100 \times 0.06 = \$6.00$
First year payment: $\$50 + \$6 = \$56$
Second year interest: $\$50 \times 0.06 = \$3.00$
Second year payment: $\$50 + \$3 = \$53$

5.4   Single Payment Discrete Cash Flow Compounding

To establish the theoretical basis for single payment discrete cash flow compounding, we will first consider simple interest. **Simple interest** is the interest that is computed on only the original amount borrowed or loaned.

$$\text{Total interest earned.} i_s = P \times i \times n$$

where i_s = simple interest and P, i, and n are as defined previously. For example, if $i = 8.0\%$, $P = \$100$, and $n = 2$,

$$i_x = \$100 \times 0.08/\text{period} \times 2\text{periods} = \$16$$

Under simple interest, the future amount of money due at the end of a loan is

$$F = P + P \times i \times n = P(1 + i \times n)$$

For the example above

$$F = \$100(1 + 0.08 \times 2) = \$116$$

Example 5.3

Given a simple interest rate of 10.0% and an initial loan of $5000, what interest paid and the future amount to be repaid at the end of a five-year loan.

$$P = \$5000 \quad i = 0.10/\text{year} \quad n = 5\text{years} \quad F =?$$

Total interest earned: $i_S -= P \times i \times n = \$5000 \times 0.10/\text{year} \times 5 \text{ year} = \$2500$.

Future amount due: $F = P(1 + i \times n) = \$5000(1 + 0.10/\text{year} \times 5 \text{ year}) = \$7500$.

Two questions arise with simple interest. Would you make a loan with simple interest terms? Would a bank? The answer is "No," because in effect you make an interest free loan for one year on the $8.00 interest earned at the end of the first year.

Year	Beg Amt	Interest	End Amt
0			$100.00
1	$100.00	($8.00)	$108.00
2	$108.00	$8.00	$116.00

Compound interest is computed on the unpaid debt and the unpaid interest due.

Total interest earned: $i = P(1+ i)^n – P$
Continuing with the above example,

$$i = P(1 + i)^n - P = \$100(1 + 0.08)^2 - \$100 = \$16.64$$

Year	Beg Amt	Interest	End Amt
0			$100.00
1	$100.00	$8.00	$108.00
2	$108.00	($8.64)	$116.64

Under compound interest, the future value, or amount of money due at the end of a loan, is

$$F = P(1 + i)_1(1 + i)_2 \ldots (1 + i)_n \text{ or } F = P(1 + i)^n$$
$$F = \$100(1 + 0.08)^2 = \$116.64$$

Example 5.4

Re-compute the total interest and future amount due for the loan in Example 5.3 using compound interest.

$$P = \$5000 \quad i = 0.10/\text{year} \quad n = 5 \text{ years} \quad F = ?$$

Total interest earned: $i = P(1 + i)^n - P = \$5000(1 + 0.10/\text{year})^5 - \$5000.$

$= \$3052.55$

Future amount due: $F = P(1 + i)^n = \$5000(1 + 0.10/\text{year})^5 = \$8052.55.$

To understand equivalence, the underlying interest formulas must be analyzed. Assume that a present sum of money P is invested for one year at an interest rate i. At the end of the year, we should get back the initial principal P plus the interest earned iP, or $F = P + iP = P(1 + i)$. If the same principal P is invested for n years, at the end of the investment period, we should get back.

Year	Beg amt	Interest	F ending amt
0			P
1	P	iP	$P(1 + i)$
2	$P(1 + i)$	$iP(1 + i)$	$P(1 + i)^2$
3	$P(1 + i)^2$	$iP(1 + i)^2$	$P(1 + i)^3$
N	$P(1 + i)^{n-1}$	$iP(1 + i)^{n-1}$	$P(1 + i)^n$

Thus, the **single payment compound amount** is defined as:

$$F = P(1 + i)^n \tag{5.3}$$

with functional notation

$$F = P(F/P, i, n) \tag{5.4}$$

Values of $(F/P, i, n)$ can be calculated in the Microsoft® Excel® Compound Interest Calculator Spreadsheet for various values of interest rates i and periods n. Reference Appendix A is for directions on the use of the Microsoft® Excel® Compound Interest Calculator Spreadsheet.

Example 5.5

Assume that $500 is deposited in an account that earns 6.0% per year for three years. Estimate its future value.

$$F = P(1 + i)^n = \$500(1 + 0.06)^3 = \$500(1.191) = \$595.50$$

or from the Microsoft® Excel® Compound Interest Calculator Spreadsheet, enter $i = 6.00\%$ and read from the Single Payment F/P column,

4		**Discrete Compounding**			
5		Interest Rate		i	6.00%
6					
7		Single Payment		Uniform Pa	
8	n	F/P		P/F	A/F
9	1	1.06000		0.94340	1.00000
10	2	1.12360		0.89000	0.48544
11	3	1.19102		0.83962	0.31411
12	4	1.26248		0.79209	0.22859

$$F = \$500(F/P, 6.0\%, 3) = \$500(1.191) = \$595.50$$

Graphically, the present amount P is multiplied three times by 1.06.

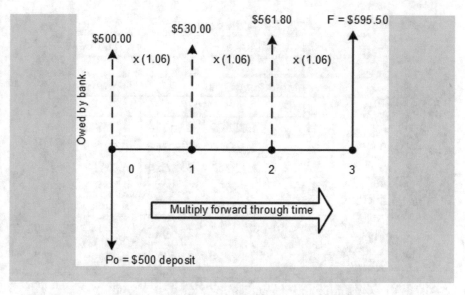

If we take $F = P(1 + i)^n$ and solve for P, we get

$$P = F(1 + i)^{-n} \tag{5.5}$$

with functional notation

$$P = F(P/F, i, n) \tag{5.6}$$

which is the **single payment present worth amount**. Values of $(P/F, i, n)$ can be calculated in the Microsoft® Excel® Compound Interest Calculator Spreadsheet for various values of interest rates i and periods n.

Example 5.6

Suppose that you need $800 at the end of four years. What amount P must be deposited now in an account that earns 5.0% per year for four years?

$$P = F(1 + i)^{-n} = \$800(1 + 0.05)^{-4} = \$800(0.8227) = \$658.16$$

or from the Microsoft® Excel® Compound Interest Calculator Spreadsheet, enter $i = 8.00\%$ and read from the Single Payment P/F column,

4		**Discrete Compounding**			
5		Interest Rate		i	5.00%
6					
7		Single Payment		Uniform Pa	
8	n	F/P	P/F	A/F	
9	1	1.05000	0.95238	1.00000	
10	2	1.10250	0.90703	0.48780	
11	3	1.15763	0.86384	0.31721	
12	(4)	1.21551	0.82270	0.23201	
13	5	1.27628	0.78353	0.18097	

$$P = \$800(P/F, 5.0\%\cdot, 4) = \$800(0.8227) = \$658.16$$

Graphically, the future amount F is divided four times by 1.05.

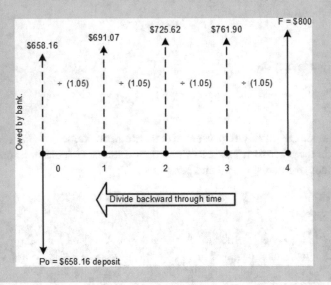

Since cash flows are linear, by the superposition theorem, a net cash flow is the algebraic sum of the individual cash flows at any given point in time. Therefore, when there is more than one cash flow, each cash flow can be discounted or compounded to a point in time (end of an interest period) and the cash flows summed to yield the net cash flow. This principle is illustrated in Example 5.7.

Example 5.7

You are offered a contract in which you will loan an amount of money and will repay $400 at the end of year 3 plus $600 at the end of year 5 both at a 12% interest rate. What amount are you willing to loan?

$$P = F_3(1 + 0.12)^{-3} + F_5(1 + 0.12)^{-5}$$
$$P = \$400(1 + 0.12)^{-3} + \$600(1 + 0.12)^{-5}$$
$$P = \$400(P/F, 12\%, 3) + \$600(P/F, 12\%, 5)$$
$$P = \$400(0.7118) + \$600(0.5674)$$
$$P = \$625.16$$

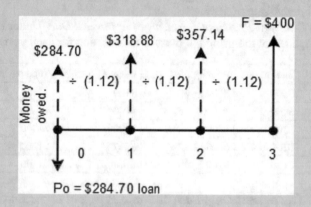

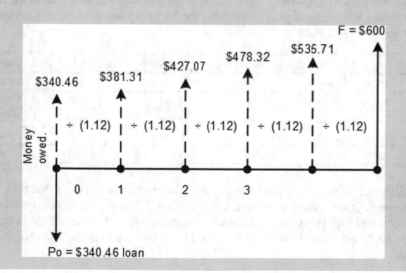

$$P(\text{loan}) = \$284.70 + \$340.46 = \$625.16$$

5.5   Four Methods of Debt Repayment

Now that we have established the basis for discounted (present value) and compounded (future value) cash flows, we can consider the four fundamental ways of repaying a loan. To focus on the repayment methods rather than interest equivalence, assume that a $5000 loan is to be repaid over a five-year period (interest period = 1 year) at an interest rate of 10%.

Method 1: *Constant Principal Payments Plus Interest Due*. Under this plan, the borrower repays $1/n$ of the principal plus the interest due for each year.

Year	Beg. bal	Interest due	Principal payment	Interest payment	Total payment	End bal
1	$5000	$500	$1000	$500	$1500	$4000
2	$4000	$400	$1000	$400	$1400	$3000
3	$3000	$300	$1000	$300	$1300	$2000
4	$2000	$200	$1000	$200	$1200	$1000
5	$1000	$100	$1000	$100	$1100	$0

Method 2: *Pay Interest Only Each Year and Principal at the End of the Loan*. Under this plan, the borrower repays all interest due for each year, and in the final year repays the principal plus interest due for the final year. This method is used in the repayment of a bond.

Year	Beg. bal	Interest due	Principal payment	Interest payment	Total payment	End bal
1	$5000	$500	$0	$500	$500	$5000
2	$5000	$500	$0	$500	$500	$5000
3	$5000	$500	$0	$500	$500	$5000
4	$5000	$500	$0	$500	$500	$5000
5	$5000	$500	$5,000	$500	$5500	$0

Method 3: *Equal Annuity Payments Each Interest Period*. Under this plan, the borrower pays an equal payment each interest period. The payment covers declining principal and interest amounts due each interest period such that the final payment draws the balance due to zero. This method is used to repay house and automobile loans.

Year	Beg. bal	Interest due	Principal payment	Interest payment	Total payment	End bal
1	$5000	$500	$819	$500	$1319	$4181
2	$4181	$418	$901	$418	$1319	$3280
3	$3280	$328	$991	$328	$1319	$2289
4	$2289	$229	$1090	$229	$1319	$1199
5	$1199	$120	$1199	$120	$1319	$0

Method 4: Accumulate Interest Owed and Pay Principal Plus Accumulated Interest at the End of the Loan. Under this plan, the borrower pays nothing in each interest period. Interest due accumulates until the end of the loan, and the final payment covers the principal plus accumulated interest. This method is often used on large, fixed-price projects where the contractor receives a final fixed payment for all or a major portion of the amount due. This repayment method is known as a "balloon loan."

Year	Beg. bal	Interest due	Principal payment	Interest payment	Total payment	End bal
1	$5000	$500	$0	$0	$0	$5500
2	$5500	$550	$0	$0	$0	$6050
3	$6050	$605	$0	$0	$0	$6655
4	$6655	$666	$0	$0	$0	$7321
5	$7321	$732	$5000	$3053	$8053	$0

Although the methods differ in cash flow patterns, *each method is equivalent for the principal owed at the stated interest rate.* We can see this equivalence by dividing the accumulated interest paid by the accumulated principal owed over the life of the loan.

Year	Method 1	Method 2	Method 3	Method 4
1	$5000	$5000	$5000	$5000
2	$4000	$5000	$4181	$5500
3	$3000	$5000	$3280	$6050
4	$2000	$5000	$2289	$6655
5	$1000	$5000	$1199	$7321
Acc principal	$15,000	$25,000	$15,949	$30,526
Acc interest	$1500	$2500	$1595	$3053
% interest	10	10	10	10

5.5.1   *Payment Series Discrete Cash Flow Compounding*

Method 3, equal annual payments, is widely used in commercial, industry, and private loans. The equal annual payments method is termed "**uniform payment series**" and provides nice properties when working with annuities A.

Uniform Series Compound Amount

From previous discussion, we know that the future value F of a single amount P invested at time 0 is $F = P(1 + i)^n$.

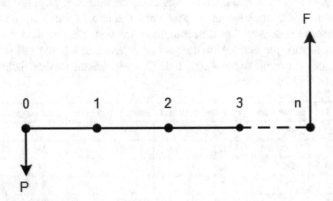

If we let $A_1 = A_2 = \cdots = A_n = P$ but invest A_i exactly at the end of each period, then F is the sum of the compounded A_i amounts.

$$F = A_1(1 + i)^{n-1} + A_{12}(1 + i)^{n-2} + \cdots + A_{n-1}(1 + i)^1 + A_n$$

Dropping the subscripts $(A_1 = A_2 = \cdots = A_n = A)$ and multiplying both sides by $(1 + i)$,

$$(1 + i)F = A(1 + i)^n + A(1 + i)^{n-1} + \cdots + A(1 + i)^2 + A(1 + i)$$

Subtracting the first equation and canceling like negative intermediate terms yields,

$$-(F = A_1(1 + i)^{n-1} + A_{12}(1 + i)^{n-2} + \cdots + A_{n-1}(1 + i)^1 + A_n)$$

$$iF = A(1 + i)^n - A$$

$$F = A\left[\frac{(1 + i)^n - 1}{i}\right] = A(F/A, i, n) \tag{5.7}$$

Equation (5.7) is termed the **uniform series compound amount factor**.

Example 5.8

If an individual invests $1000 in a certificate of deposit paying a 4.0% annual rate every quarter for five years and interest is paid quarterly, how much will the individual have in certificates deposit at the end of the five years?

$$A = \$1000/\text{qtr} \quad n = 5 \text{ years} \quad i = 4\%/\text{year} \quad F = ?$$
$$i_q = 4.0\%/4 \text{ qtr/year} = 1.0\%/\text{qtr}$$
$$n_q = 4 \text{ qtr/year} \times 5 \text{ years} = 20 \text{ qtr}$$
$$F = \$1000 \left[(1 + 0.01)^{20} - 1 \right]/0.01$$
$$= \$1000(22.019)$$
$$= \$22,019$$
$$F = \$500(F/A, 1\%, 20)$$
$$= \$1000(22.019)$$
$$= \$22,019$$

Uniform Series Sinking Fund Amount

If we solve for A using the uniform series compound amount formula, we get

$$A = F \left[\frac{i}{(1 + i)^n - 1} \right] = F(A/F, i, n) \tag{5.8}$$

the **uniform series sinking fund factor**.

Example 5.9

An engineering manager needs $10,000 at the end of the new fiscal year to overhaul a catalytic converter in an exhaust manifold of an air scrubber. Due to other unplanned corrective maintenance activities in this the first fiscal month of the year, she will not have sufficient funds in her budget to cover the entire cost and is considering investing in a uniform amount in a savings account that pays 6.0% annually but compounded monthly. How much does she have to deposit each month to have the $10,000 at the end of the fiscal year?

$$F = \$10,000 \quad n = 1 \text{ year or 12 months} \quad i = 6\%/\text{year or}$$
$$i_m = 6.0\%/12 = 0.5\%/\text{mon} \quad A = ?$$

$$A = \$10,000 \left(0.005/\left[(1 + 0.005)^{12} - 1 \right] \right)$$

$$= \$10,000(0.0811)$$
$$= \$810.66/\text{month}$$

$$F = \$10,000(A/F, 0.5\%, 12)$$
$$= \$10,000(0.0811) \qquad .$$
$$= \$811 \text{ per month}$$

Uniform Series Capital Recovery Amount

If we substitute $F = P(1 + i)^n$ in the uniform series sinking fund factor formula, we get

$$A = P(1+i)^n \left[\frac{i}{(1+i)^n - 1} \right]$$

$$A = P \left[\frac{i(1+i)^n}{(1+i)^n - 1} \right] = P(A/P, i, n) \qquad (5.9)$$

the **uniform series capital recovery factor**.

Example 5.10

A new more efficient IC imaging unit costs \$50,000 and has a life of four years. If the MARR for this unit is 8.0%, what must the efficiency savings be each year to recover the unit's cost?

$$P = \$50,000 \quad n = 4 \quad i = 8.0\% \quad A = ?$$

$$A = \$50,000 \left[0.08\,(1+0.08)^4/((1+0.08)^4 - 1) \right]$$
$$= \$50,000\,[0.30192]$$
$$= \$15,096.04$$

$$A = \$50,000\,(A/P, 0.08, 5)$$
$$= \$50,000\,(0.3019) \qquad .$$
$$= \$15,095$$

Uniform Series Present Worth Amount

If we solve for P in the uniform series capital recovery factor, we get

$$P = A\left[\frac{(1+i)^n - 1}{i(1+i)^n}\right] = A(P/A, i, n) \tag{5.10}$$

the **uniform series present worth factor**.

Example 5.11

A bank holds a note on purchase of a backhoe-loader used by a construction firm. The bank will receive $1,244.25 payment per month for four years at an annual interest rate of 8.0%. How much did the bank loan the construction company?

$$A = \$1,244.25 \quad n = 4 \text{ years or 48 months}$$
$$i = 9.0\%/\text{year or } 0.75\%/\text{mon} \quad P = ?$$

$$P = \$1,244.25\left[(1 + 0.0075)^{48} - 1/0.0075(1 + 0.0075)^{48}\right]$$
$$= \$1,244.25[40.1848]$$
$$= \$49,999.91$$

$$P = \$1,244.25(A/P, 0.08, 5)$$
$$= \$1,244.25(40.185)$$
$$= \$50,000$$

Linear Interpolation of Compounding and Discounting Factors

We may apply linear interpolation to estimate factor values between interest rates for the same number of periods and between periods for the same interest rate. Linear interpolation is a method of estimating missing values between two known values or of curve fitting using linear polynomials to construct new data points within the range of a discrete set of known data points. If the two known points are given by the coordinates (x_0, y_0) and (x_1, y_1), the unknown y given a known x is the linear interpolant straight line between these points.

$$\frac{y - y_0}{x - x_0} = \frac{y_1 - y_0}{x_1 - x_0} \tag{5.11}$$

Solving for y, the linear interpolation equation becomes

$$y = y_0 + (x - x_0)\frac{y_1 - y_0}{x_1 - x_0} \tag{5.12}$$

Example 5.12

A bank holds a note issued for \$40,900 on a new automobile. The bank is paid \$563 payment per month for seven years plus one month. What interest rate is the bank earning?

$$P = \$40,900 \quad A = \$563 \quad n = 7 \text{ years} \times 12 \text{ mon/year}$$
$$= 84 \text{ months} \quad i = ?$$

$$P = A(P/A, i,)$$
$$(P/A, i, 84) = P/A$$
$$(P/A, i, 84) = \$40,900 / \$563$$
$$(P/A, i, 84) = 72.647$$
$$(P/A, 0.25\%, 84) = 75.682 \quad (P/A, 0.50\%, 84) = 68.453$$

$$b\left\{ \begin{matrix} a\left\{ \begin{matrix} 0.25\% & 75.682 \\ i & 72.647 \end{matrix} \right\}c \\ 0.50\% & 68.453 \end{matrix} \right\}d$$

$$\text{Ratio} = \frac{a}{b} = \frac{c}{d}$$

$$\frac{i - 0.25\%}{0.50\% - 0.25\%} = \frac{72.647 - 75.682}{68.453 - 75.682}$$

$$i = 0.25\% + (0.50\% - 0.25\%)\left(\frac{72.647 - 75.682}{68.453 - 75.682}\right) = 0.355\%$$

From the property of similar triangles or the two-point equation of a line, the graphical solution is shown below.

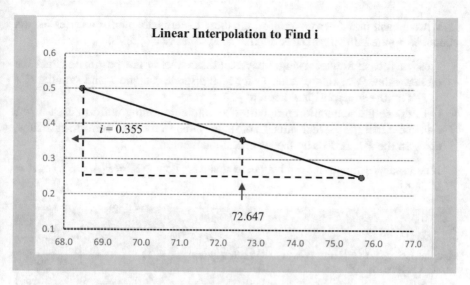

Cash Flows That Do Not Match Uniform Series Patterns

To use the uniform series factors, the cash flows must have equal annual cash flows from period one to period n with a P at time 0 or F at time n as illustrated below.

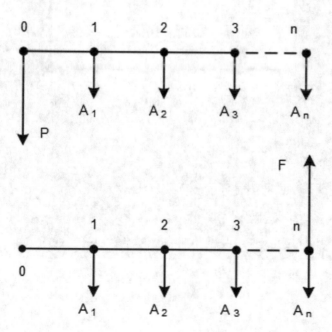

When a cash flow does not conform to these patterns, the uniform series factors cannot be applied directly. The economic analyst has two options.

- Treat each cash flow as a single payment discrete cash flow, estimate either the present value P_i or future value F_i for each discrete cash flow, and sum the P_i's or F_i's by the superposition theorem.
- Decompose the nonconforming cash flow into conforming uniform cash flows, estimate either the present value P_i or future value F_i for each uniform cash flow, and sum the P_i's or F_i's by the superposition theorem.

If necessary, estimate the equivalent annual cash flow as $A = P(A/P, i, n)$ or $A = F(A/F, i, n)$.

Example 5.13

Find F for the following cash flow pattern when $i = 15\%$.

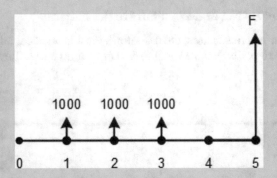

Method 1: Solve by setting each $A = P$, multiplying by $(P/F, 15\%, j)$, and summing.

$$F = \$1000\,(1 + 0.15\,)^4 + \$1000\,(1 + 0.15\,)^3 + \$1000\,(1 + 0.15\,)^2$$
$$= \$1000\,(1.749) + \$1000\,(1.521) + \$1000\,(1.322)$$
$$= \$4592.00$$

Method 2: Solve for $F(3)$ and multiply by $(F/P, 15\%, 2)$.

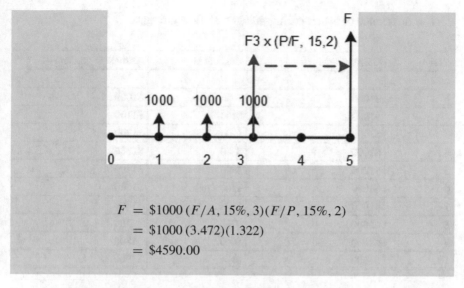

$$F = \$1000 \, (F/A, 15\%, 3)(F/P, 15\%, 2)$$
$$= \$1000 \, (3.472)(1.322)$$
$$= \$4590.00$$

Payment Series Discrete Cash Flow Compounding—Arithmetic Gradient

Often an observed cash flow increases or decreases rather than remaining at a constant value A. The first case we will consider is when a cash flow increases in uniform steps or by an arithmetic gradient.

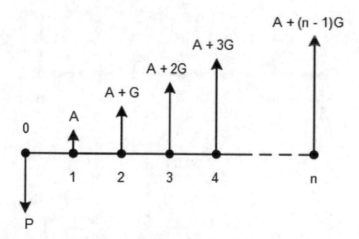

An arithmetic gradient is composed of

- A uniform amount A at the end of time period one.
- A uniform increasing amount G added to A beginning at the end of time period two.
- The same uniform increasing amount G added at the end of each subsequent time period up to time period n.

- G = difference between each gradient cash flow amount.

Year	Uniform amount A	Gradient G	Arithmetic gradient
0			
1	$1000	$0	$1000
2	$1000	$100	$1100
3	$1000	$200	$1200
4	$1000	$300	$1300
5	$1000	$400	$1400
6	$1000	$500	$1500
7	$1000	$600	$1600
8	$1000	$700	$1700
9	$1000	$800	$1800
10	$1000	$900	$1900

The arithmetic gradient cash flow may be resolved into a uniform cash flow plus a gradient cash flow with

$$P = P_A + P_G = A(P/A, i, n) + G(P/G, i, n)$$

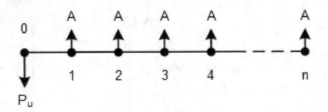

PLUS

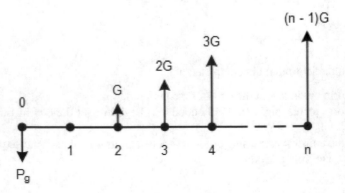

We already know how to solve P_A given A. To find G, we can proceed as we did in finding the uniform series compound amount factor F.

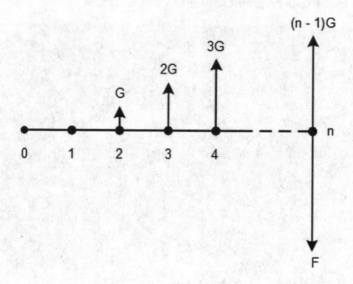

$$F = G(1+i)^{n-2} + 2G(1+i)^{n-3} + \cdots + (n-2)G(1+i) + (n-1)G$$

Multiplying both sides by $(1+i)$ and factoring out G:

$$(1+i)F = G[(1+i)^{n-1} + 2(1+i)^{n-2} + \cdots + (n-2)(1+i)^2 + (n-1)(1+i)]$$

Subtracting the first equation from the second yields

$$iF = G[((1+i)^n - 1)/i] - nG$$

$$F = G[((1+i)^n - 1 - in)/i^2]$$

Setting $F = P(1+i)^n$:

$$P(1+i)^n = G[((1+i)^n - 1 - in)/i^2]$$

$$P = G[((1+i)^n - 1 - in)/i^2(1+i)^n] = G(P/G, i, n) \qquad (5.13)$$

This is the **arithmetic gradient present worth amount**. Setting $F = A\,[((1+i)^n - 1)/i]$:

$$A[((1+i)^n - 1)/i] = G[((1+i)^n - 1 - in)/i^2]$$

$$A = G\big[\big((1+i)^n - 1 - in\big)/\big(i(1+i)^n - i\big)\big] = G(A/G, i, n) \qquad (5.14)$$

This is the **arithmetic gradient uniform series amount**.

Example 5.14

An engineering manager is preparing a purchase order for a new automated arc welder with a five-year useful life. The vendor offers a five-year maintenance contract for $8000. Records for similar model arc welder indicate that it will cost $1200 the first year for maintenance and increase by $300 per year for years 2–5. The arc welding process MARR is 5.0%. Should the engineering manager purchase the maintenance contract?

$$A = \$1200 \quad G = \$300 \quad i = 5.0\% \quad P = ?$$

Uniform series present worth:

$$P_A = A(P/A, 5\%, 5) = \$1200(4.329) = \$5{,}194.80$$

Arithmetic gradient present worth:

$$P_G = G(P/G, 5\%, 5) = \$300(8.237) = \$2471.10$$

$P = PA + PG = 7665.90$. The present worth of incurring the annual maintenance expenses is less than the $8000 present worth of the contract cost. Do not purchase the maintenance contract.

Example 5.15

A logistics manager just purchased a new semi-truck with a three-year maintenance agreement as part of the purchase price. Analysis of maintenance cost data for similar semi-trucks indicates that year 4 maintenance cost should be $1500 and increase each year thereafter by $500. The semi-truck has a useful life of seven years. The MARR for trucking equipment is 10%. For budgeting purposes, estimate the present worth and equivalent annual worth of the future maintenance cash flows.

The cash flows are shown in the following table. Note that we set up a secondary timeline for the maintenance cash follows with year 0 = year 3 of the budgeting period.

Year	Cash flow
0	P
1	0
2	0
3 (0)	$F = P_3 = P_U + P_G$
4 (1)	$1500
5 (2)	$2000
6 (3)	$2500
7 (4)	$3000

$$A_4 = \$1500 \quad G = \$500 \quad i = 10.0\% \quad P_0 = ?A = ?$$

Arithmetic gradient present worth, year 3:

$$P_A = A(P/A, 10\%, 4) = \$1500(3.170) = \$4755$$

$$P_G = G(P/G, 10\%, 4) = \$500(4.378) = \$2189$$

$$P_3 = \$4755 + \$2189 - \$6944$$

Present worth at time period 0:

$$P_0 = P_3(P/F, 10\%, 3) = \$6944(0.7513) = \$5217.03$$

The equivalent uniform annual worth is:

$$A = P_0(A/P, 10\%, 7) = \$5217.03(0.2054) = \$1,071,58.$$

Payment Series Discrete Cash Flow Compounding—Geometric Gradient

Alternative to increasing in uniform steps, cash flows may also increase in a constant proportion in each successive period. Proportional increases at discrete times is termed as a **geometric gradient series**. The initial amount A_1 increases at a uniform rate g. The initial amount occurs in year 1; hence, the amount in year n is

$$A_n = A_1(1 + g)^{n-1}$$

Since any A_n can be considered as a future value F, we can multiply any A_n by $(P/F, i, n) = (1 + i)^{-n}$ to compute the present value.

$$P = A_1(1 + g)^{n-1}(1 + i)^n$$

Rewriting as

$$P = A_1(1 + i)^{-1} \sum_n [(1 + g)/(1 + i)]^{x-1}$$

Expanding,

$$P = A_1(1 + i)^{-1}$$
$$\left[1 + [(1 + g)/(1 + i)] \quad + [(1 + g)/(1 + i)]^2 + \cdots + [(1 + g)/(1 + i)]^{n-1}\right]$$

Let $x = A_1(1 + i)^{-1}$ and $y = ((1 + g)/(1 + i))$,

$$P = x\left[1 + y + y^2 + y^3 + \cdots + y^{n-1}\right]$$

Multiplying both sides by y:

$$yP = x\left[y + y^2 + y^3 + \cdots + y^{n-1} + y^{n-1}\right]$$

Subtracting the second equation from the first yields:

$$(1 - y)P = x\left[1 - y^n\right]$$

$$P = x\left[1 - y^n\right]/(1 - y)$$

Substitution back $x = A_1(1 + i)^{-1}$ and $y = ((1 + g)/(1 + i))$, yields

$$P = A(1 + i)^{-1}\left[1 - [(1 + g)/(1 + i)]^n\right]/(1 - [(1 + g)/(1 + i)])$$

$$P = \frac{\left[1 - [(1 + g)/(1 + i)]^n\right]}{(1 + i) - [(1 + g)/(1 + i)](1 + i)}$$

$$P = A_1\frac{1 - ((1 + g)/(1 + i))^n}{i - g} = A_1(P/A, g, i, n) \qquad (5.15)$$

where $i \neq g$. This is termed the **geometric gradient present worth factor**. In the case of $i =$ Eq. (5.15)becomes

$$P = A_1(1 + i)^{-1} \qquad (5.16)$$

Example 5.16

A quality manager is preparing a five-year budget. The cost for external calibration laboratory service was $9615.38 this year. Over the prior five-year period,
the cost for calibration services has shown a steady average annual growth of
4.0%. The MARR for calibration services is 6.0%. What is the present worth
and equivalent uniform annual cash flow cost of calibration services over the
next five-year budget planning period?

The estimated cost of the first year calibration service for the five-year
budget period is $A_1 = \$9615.38(1.04) = \$10{,}000$. The present worth of
calibration service is

$$P = \$10,000\frac{1 - ((1 + 0.04)/(1 + 0.06))^5}{0.06 - 0.04} = \$45,423.09$$

$$A = \$45,423.09(A/P, 6.0\%, 5) = \$45,423.09(0.2374) = \$10,783,44.$$

5.6 Compounding Periods and Payment Periods Differ

When the compounding and payment periods differ, we must adjust the cash flows,
so that they are in one of the standard forms. Then, we convert from one standard
form cash flow into the desired standard form cash flow.

Example 5.17

An accounting manager invests $25,000 in a money market fund that pays 3%
nominal annual interest rate compounded monthly. The accounting manager
seeks to withdraw an equal annual amount at the end of each year for five years
to cover the cost of the annual Holiday Party. How much will he be able to
withdraw each year?

First compute the equivalent monthly A for each of the 12 mon/year $\times$ 5 year
= 60 months.

$$i_m = 3\%/12 \text{ month} = 0.25\% \text{ per month}$$
$$A = \$25,000(A/P, 0.25\%, 60)$$
$$A = \$25,000(0.0179687)$$
$$A = \$449.22$$

Now, convert from the monthly A to the end-of-year F withdrawals.

$$F = \$449.22\,(F/A,\ 0.25\%,\ 12)$$
$$F = \$449.22\,(12.16638)$$
$$F = \$5465.35$$

5.7   Nominal and Effective Interest Rates

In repayment methods 3 and 4 where interest is charged on interest, the stated annual nominal interest rate is not the rate actually paid. The **effective annual interest rate** is the actual interest rate charged over the life of the loan. There are three interest rates that apply to repayment methods 3 and 4.

Nominal interest rate/year, i_n: the stated annual interest rate on a loan without considering the effect of interest compounding (i.e., 12%/year).

Number of compounding subperiods, m: the number of compounding subperiods per year (i.e., month, quarter, semi-annual).

Effective interest rate/period, i: the nominal interest rate/year divided by the number of interest compounding periods (i.e., 12%/year/12 months/year = 1%/month)

Effective annual interest rate, i_a: the annual interest rate accounting for the effect of compounding interest over the number of interest compounding periods (i.e., 12%/year or 1%/month has an effective annual interest rate of 12.68%/year).

Using these definitions, we can use the single payment future amount formula to restate the nominal annual interest rate to its effective annual interest rate equivalent.

$$F = P(1 + i_n)^n$$

$$F = P\left(1 + \frac{i_n}{m}\right)^{nm}$$

Using the later interest representation, the effective annual interest rate is:

$$i_a = \left(1 + \frac{i_n}{m}\right)^{nm} - 1 \tag{5.17}$$

Example 5.18

Depending on macroeconomic and personal factors, the typical credit card charges a nominal annual interest rate of low of 15%, median 18%, to a high of 21%. What is the effective annual interest rate for each nominal annual interest rate?

For $i_n = 15.0\%$,

Effective interest rate/subperiod,$i = 0.15/12 = 0.0125$

Effective annual interest rate,$i_a = (1 + 0.0125)^{12} - 1 = 0.1608$ or 16.08%
   For $i_n = 18.0\%$,

Effective interest rate/subperiod,$i = 0.18/12 = 0.015$

Effective annual interest rate,$i_a = (1 + 0.015)^{12} - 1 = 0.1956$ or 19.56%.
   For $i_n = 21.0\%$,

Effective interest rate/subperiod,$i = 0.21/12 = 0.0175$

Effective annual interest rate,$i_a = (1 + 0.0175)^{12} - 1 = 0.2314$ or 23.14%.

Example 5.19

Example 5.17 revisited. We found that the accounting manager can withdraw $5475.15 on $25,000 in a money market fund that pays 3% nominal annual interest rate compounded monthly. We can use the effective annual interest rate to verify the withdrawal rate.

$$I_a = (1 + 0.0025)^{12} - 1 = 0.030416$$
$$P = A\,(P/A,\ 3.0415\%,\ 5) = \$5465.35\,(4.574273) = \$25,000.01^{\cdot}$$

5.8   Continuous Cash Flow Compounding

Continuous compounding is the mathematical limit of compound interest if it is calculated and re-invested into an account's balance over a theoretically infinite number of periods as $\Delta t \to 0$. While compound interest has limited use in practice, the concept of continuously compounded interest is important in finance. As an example, an international company on the United States west coast could create a contract with a bank in Japan (eight time zone difference) to deposit its working cash into an continuous interest-bearing account at 5:00 PM west coast time (8:00

AM Japanese time) for 8 h. It could create a second continuous interest-bearing account with a bank in Germany (eight time zone difference from Japan) for the Japanese bank to transfer the cash balance plus accrued interest to the German bank at Japan 5:00 PM (Germany 8:00 AM) time. The German bank would then transfer cash balance plus accrued interest back into the US company's cash account at 5:00 PM (8:00 AM US west coast time). The US west coast company earns continuous compounding on its working cash over night when it is not in use, and each bank adds the cash balance as a cash asset available for loans locally.

We can use the effective annual interest rate formula to develop the formula for continuous compounding.

$$\left(1 + \frac{i_n}{m}\right)^{nm}$$

For continuous compounding, $\Delta t \to 0$ as $m \to \infty$

$$\lim_{m \to \infty} \left(1 + \frac{i_n}{m}\right)^{nm} = e^{i_n} \tag{5.18}$$

$$\lim_{m \to \infty} (1 + i_n) = e^{i_n} \tag{5.19}$$

$$i_a = e^{i_n} - 1 \tag{5.20}$$

Single Payment Continuous Compounding and Present Worth Amounts

$$F = P(1 + i_n)^n = Pe^{i_n n} \tag{5.21}$$

$$P = F(1 + i_n)^{-n} = Fe^{-i_n n} \tag{5.22}$$

Uniform Series Continuous Compounding

Substituting $i_a = e^{i_n} - 1$:

Continuous Compounding Sinking Fund Amount

$$(A/F, i, n) = \left[\frac{i}{(1 + i)^n - 1}\right] = \frac{\left(e^{i_n} - 1\right)}{\left(e^{i_n n} - 1\right)} \tag{5.23}$$

Continuous Compounding Series Amount

$$(F/A, i, n) = \left[\frac{(1 + i)^n - 1}{i}\right] = \frac{\left(e^{i_n n} - 1\right)}{\left(e^{i_n} - 1\right)} \tag{5.23}$$

Continuous Compounding Capital Amount

$$(A/P, i, n) = \left[\frac{i(1+i)^n}{(1+i)^n - 1}\right] = \frac{e^{i_n n}(e^{i_n} - 1)}{(e^{i_n n} - 1)} \tag{5.24}$$

Continuous Compounding Series Present Worth Amount

$$(A/P, i, n) = \left[\frac{i(1+i)^n}{(1+i)^n - 1}\right] = \frac{(e^{i_n n} - 1)}{e^{i_n n}(e^{i_n} - 1)} \tag{5.25}$$

Example 5.19

A deposit of \$1000 per year (end of year) is made into a retirement account that pays a 7.0% nominal interest rate. Estimate how much is in the account at the end of 50 years. Use both discrete compounding and continuous compounding.

Discrete compounding:

$$F = \$1000(F/A, 7.0\%, 50) = \$1000(406.530) = \$406,530$$

Continuous compounding:

$$F = A\frac{(e^{i_n n} - 1)}{(e^{i_n} - 1)} = \$1000\frac{(e^{0.07 \times 50} - 1)}{(e^{0.07} - 1)} = \$1000(442.922) = \$442,922$$

5.9 Spreadsheets for Economic Analysis

In practice, engineering departments and projects require large quantities of economic data. It is impractical to analyze the required data manually. Spreadsheets have become the standard for acquiring, managing, and analyzing engineering managerial economic data. In many engineering firms, data is structured and automatically analyzed in multiple interrelated workbooks of spreadsheets. Common spreadsheet analyses include:

- Structuring tables of cash flows.
- Manually inputting data or developing scripts and macros to import data from databases, Internet sources, or other spreadsheets.

- Structuring worksheet economic functions to automatically update and model discounted or compound cash flows of ongoing engineering operations and projects.
- Structuring output reports and graphics to support engineering managerial decisions.
- Verifying and validating spreadsheet economic models.
- Revising workbook spreadsheet model structure to reflect changing organizational, process, and product economics.

Spreadsheets are used in engineering managerial economic analyses to:

- Construct operations and project cash flow.
- Calculate P, F, A, n, or i using annuity functions.
- Find the present worth or internal rate of return of cash flows using a block of functions.
- Make graphs to support visual analysis.
- Conduct "what-if" scenario analyses to understand the ranges over which differing decisions are optimal.

Basic Spreadsheet Operations and Arithmetic

The most basic spreadsheet operation is cell referencing. Spreadsheets use absolute, relative, and mixed references. In cell referencing, the active cell formula points the data value in another cell. A formula with an *absolute reference* points to the data in another cell by its fixed position in the current or another worksheet. If the formula is copied to another cell, the formula will still point to the original cell. A formula with a *relative reference* points to the data in another cell by the relative number column and row differences. If the formula is copied to another cell, the formula will point to data in a new cell that is the same relative number column and row differences. A formula with a *mixed reference* either sets an absolute reference to a fixed column or row and leaves the other a relative reference. If the formula is copied to another cell, the absolute column or row references will remain fixed, but the relative column or row reference will be to a new cell the same number of relative rows or column away. Typically, a "$" in front of the column or row designation to make it an absolute reference. The first four rows of Table 5.1 illustrate absolute, relative, and mixed designations of a formula that points to data in cell B3 (column B, row 3). The basic arithmetic operations of addition, subtraction, multiplication, division, and exponentiation are performed in an active cell's formula as illustrated in the last five rows of Table 5.1.

Table 5.1 Basic spreadsheet arithmetic operations

Operation	Symbol	Example	Comments
Relative	$	=B3	Column and row relative
Fixed column	$	=$B3	Column fixed and row relative
Fixed row	$	=B$3	Column relative and row fixed
Absolute	$	=$B$3	Column fixed and row fixed
Addition	+	=A3 + B3	=SUM(A3, B3) or =SUM(A3:B3)
Subtraction	−	=A3 − B3	=SUM(A3, −B3)
Multiplication	*	=A3 * B3	=PRODUCT(A3, B3)
Division	/	=A3/B3	No worksheet function
Exponentiation	^	=B3^2	=POWER(B3, 2)

Table 5.2 Spreadsheet financial functions

Argument	Description	Function
P: present value	Value of cash flows today	$-PV(i, n, A, F, \text{type})$
F: future value	Value of cash flows at an end time	$-FV(i, n, A, P, \text{type})$
A: annual cash flow	Equal periodic cash flows	$-PMT(i, n, P, F, \text{type})$
n: time periods	Number of time periods in the term	$NPER(i, A, P, F, \text{type})$
i: interest rate	Constant interest rate	$RATE(n, A, P, F, \text{type})$
Type	Timing of cash flows	0 (default)—end of period

Financial Analysis Functions

Spreadsheet financial functions allow common business calculations without having to construct long, complex formulas. The basic financial functions are listed in Table 5.2. Note that the financial functions accept similar arguments.

Example 5.20

A small architectural engineering firm purchases a new truck with no down payment. The truck costs $35,732 including taxes and registration fees. Monthly payments are $645 and the financing period is six years. What is the monthly, nominal, and effective annual interest rate?

$$P = \$35,732 \quad A = \$645 \quad n = 72 \text{ months} \quad i = ? \quad i_n = ? \quad i_a = ?$$

F4	▼	⋮	✕ ✓	f_x	=RATE($C$4,$C$5,$C$6)

◢	A	B	C	D	E	F
1	*Spreadsheets Interest Rate Estiamte*					
2						
3	Inputs:				Analysis:	
4	Periods: n		72		i =	0.754%
5	Monthly Payment: A	($645.00)			i(n) =	9.05%
6	Total Loan: P	$35,732.00			i(a) =	9.44%

F6	▼	⋮	✕ ✓	f_x	=((1+$F$4)^12)-1

◢	A	B	C	D	E	F
1	*Spreadsheets Interest Rate Estiamte*					
2						
3	Inputs:				Analysis:	
4	Periods: n		72		i =	0.754%
5	Monthly Payment: A	($645.00)			i(n) =	9.05%
6	Total Loan: P	$35,732.00			i(a) =	9.44%

Financial Block Functions

The present value or rate of return of cash flows across multiple periods (spreadsheet columns or rows) can be found using the NPV and IRR block functions (Table 5.3).

Table 5.3 Spreadsheet financial block functions

Argument	Description	Function
NPV: net present value	Net Present Value	NPV(i,values)
IRR: internal rate of return	Internal Rate of Return	RATE(n,A,P,F,type)

Example 5.21

A new production line will cost $500,000 to design and $7,000,000 to install. Revenue for the first year of product sales is estimated to be $2,500,000 and to increase by $250,000 for years 2 through 4, level at $3,250,000 for years 5 and 6, and decline by $100,000 per year for years 7 and 8. Operating and maintenance expenses are expected to be $1,000,000 the first year, increase by 10% in years 2 through 4, and remain steady at the year 4 level for years 5 through 8. Allocated overhead will be $250,000 each year, and the equipment will be depreciated by straight line at $937,500. The company requires 8.0% MARR on this type of project and pays combined 30% state and federal taxes. Estimate the net present value and internal rate of return.

Inputs:									
MARR		8.00%			Revenue				
Equipment cost						Yr 1	$2,500,000		
	Design	$500,000				Yr 2-4		$250,000 increase/yr	
	Purchase	$7,000,000				Yr 5-6	$3,250,000		
	SL Depr.	$937,500 /year				Yr 7-8		$100,000 decline/yr	
	Overhead		$250,000		O&M	Yr 1	$1,000,000		
	Depr		$937,500 / yr			Yr 2-4		10% increase/yr	
	Taxes		30%			Yr 5-8	= yr 4		

Output									
Year	0	1	2	3	4	5	6	7	8
Equip Cost	($7,500,000)								
Revenue		$2,500,000	$2,750,000	$3,000,000	$3,250,000	$3,250,000	$3,250,000	$3,150,000	$3,050,000
O&M Exp		($1,000,000)	($1,100,000)	($1,210,000)	($1,331,000)	($1,331,000)	($1,331,000)	($1,331,000)	($1,331,000)
Depr		($937,500)	($937,500)	($937,500)	($937,500)	($937,500)	($937,500)	($937,500)	($937,500)
Overhead		($250,000)	($250,000)	($250,000)	($250,000)	($250,000)	($250,000)	($250,000)	($250,000)
Earning bf taxes		$312,500	$462,500	$602,500	$731,500	$731,500	$731,500	$631,500	$531,500
Comb taxes		($93,750)	($138,750)	($180,750)	($219,450)	($219,450)	($219,450)	($189,450)	($159,450)
Net income		$218,750	$323,750	$421,750	$512,050	$512,050	$512,050	$442,050	$372,050
Cash flows	($7,500,000)	$1,156,250	$1,261,250	$1,359,250	$1,449,550	$1,449,550	$1,449,550	$1,379,550	$1,309,550
NPV	$208,864								
IRR	8.71%								

5.10 Summary

Risk Uncertainty about the future causes a decline in the value of money. Risk increases into future times. As risk increases, the value of money decreases.

Inflation The general price increase in an economy. As prices increase, the value of a monetary unit decreases.

Liquidity Investors have a ***preference for liquidity***, i.e., they prefer money today to the promise of future money. Investments in monetary returns generating assets are the same as a promise of future money. Investment must be made in the assets today for future returns.

The *fundamental relationship between risk and return*—the return on an investment should be proportional to the risk involved.

There are two components to risk volatility:

Structural   difference between long run average realized return and the expected return.

Random   random error difference between realized return and realized long run average return.

Equivalence of cash flows—any set of two or more cash flows are equivalent at interest rate i if we can convert one cash flow into the equivalent value of the other(s) by using the proper interest factors.

Principal—the initial amount money for a loan, bond, stock offering, or asset purchase.

Interest rate—the proportion cost of borrowed money per period of time.

Interest period—the time over which interest is paid.

Number of interest periods—duration of the loan, bond, stock offering, or asset purchase.

Receipts or disbursements schedule—the cash flow pattern with interest over the number of interest periods.

Future amount of money—cumulative effect of principal and interest over the number of interest periods.

Present amount of money—the current value of a set of future cash flows discounted for interest costs.

A_n = a discrete payment or receipt of money occurring at the end of an interest period.

i = interest rate per period.

N = number of interest periods.

P = present value or worth of a set of future cash flows discounted for interest costs.

F = single amount that is equivalent to the cumulative effect of principal and interest over the number of interest periods.

A = an end of period payment or receipt of money in a uniform series that continues for n periods; $A_1 = A_2 = \ldots = A_n$

Simple interest:

Total interest earned: $i_s = P \times i \times n$

Future amount due: $F = P + P \times i \times n = \underline{i}(1 + i \times n)$

Compound interest:

Single payment compound amount:

$$F = P(1+i)^n = P(F/P, i, n)$$

Single payment present worth amount:

$$P = F(1+i)^{-n} = F(P/F, i, n)$$

Uniform series compound amount:

$$F = A\left[\frac{(1+i)^n - 1}{i}\right] = A(F/A, i, n)$$

Uniform series sinking fund amount:

$$A = F\left[\frac{i}{(1+i)^n - 1}\right] = F(A/F, i, n)$$

Uniform series capital recovery amount:

$$A - P\left[\frac{i(1+i)^n}{(1+i)^n - 1}\right] = P(A/P, i, n)$$

Uniform series present worth amount:

$$P = A\left[\frac{(1+i)^n - 1}{i(1+i)^n}\right] = A(P/A, i, n)$$

Arithmetic gradient present worth amount:

$$P = G\left[\left((1+i)^n - 1 - in\right)/i^2(1+i)^n\right] = G(P/G, i, n)$$

Arithmetic gradient uniform series amount:

$$A = G\left[\left((1+i)^n - 1 - in\right)/\left(i(1+i)^n - i\right)\right] = G(A/G, i, n)$$

Geometric gradient present worth amount:

$$P = A_1 \frac{1 - ((1+g)/(1+i))^n}{i - g} = A_1(P/A, g, i, n)$$

where $i > g$.

$$P = A_1 n (1+i)^{-1}$$

where $i = g$.

Nominal and Effective Interest Rates.

Nominal interest rate/year, i_n: the stated annual interest rate on a loan without considering the effect of interest compounding.

Number of compounding subperiods, m: the number of compounding subperiods per year (i.e., month, quarter, semi-annual).

Effective interest rate/period, i: the nominal interest rate/year divided by the number of interest compounding periods.

Effective annual interest rate, i_a: the annual interest rate accounting for the effect of compounding interest over the number of interest compounding periods.

$$i_a = \left(1 + \frac{i_n}{m}\right)^{nm} - 1$$

Continuous Cash Flow Compounding

Single Payment Continuous Compound Amount

$$F = P(1 + i_n)^n = P e^{i_n n}$$

Single Payment Present Worth Amount

$$P = F(1 + i_n)^{-n} = F e^{-i_n n}$$

Continuous Compounding Sinking Fund Amount

$$(A/F, i, n) = \left[\frac{i}{(1+i)^n - 1}\right] = \frac{\left(e^{i_n} - 1\right)}{\left(e^{i_n n} - 1\right)}$$

Continuous Compounding Series Amount

$$(F/A, i, n) = \left[\frac{(1+i)^n - 1}{i}\right] = \frac{\left(e^{i_n n} - 1\right)}{\left(e^{i_n} - 1\right)}$$

Continuous Compounding Capital Recovery Amount

$$(A/P, i, n) = \left[\frac{i(1+i)^n}{(1+i)^n - 1} \right] = \frac{e^{i_n n}(e^{i_n} - 1)}{(e^{i_n n} - 1)}$$

Continuous Compounding Series Present Worth Amount

$$(A/P, i, n) = \left[\frac{i(1+i)^n}{(1+i)^n - 1} \right] = \frac{(e^{i_n n} - 1)}{e^{i_n n}(e^{i_n} - 1)}$$

5.11   Key Terms

Arithmetic gradient
Cash flow
Compound interest
Compounding period
Continuous compounding
Effective interest rate
Equivalence
Geometric gradient
Interest rate per period
Internal rate of return
Nominal interest rate
Payment period
Simple interest
Spreadsheet
Uniform cash flow series.

Problems

1. A proprietor borrowed $5000 from family members to launch his organic farm. His family members agreed to repayment at the end of three (3) years at annual simple interest of 6%. What is the final repayment?
2. A $5000 loan was repaid at annual 5% simple interest. The amount repaid was $6,250. How many years was the loan made?
3. How long will it take for an investment to double in value at 8.0% annual interest rate? (a) Using simple interest. (b) Using compound interest.
4. What sum of money at the present time is equivalent to $5627.50 at the end of two years if the interest rate is 6.0%, compounded every 6 months?
5. An engineering manager estimates that the replacement cost for a certain piece of equipment will cost $250,000 in five years. How much should be deposited today at 8.0% to pay for the equipment? (a) Use 8.0% annual interest rate. (b) Use 8.0% annual interest rate compounded quarterly.

6. Suppose that $2000 is deposited into an account that earns an annual 10% interest. (a) How much is in the account in 1, 2, 3, 4, and 5 years? (b) How much is in the account in 1, 2, 3, 4, and 5 years if annual interest is 10% compounded semi-annually?

7. A proprietor borrows $5000 from a bank to start his plumbing service. He is to repay $6000 to cover the principal and interest. If he repays the loan in (a) 2 years, (b) 3 years, or (c) 5 years, what is the corresponding interest rate?

8. Quickie Loans charges its clients a mere 1.0% interest per week for cash advance loans. What is the effective annual interest rate?

9. A local credit unit advertises that it pays 5.0% annual interest, compounded daily, on money deposited in a savings account provided the money is left in the account for five years. What is the effective annual interest rate? If $1,000 is deposited, how much is in the account at the end of five years?

10. A recently graduated engineer purchased a used automobile for transportation to her new engineering position with a local manufacturing firm. She paid $20,000 with no down payment. The loan is for three years with nominal annual interest of 9.0% compounded monthly. What is her monthly payment?

11. Compute the present and future values of the following cash flows over a five-year planning period. Use an annual interest rate $r = 5\%$.

Year	(a)	(b)	I
1	$100	$100	$150
2	$100	$100	$100
3	$100	$100	$50
4	$100	$0	$100
5	$100	$0	$150

12. A company purchases a machine for $20,500 on which it will make five equal annual payments at an annual nominal interest rate of 6%. The loan contract allows the company to pay off the loan at the end of any year when payment is due without penalty. What is the company's payoff amount for each year of the loan contract?

13. Given $(F/P, 0.015, 25) = 32.919$ and $(F/P, 0.18, 25) = 62.669$, use linear interpolation to determine the multiplier for $(F/P, 0.17, 25)$. What is the computed value using the formula for the single payment compound amount formula? Explain any difference in the estimates.

14. A natural gas company is installing solar powered, Wi-Fi meter readers to replace manual reading. The readers will be installed in equal stages in the company's operating region over the next ten years. The readers will save $25,000 in the first year, increasing $25,000 in each subsequent year. The natural gas company requires 7.0% return on cost-savings projects. What is the present worth and equivalent uniform annual worth of the project?

15. Snappy Lawn Service is considering the purchase of a new electric riding mower. As a sales incentive, the manufacturer provides the mower with a two-year warranty on service and parts. For this mower, online data indicates that the annual maintenance expense for the mower in its third year of use will be $160 and increase by $80 per year in years 4 through 7 of its useful life. How much money should Snappy deposit in an interest-bearing account at 4% today to pay for the mower's maintenance expenses over its useful life? What is Snappy's equivalent uniform annual cost of maintenance at the 4% interest rate?

16. A maintenance manager is considering is whether to install a new high-efficiency electric blower motor on a roof HVAC unit. The motor cost $2000 is reports energy saving of $400 the first year of use. The motor qualifies for an energy efficiency discount (additional savings) of 3.0% per year beginning with the second year of life and increasing at 3.0% per year until the end of the motor's useful life of five years. This type of project requires a 7.0% rate of return. Should the motor be installed?

Chapter 6
Measures of Investment Worth

Abstract Investing in productive assets requires not only the availability of money but also sound, relevant decision criterion. Equivalence provides the basis on which cash flows can be adjusted to account for the interest that must be paid on borrowed or invested money to compensate for the riskiness each alternative. In this chapter, we will explore two criteria used primarily for short-term investments and the four primary criteria used for medium-term and long-term investments.

Short-term investment:

- Undiscounted payback.
- Discounted payback.

  Medium-term and long-term investment:

- Net present worth.
- Equivalent uniform annual worth.
- Net future worth.
- Internal rate of return.

6.1 Investment Time Periods

Understanding the revenue and expense cash flows of an investment over its lifespan is key to building a budget. This is especially true when the asset is high risk and high value incurring higher interest and return on investment costs.

Definition: A *budget period* is the management equivalent of an accounting period (i.e., month, quarter, semi-annual, year, or multi-year) tied to the product life cycle.

© The Author(s), under exclusive license to Springer Nature Switzerland AG 2022   173
T. S. Cotter, *Engineering Managerial Economic Decision and Risk Analysis*,
Topics in Safety, Risk, Reliability and Quality 39,
https://doi.org/10.1007/978-3-030-87767-5_6

Budget periods correspond to specific future periods of time, such as a month, an accounting year, product life cycle, or a natural business cycle. Typical management budgeting periods are the zero-period budget for the current fiscal year, the life-cycle budget, and the long-term business-cycle budget. The *zero-period budget* delineates the short-term budget and sets controls for current fiscal year revenues and expenses. A *life-cycle budget* is a medium-term budget greater than one year that is an estimate of all costs to design, develop, install productive assets, market, produce revenues, and retire a product. A life-cycle budget is tied directly to an organization's product life cycle and may range from two to ten or fifteen years. A *business-cycle budget* is a long-term budget across multiple product life-cycle budgets that predict the investments needed in facilities and equipment to support a chain of products.

The conventional stages of product development include *embryonic*, *growth*, *maturity*, and *decline*. The *embryonic* stage covers the first four product life cycle phases (needs assessment and justification, conceptual design, detailed design, and production/construction) plus product introduction in the early operational phase. Sales growth tends to be slow because of buyer unfamiliarity with the product and the need to build product acceptance. This stage requires heavy start-up investments in research, manufacturing, and marketing. It tends to be unprofitable and often involves high risks and a negative cash-flows. In the *growth* stage, demand for the product or service is strong with sales growing at increasing rates. As sales grow more rapidly, the product begins to generate profits. In the *maturity* stage, market penetration saturates resulting in demand and corresponding sales stabilizing. Inefficient competitors are eliminated from the industry, and few, if any, entrants are attracted. Profits and cash flow peak in the maturity stage. In the *decline* stage, product sales decline as new technological competitors replace it. Management's goal is to close out the product with minimum loss and transition to its replacement.

The analysis period for an engineering managerial economic analysis depends on the project situation. Some engineering economic analyses may directly support a new product and be tied to its life cycle. Other engineering economic analyses may be for equipment upgrade or replacement and be independent of the products they support. Other engineering economic analyses may be for equipment required to fulfill legal or regulatory requirements and be independent of all products. In this chapter, we will begin with the assumption that the alternatives project lives are equal and equal to the project life. We will then relax this assumption and evaluate more realistic situations in which alternatives lives are not equal and not equal to the project life.

6.2 Undiscounted Payback Period

Definition: *Undiscounted payback period* is number of time periods required to recover the initial cost cash flow(s) of an investment from the net cash flows (profit or other benefits) produced by that investment for an interest rate of zero.

Criterion: Select the alternative with the minimum payback period.

If CFt = cash flow (cost or benefit) in period t, the payback period is defined as the smallest value of n that satisfies

$$\sum_n CF_t = 0 \tag{6.1}$$

Payback period is a popular method of selecting among investments in productive assets, because it is simple to estimate and indicates when the asset will pay off itself restoring lenders and stockholders to their initial financial position.

Example 6.1

An engineering manager seeks to select between two automatic labeling devices to apply shipping labels for a one-time, special-order 4-month contract. Both devices have a three-year technological life and will be sold at the end of the contract. Each labeling device has the following initial cost and after-tax net cash flow benefits. Use the undiscounted payback period to select between the two devices.

Month	A labels	B labels
0	(−$1200)	(−$1400)
1	$400	$400
2	$500	$400
3	$500	$500
4	$500 + $600	$600 + $1000

Payback A:

$$0 = \sum_0^4 CF_t = \left(-\$1200\right) + \$400 + \$500 + f\left(\$300/\$500\right)$$

$$= 2.6 \text{ months Select A}$$

Payback B:

$$0 = \sum_{0}^{4} CF_t = (-\$1400) + \$400 + \$400 + \$500 + f(\$100/\$600)$$

$$= 3.17 \text{ months}$$

Payback period may be visualized by the balance outstanding.

Month	A Labels	A Bal-ance	B Labels	B Balance
0	(-$1,200)	(-$1,200)	(-$1,400)	(-$,400)
1	$400	(-$800)	$400	(-$1,000)
2	$500	(-$300)	$400	(-$600)
3	$500	$200	$500	(-$100)
4	$500 + $600		$600 + $1,000	$500 + $1,000

6.3   Discounted Payback Period

Definition: *Discounted payback period* is the number of time periods required to recover the initial cost cash flow(s) of an investment from the net cash flows (profit or other benefits) produced by that investment at the MARR interest rate.

Criterion: Select the alternative with the minimum discounted payback period.

If CF_t = cash flow (cost or benefit) in period t, the discounted payback period is defined as the smallest value of n that satisfies

$$\sum_{n} CF_t(1+i)^{-t} = 0 \qquad (6.2)$$

Example 6.2

Repeat the payback period analysis of Example 6.1 using discounted payback for MARR = 5.0%.

Month	A labels	B labels
0	(−$1200)	(−$1400)
1	$400	$400
2	$500	$400
3	$500	$500
4	$500 + $600	$600 + $1,000

Payback A:

$$\sum_0^4 CF_t = (-\$1200) + \$400\,(P/F, 5\%, 1) + \$500\,(P/F, 5\%, 2)$$
$$+ f\big(\$500\,(P/F, 5\%, 3)\big)$$

$$\sum_0^4 CF_t = (-\$1200) + \$400\,(0.9524) + \$500\,(0.9070)$$
$$+ f\big(\$500\,(0.8638)\big)$$

$$\sum_0^4 CF_t = (-\$1200) + \$381 + \$454 + f\big(\$365/\$432\big)$$

$$f = \$365/\$432 = 0.85$$

Payback period = 2.85 months Select A.

Payback B:

$$\sum_0^4 CF_t = (-\$1400) + \$400\,(P/F, 5\%, 1) + \$400\,(P/F, 5\%, 2)$$
$$+ \$500\,(P/F, 5\%, 3) + f\big(\$600(P/F, 5\%, 4)\big)$$

$$\sum_0^4 CF_t = (-\$1400) + \$400\,(0.9524) + \$400\,(0.9070)$$
$$+ \$500\,(0.8638) + f\big(\$600(0.8227)\big)$$

$$\sum_0^4 CF_{tt} = (-\$1400) + \$381 + \$363 + \$319 + f\big(\$337/\$494\big)$$

$$f = \$337/\$494 = 0.68$$

Payback period = 3.68 months.

Problems with Payback Period

Although undiscounted and discounted payback are simple to apply and popular, both have significant deficiencies relative to the criteria of equivalence.

- Approximate economic analysis method.
- Prior to payback, the effect of timing is ignored when applying undiscounted payback.
- After payback, all economic consequences are ignored. In Example 6.2, the total cash flows for device A are $(-\$1,200) + \$400 + \$500 + \$500 + (\$500 + \$600) = \$1,300$, and the total cash flows for device B are $(-\$1400) + \$400 + \$400 + \$500 + (\$600 + \$1000) = \$1,500$ reversing the decision and favoring selection of device B.
- Will not necessarily produce a recommended alternative consistent with equivalent worth and rate of return methods.

The most serious deficiencies are:

- Undiscounted payback does not consider the time value of money.
- Neither consider the consequences of cash flows (investments or benefits) prior to or after the payback period and the expected life of the investment.

However, organizations often apply payback to the selection of alternatives for multi-year life-cycle and business-cycle budgeting. Given the above deficiencies, applying payback period to multi-year alternatives will not meet criteria of equivalence. Rather, payback period is applicable to small asset investments that can pay off initial cost with net cash flows during the zero-budget period or 12–18 months.

Example 6.3

Consider the following three alternatives, each using undiscounted payback period of three periods and MARR = 5.0%.

Month	A	B	C
0	(-$1,500)	(-$1,500)	(-$1,500)
1	$600	$400	$500
2	$500	$500	$500
3	$400	$600	$500
4	$200	$1,000	$0
5	$200	$2,000	$0
6	$200	$3,000	$0
Payback	3 months	3 months	3 months
NPV @ 5%	$341	$5,227	(-$138)

> By undiscounted payback period, all three alternatives are equally desirable.
> By NPV at MARR = 5.0%, prefer alternative B.

6.4   Net Present Worth

The net present worth criterion addresses the management question, "If we install asset X today for $1,000,000 (or any other initial cost), what is the net worth today of future cash revenues and expenses at the required MARR?" To answer this question, we will use the following notation.

i = Interest rate, MARR.

C_0 = Initial cost at time 0; a negative amount.

C_n = Cost or expense at the end of period n.

R_n = Revenue or income at the end of period n.

F_t = Net cash flow at the end of period n. $F_n = R_n - C_n$.

N = Project life.

n — Time measured in discrete compounding periods.

The *net present value* or *net present worth* of a project, NPV or NPW, is defined as the difference between discounted revenues and expenses at the MARR interest rate.

$$\text{NPW} = -C_0 + \sum_{t=1}^{N} \frac{R_n - C_n}{(1+i)^n} = -C_0 + \sum_{n=1}^{N} \frac{F_n}{(1+i)^n} \tag{6.3}$$

Criterion: Select the alternative that maximizes net present worth.

The procedure for applying the net present worth criterion (and equivalent uniform annual worth) is:

1. Determine the MARR interest rate applicable to the alternative or project.
2. Estimate the required alternative life.
3. Identify and estimate expected revenue or benefit cash flows and associated expense cash flows for each period over the required life.
4. Determine the net cash flows using Eq. (6.3).
5. Find the present worth of each net cash flow.
6. Sum the net present worth values to estimate the net present worth.
7. If the alternative or project net present worth >$0, the alternative or project is acceptable for consideration.

8. Among alternatives with equivalent lives, select the alternative with the maximum net present worth.

Example 6.4

A new product will cost $1,000,000 to develop and install manufacturing equipment and will generate the follow revenues and expenses ($\times$ $1000) including taxes over its five-year life. This product category must earn MARR $= 10.0\%$. Estimate its net present worth.

Time	0	1	2	3	4	5
Revenue, r_n	$0	$550	$550	$550	$550	$550
Expense, c_n	$1000	$125	$175	$225	$275	$325
Net, F_n	(−$1000)	$425	$375	$325	$275	$275

Using steps 4 and 5 of the net present worth procedure:

$$\text{NPW} = -\$1,000 + \frac{\$425}{(1+0.1)^1} + \frac{\$375}{(1+0.1)^2} + \frac{\$325}{(1+0.1)^3} + \frac{\$275}{(1+0.1)^4}$$
$$+ \frac{\$225}{(1+0.1)^5}$$

$$\text{NPW} = +\$267.99$$

We can also view the net cash flows as an arithmetic gradient:

$$\text{NPW} = -\$1000 + (\$425(P/A, 10\%, 5) - \$50(P/G, 10\%, 5))$$

$$\text{NPW} = -\$1000 + (\$425(3.791) - \$50(6.862))$$

$$\text{NPW} = +\$268.08$$

6.5 Equivalent Uniform Annual Worth

Equivalent uniform annual worth addresses two investment questions.

1. For lenders, will this project generate sufficient cash flows to make its loan payments and to generate sufficient after-tax cash flows for re-investment in future products?

2. For management, what are the expected annual after-tax cash flows discounted for the MARR?

The equivalent uniform annual worth criterion provides the basis for determining discounted equal annual cash flows. The *equivalent uniform annual worth*, EUAW, of an alternative is defined as its annualized Net Present Worth.

$$\text{EUAW} = \text{NPW}\left[\frac{i(1_- + i)^N}{(1 + i)^N - 1}\right] = \text{NPW}(A/P, i, N) \qquad (6.4)$$

Criterion: Select the alternative that maximizes equivalent uniform annual worth.

The *equivalent uniform annual worth* of an alternative is the difference between the *equivalent uniform annual benefit* and the *equivalent uniform annual cost*; EUAW = EUAB − EUAC. For example, the EUAW of the project in Example 6.4 is estimated in Example 6.5.

Example 6.5

Estimate the EUAW of the after-tax cash flows for the alternative in Example 6.4.

$$\text{EUAW} = +\$268.08(A/P, 10\%, 5) - \$268.08(0.2638) - \$70.72$$

Interpretation: A lender is assured that the alternative will be able to repay a loan at MARR = 10% plus yield EUAW = \$70,720 per year for reinvestment in new products. Management should budget and additional EUAW = \$70,720 after-tax cash flow per year from this alternative. At MARR = 10%, both the lender and management are indifferent to the actual net cash flows F_n or to the EUAW.

6.6 Future Worth

Net present worth measures the discounted cash flow worth of an alternative at time 0. Correspondingly, equivalent uniform annual worth measures the discounted cash flow worth of an alternative at the end of each time period t of its life. Net *future worth* measures the compounded cash flows of an alternative at the end of any time period t in the future or at the end of the alternative's useful life t_n. The net future worth is most useful when we need to compare investments that must be made over time such as building a new facility or adding a new production line, which each may take two or three years. Some large-scale projects such as building a new power plant may take seven to ten years due to the complexities of design and meeting governmental regulations. Governmental projects such as building a new or upgrading and existing interstate highway or an airport runway may take up to 20 years of development engineering and construction.

The *future value* or *future worth* of a project, FV or FW, is defined as the difference between compounded revenues and expenses at the MARR interest rate at the end of an alternative's life.

$$FW = C_0(1 + i)^N + \sum_{n=1}^{N}(r_n - c_n)(1 + i)^{N-n}$$

$$= C_0(F/P, i, N) + \sum_{n=1}^{N}(r_n - c_n)\left(\frac{F}{P}, i, N - n\right) \tag{6.5}$$

Criterion: Select the alternative that maximizes future worth.

The simplest approach to estimating a future worth is to first estimate the net present worth or equivalent uniform annual worth from an alternative's cash flows and us the appropriate F/P or F/A factor to estimate the compounded future worth.

$$F = P(F/P, i, n) \tag{6.6}$$

$$F = A(F/A, i, n) \tag{6.7}$$

Example 6.6

Estimate the FW of the after-tax cash flows for the alternative in Example 6.4.

Time	0	1	2	3	4	5
Revenue, r_n	$0	$550	$550	$550	$550	$550
Expense, c_n	$1000	$125	$175	$225	$275	$325
Net, F_n	(−$1,000)	$425	$375	$325	$275	$275
EUAW	$0	$70.72	$70.72	$70.72	$70.72	$70.72

$$FW = -\$1000(1 + 0.1)^5 + \$425(1 + 0.1)^4 + \$375(1 + 0.1)^3$$
$$+ \$325(1 + 0.1)^2 + \$275(1 + 0.1)^1 + \$225 = \$431.61$$

$$FW = P(F/P, i, n) = \$268(F/P, 10\%, 5) = \$268(1.611) = \$431.75$$

$$FW = A(F/A, i, n) = \$268(F/A, 10\%, 5) = \$70.72(6.105) = \$431.75$$

Example 6.7

A west coast robotics firm has decided to build a production facility on the US east coast to manufacture an innovative new robot for sale there and into the European market. The project manager has identified two options. She has located an existing vacant building in Maryland that can be purchased for $9,500,000 and remodeled. She has also located land in central Virginia that can be purchased for $850,000. With production scale-up, the project will require four years engineering and construction before full release to manufacturing. The expected costs per year are set forth in the following table. The MARR for this project is 12%. Estimate the time 0 initial cost at the end of year 4 for release to manufacturing. Estimate the project present worth and equivalent uniform annual worth to the firm

Engineering, Construction, Scale-Up:

Year	Vacant building		New building	
0	Purchase building	$9,500,000	Purchase land	$850,000
1	Design engineering	$3,500,000	Develop land and design	$4,000,000
2	Remodel	$4,500,000	Construction	$21,600,000
3	Production equip.	$2,500,000	Production equip.	$2,000,000
4 (0)	Scale-up revenue	$500,000	Scale-up revenue	$2,750,000
	Scale-up expenses	$1,900,000	Scale-up expenses	$2,300,000

Operational Revenues and Expenses (including depreciation and taxes):

	Vacant building		New building	
Year	Revenue	Expenses	Revenue	Expenses
5 (1)	$10,000,000	$4,400,000	$10,000,000	$4,000,000
6 (2)	$18,000,000	$11,000,000	$18,000,000	$10,000,000
7 (3)	$31,000,000	$19,800,000	$31,000,000	$18,000,000
8 (4)	$49,000,000	$34,100,000	$49,000,000	$31,000,000
9 (5)	$47,000,000	$34,100,000	$47,000,000	$31,000,000
10 (6)	$45,000,000	$34,100,000	$45,000,000	$31,000,000

Future Worth: Engineering, Construction, Scale-Up—Vacant Building

$$FW = \$9.5(F/P, 12\%, 4) + \$3.5(F/P, 12\%, 3)$$
$$+ \$4.5(F/P, 12\%, 2) + \$2.5(F/P, 12\%, 1) + \$1.4$$

$$FW = \$9.5(1.574) + \$3.5(1.405) + \$4.5(1.254)$$
$$+ \$2.5(1.120) + \$1.4 = \$29,713,500$$

Future Worth: Engineering, Construction, Scale-Up—New Building

$$FW = \$0.85(F/P, 12\%, 4) + \$4.0(F/P, 12\%, 3) + \$21.6(F/P, 12\%, 2)$$
$$+ \$2.0(F/P, 12\%, 1) - \$0.45$$

$$FW = \$0.85s(1.574) + \$4.0(1.405) + \$21.6(1.254) + \$2.0(1.120)$$
$$- \$0.45 = \$35,834,300$$

Present Worth: Operations − Vacant Building

NPW = (−\$29.7135) + \$5.6(P/F,12%,1) + \$7.0(P/F,12%,2) + \$11.2(P/F,12%,3) + \$14.9(P/F,12%,4) + \$12.9(P/F,12%,5) + \$10.9(P/F,12%,6)

NPW = (−\$29.7135) + \$5.6(0.8929) + \$7.0(0.7972) + \$11.2(0.7118) + \$14.9(0.6355) + \$12.9(0.5674) + \$10.9(0.5066) = \$11.14965 or \$11,149,650

Present Worth: Operations − New Building

PW = (\$35.8343) + \$6.0(P/F,12%,1) + \$8.0(P/F,12%,2) + \$13.0(P/F,12%,3) + \$18.0(P/F,12%,4) + \$16.0(P/F,12%,5) + \$14.0(P/F,12%,6)

PW = (\$35.8343) + \$6.0(0.8929) + \$8.0(0.7972) + \$13.0(0.7118) + \$18.0(0.6355) + \$16.0(0.5674) + \$14.0(0.5066) = \$12.7639 or \$12,763,900

6.7 Internal Rate of Return

The internal rate of return criterion addresses stockholders' question of return on their investment. Stockholders fundamental question is, "Will management's proposed portfolio of new investments yield a weighted return on investment ≥MARR?" The *internal rate of return* is the interest rate at which the equivalent benefits are equal to the equivalent costs.

$$PW(\text{Benefits}, i) = PW(\text{Costs}, i)$$
$$EUAB(i) = EUAC(i)$$
$$FW(\text{Benefits}, i) = FW(\text{Costs}, i)$$

To calculate a rate of return:

1. Convert benefits and costs into cash flows.
2. Solve the cash flows for the rate of return that equates equivalent benefits and equivalent costs.

$$IRR = i * -\text{interest rate at which NPW} = \$0 \text{ or EUAW} = \$0$$

- PW Benefit – PW Cost = $0.
- NPW = $0.
- EUAW = EUAB – EUAC = $0.
- FW Benefit – FW Cost = $0.

Criterion: Select the alternative that maximizes IRR $\geq$ MARR.

Example 6.8

Estimate the internal rate of tension of the after-tax cash flows for the alternative in Example 6.4.

Example: MARR 10%. Cash flows $\times$ $1000. IRR = ?

Time	0	1	2	3	4	5
Revenue, r_n	$0	$550	$550	$550	$550	$550
Expense, c_n	$1000	$125	$175	$225	$275	$325
Net, F_n	(−$1000)	$425	$375	$325	$275	$225

$0 = PW Benefit – PW Cost

$$\$0 = \$425(P/A, i, 5) - \$50(P/G, i, 5) - \$1000$$

Try $i = 20\%$:

$$\$0 = \$425(P/A, 20\%, 5) - \$50(P/G, 20\%, 5) - \$1000$$

$$\$0 = \$425(2.991) - \$50(4.906) - \$1000$$

$$\$0 \neq \$25.88$$

Try $i = 25\%$:

$$\$0 = \$425(P/A, 25\%, 5) - \$50(P/G, 25\%, 5) - \$1000$$

$$\$0 = \$425(2.689) - \$50(4.204) - \$1000$$

$$\$0 \neq (-\$67.38)$$

Linear interpolation:

$$\left\{ a \begin{cases} 20\% & \$25.88 \\ \text{IRR} & \$0 \\ 25\% & -\$67.38 \end{cases} c \right\} d$$

$$\text{Ratio} = \frac{a}{b} = \frac{c}{d}$$

$$\frac{\text{IRR} - 20\%}{25\% - 20\%} = \frac{\$0 - \$25.88}{-\$67.38 - \$25.88}$$

$$i = 20\% + (25\% - 20\%)\left(\frac{\$0 - \$25.88}{-\$67.38 - \$25.88} \right) = 21.388$$

Example 6.9

A newly graduated engineer obtains his first engineering position and wants to invest \$5000 at the end of every year for 40 years toward his retirement. If he desires \$1,000,000 to be in the investment when he retires, what average interest rate must the investment earn?

$F = \$1,000,000\ A = \$5000\ \text{IRR} = ?$

$\text{EUAB} = \$1,000,000\ (A/F, i, 40) = \$5000 = \text{EUAC}$

$(A/F, i, 40) = \$5000/\$1,000,000$

$(A/F, i, 40) = 0.00500$

From the compound interest tables

i	$(A/F, i, 40)$
6.0%	0.00646
7.0%	0.00501
8.0%	0.00386

Linear interpolation is not needed.

6.8 Investment Worth Metrics Under Differing Project Life Analysis Periods

In measuring interest equivalence, we must carefully consider the analysis period or planning horizon. Differing combinations of alternatives and project lives include:

- Useful life of each alternative equals each other and the analysis period.
- Alternatives have useful lives different from each other and from the analysis period.

  - *Least common multiple life analysis.*
  - *Project life analysis.*

- Special case, EUAW of a continuing requirement.
- The analysis period is infinite, $n = \infty$.

For the second case of unequal lives, we must also consider how end of alternative life cash flows terminate. One of two cases may occur. Either the asset life will be longer than the project life, or the asset life will be shorter than the project life. In the first case, we must use the assets market value at the end of the project life. In the second case, we must use the assets salvage value at the end of its life and re-purchase the same asset for use until the end of the project life. At the end of the project life, we use the market value of the replacement asset to value the asset's cash value.

Cash flow equation for asset disposal *during* its useful life:

$$ NPW = P_0 + \sum_1^t \left(\text{Net Cash Flow}_j \right) (P/F, i, j) + \text{Market Value}_t (P/F, i, t)a $$

Cash flow equation for asset disposal *at the end* of its useful life:

$$ NPW = P_0 + \sum_1^n \left(\text{Net Cash Flow}_j \right) (P/F, i, j) + \text{Salvage Value}_n (P/F, i, n) $$

The above representations can be made equivalent to annual and future cash flows, since $EUAW = NPW(A/P, i, n)$ and $FW = NPW(F/P, i, n)$.

Example 6.10

A development engineering manager must decide between two vendors for a new automated fluorescence metallurgical microscope. The microscope is required to examine the grain structure of a new alloy in the quality control laboratory. The microscope will reduce false rejects of required grain size distribution relative to specification and result decreased rework costs (uniform annual benefit). The new alloy is expected to have a six-year technological life before competitors can duplicate its properties. At MARR = 8.0%, which microscope should be purchased?

Cash flow	Nixon	Zike
Initial cost	$20,000	$30,000
Net uniform annual benefit	$4500	$6000
Salvage value	$1000	$7000
Useful life	6 years	6 years

Since each alternative's useful life equal each other's and the project life, terminate the asset's cash flows with its salvage value. The cash flow equation for each alternative is,

$$PW = -P_0 + A(P/A, 8\%, 6) + S(P/F, 8\%, 6)$$

Nixon:

$$PW = (-\$20,000) + \$4500(P/A, 8\%, 6) + \$1000(P/F, 8\%, 6)$$
$$PW = (-\$20,000) + \$4500(4.623) + \$1000(0.6302)$$
$$PW = \$1433.70$$

Zike:

$$PW = (-\$30,000) + \$6000(P/A, 8\%, 6) + \$7000(P/F, 8\%, 6)$$
$$PW = (-\$30,000) + \$6000(4.623) + \$7000(0.6302)$$
$$PW = \$2149.40$$

Decision: Select the Zike microscope to maximize NPW at MARR = 8.0%.

In our application of equivalence to this point, we have considered only situations in which alternatives useful lives were equal to each other and to the project period. There are cases where this assumption does not hold. When the useful lives of some alternatives differ, we must select a common service period. For projects with finite lives, two methods of common service period analyses are least common multiple life and project life. As a simple example, if alternative one has a useful life of three years, alternative two a useful life of four years, and the project life is six years, the engineering economic analyst would choose a common service period of 12 years or two project lives. During the common analysis period, alternative one would be purchased four times, each purchase used to its useful life of three years at which time each would be scrapped and salvage value realized. Alternative two would be purchased three times, each purchase used to its useful life of four years at which time each would be scrapped and salvage value realized.

Example 6.11

A calibration provider needs to purchase a new high precision camera calibration system for a new robot calibration service. Two systems are under consideration, and their respective cash flows are given in the following table. The robot calibration service is expected to last for 12 years and must earn MARR = 8.0%.

Cash flow	Alt 1	Alt 2
Initial cost	$500,000	$750,000
Calibration service net cash flows	$235,000	$255,000
Salvage value	$100,000	$160,000
Useful life	4 years	6 years

Use 12 years as the least common multiple life for the common service period.

$NPW(1) = -P_0 + A(P/A, 8\%, 12) + (S_4 - P_0)(P/F, 8\%, 4) + (S_8 - P_0)(P/F, 8\%, 8) + S_4(P/F, 8\%, 12)$

$NPW(1) = (-\$500,000) + \$235,000(7.536) + (-\$400,000)(0.7350) + (-\$400,000)(0.5403) + \$100,000(0.3971)$

$NPW(1) = \$800,550$

$NPW(2) = -P_0 + A(P/A, 8\%, 12) + (S_6 - P_0)(P/F, 8\%, 6) + S_6(P/F, 8\%, 12)$

$NPW(2) = (-\$750,000) + \$255,000(7.536) + (-\$590,000)(0.6302) + \$160,000(0.3971)$

$NPW(2) = \$863,398$ Select Alternative 2.

When the alternatives lives and the project life are unequal such that a reasonable least common multiple life cannot be used, we can use the *project life analysis* given that we have an estimate of the *market value* of each alternative in its year of disposal.

Example 6.12

Reanalyze problem 6.10 using project life analysis for the following alternatives' cash flows, useful lives, and market values.

Cash flow	Alt 1	Alt 2
Initial cost	$500,000	$750,000
Calibration service net cash flows	$235,500	$255,000
Salvage value	$100,000	$160,000
Market value in year of disposal	$250,000	$180,000
Useful life	7 years	13 years

$NPW(1) = -P_0 + (S_7 - P_0)(P/F, 8\%, 7) + A(P/A, 8\%, 12) + M_5(P/F, 8\%, 12)$

$NPW(1) = (-\$500,000) + (-\$400,000)(0.5835) + \$235,000(7.536) + \$250,000(0.3971)$

$NPW(1) = \$1,136,835$

$NPW(2) = -P_0 + A(P/A, 8\%, 12) + M_{12}(P/F, 8\%, 12)$

$NPW(2) = (-\$750,000) + \$255,000(7.536) + \$180,000(0.3971)$

$NPW(2) = \$1,243,158$ Select Alternative 2.

Example 6.13

Estimate the EUAW of Problems 6.10 and 6.11.

Problem 6.10 $EUAW(1) = P(A/P, 8.0\%, 12) = \$800,500(0.1327) = \$106,226.35$

$EUAW(2) = P(A/P, 8.0\%, 12) = \$863,398(0.1327) = \$114,572.91$

Problem 6.11 $EUAW(1) = P(A/P, 8.0\%, 12) = \$1,136,835(0.1327) = \$150,858.00$

$$EUAW(2) = P(A/P, 8.0\%, 12) = \$1,243,158(0.1327) = \$164,967.07$$

EUAW of a Continuing Requirement

One exception to the common service period is the equivalent uniform annual worth estimate for a continuing requirement (i.e., an asset is needed for the long-term business-cycle budget period but has a useful life much shorter than the business-cycle budget period). A continuing requirement is an asset that is required for basic organizational functioning regardless of the product life cycle mix. Examples include the organizational information system, heating-ventilation-air-conditioning systems, utilities, logistics equipment, and equipment required to meet regulatory requirements. In the case of a continuing requirement, we just need to estimate the one-life EUAW of each alternative. The EUAW will not change over future common multiple lives of the same alternative.

Example 6.14

Two network servers are being considered for purchase to support a new process line. Whichever server is purchased, it will continue to be purchased to replace itself over the expected 60-year facility life. MARR $= 7.0\%$. Estimate the one-life EUAC for each alternative server and the EUAC for each alternative for the least common multiple life of the two servers.

Cash flow	Server A	Server B
Initial cost	$17,000	$15,000
Salvage value	$1500	$1000
Useful life	12 years	9 years

One-life EUAC estimate.

$$\begin{aligned}
EUAC(A) &= -P(A/P, i, n) + S(A/F, i, n) \\
&= (-\$17,000)(A/P, 7\%, 12) + \$1500\,(A/F, 7\%, 12) \\
&= (-\$17,000)(0.1259) + \$1,500\,(0.0559) \\
&= (-\$2,056.45) \quad \text{Select server } A.
\end{aligned}$$

$$\begin{aligned}
EUAC(B) &= -P(A/P, i, n) + S(A/F, i, n) \\
&= (-\$15,000)(A/P, 7\%, 9) + \$1000\,(A/F, 7\%, 9) \\
&= (-\$15,000)(0.1535) + \$1,000\,(0.0835) \\
&= (-\$2,219.00)
\end{aligned}$$

Least common multiple life EUAC estimate.

$$\begin{aligned}
\text{EUAC}(A) &= [-P_0 + (S_{12} - P_0)(P/F, 7\%, 12) + (S_{24} - P_0)(P/F, 7\%, 24) \\
&\quad + S(P/F, 7\%, 36)](A/P, 7\%, 36) \\
&= [(-\$17,000) + (\$1,500 - \$17,000)(0.4440) \\
&\quad + (\$1,500 - \$17,000)(0.1971) + \$1500(0.0883)](0.0768) \\
&= (-\$26,804.60)(0.0768) = (-\$2,058.59)
\end{aligned}$$

$2 difference due to rounding error.

$$\begin{aligned}
\text{EUAC}(B) &= [-P_0 + (S_9 - P_0)(P/F, 7\%, 9) + (S_{18} - P_0)(P/F, 7\%, 18) \\
&\quad + (S_{27} - P_0)(P/F, 7\%, 27) + S(P/F, 7\%, 36)](A/P, 7\%, 36) \\
&= [(-\$15,000) + (\$1,000 - \$15,000)(0.5439) \\
&\quad + (\$1,000 - \$15,000)(0.2959) \\
&\quad + (\$1,000 - \$15,000)(0.1609) \\
&\quad + \$1,000(0.0883)](0.0768) \\
&= (-\$28,921.50)(0.0768) = (-\$2,221.17)
\end{aligned}$$

$2 difference due to rounding error.

Infinite Analysis Period

In the governmental, utilities, and transportation sectors, some asset investments have project lives that span multiple decades or become permanent infrastructure. Since the compound interest values asymptotically approach their exponential limits, these projects lives can be treated as infinite. Present worth estimates for projects with essentially infinite lives are termed capitalized cost analysis.

Definition: *Capitalized cost* is the present worth of an infinite series of future cash flows, at a stated interest rate, that represents the amount of money that needs to be invested at time 0 to cover all future expenditures to maintain the service or asset.

The present value of the initial investment must never decline. For any present amount P, there can be an end of period withdrawal of $A = iP$ at the end of each period into the infinite future. This sets up the fundamental relationship

$$\text{For } n = 1 \to \infty, \quad A = iP \tag{6.8}$$

Capitalized cost is the P in Equation (6.8).

$$\text{For } n = 1 \to \infty, \quad P = A/i \tag{6.9}$$

Example 6.15

A local electric utility company needs to install power lines to a new suburb. The initial installation will cost $12,000,000 and will have an expected life of 100 years at which time the towers and lines will have to be replaced. The utility company will need to keep the power service indefinitely. Assuming a 7% interest rate for this project, how much money will the utility company have to invest at time 0 to finance the power line's initial construction and all future replacements? Explain the future cash flows.

From Example 6.13, the continuing requirement annual cost is,

$$A = P(A/P, 7\%, 100)$$
$$= \$12,000,000\,(0.07000808)$$
$$= \$840,969.18$$

Thus, the required time 0 investment is,

$$P = A/i$$
$$= \$840,969.18/0.07$$
$$= \$12,013,845.36$$

If the electric utility company invests a total of $12,013,845.36 ($12,000,000 initial cost plus $13,845.36 in a 7% interest-bearing account), the interest-bearing account will grow to $13,845.36(1 + 0.07)^{100} - \$12,013,845.36$ every 100 years allowing replacement of the old power lines.

Multiple Alternatives

Multiple alternative problems are solved using the same methods as problems with two alternatives. Analysis is determined by the cash flows.

- Useful life of each alternative equals each other and the analysis period.
- Alternatives have useful lives different from each other and from the analysis period.

  - *Least common multiple life analysis.*
  - *Project life analysis.*

- Special case, EUAW of a continuing requirement.
- The analysis period is infinite, $n = \infty$.

Example 6.16

A bridge contractor needs a water pump and pipe to remove water from an excavated hole during footing installation. Engineering estimates estimate that water seepage will be about 25 gal/min. Available matched pumps and pipe diameter options are presented in the following table. Pipe length to the adjacent river will be 90 ft. downstream to avoid back seepage. Work on the footing will be completed in four months working 24-hour days. The contractor's MARR = 9.0% for this project.

	Matching pump hp versus pipe diameter (in.)			
Pump hp–self priming	1.0	1.5	2.0	2.5
Pump initial cost	$400	$450	$490	$700
Pipe diameter	1	1¼	1½	2
Capacity (gal/min)	25	45	60	80
Pipe installed cost	$450	$475	$510	$625
Pumping cost/h	$1.20	$0.90	$0.70	$0.60

Interest rate/period = 9.0%/3 = 3.0% per 4-month period.

Estimate h/day pumping time.

1 hp pump w 1-in. pipe ~ (25 gal/min/25 gal/min) × 24 h = 24 h/day

1.5 hp pump w 1.5-in. pipe ~ (25 gal/min/45 gal/min) × 24 h = 13.33 h/day

2.0 hp pump w 2-in. pipe ~ (25 gal/min/60 gal/min) × 24 h = 10 h/day

2.5 hp pump w 2.5-in. pipe ~ (25 gal/min/80 gal/min) × 24 h = 7.50 h/day

Estimate pumping cost per month.

1 hp pump w 1-in. pipe ~ $1.20/h × 24 h/day × 30.5 days/mon = $878.40

1.5 hp pump w 1.5-in. pipe ~ $0.90/h × 13.33 h/day × 30.5 days/mon = $365.91

2.0 hp pump w 2-in. pipe ~ $0.70/h × 10 h/day × 30.5 days/mon = $213.50

2.5 hp pump w 2.5-in. pipe ~ $0.60/h × 7.5 h/day × 30.5 days/mon

= $137.25

$PW(a) = (-P_0(\text{pump}) - P_0(\text{pipe})) + (\text{pumping cost/mon})(P/A, 3.0\%, 4)$

	Matching pump hp versus pipe diameter (in.)			
	1.0	1.5	2.0	2.5
Pump hp–self priming	$400	$450	$490	$700
Pipe installed cost	$450	$470	$510	$625
1.0|$878.40 × 3.717	$3265			
1.5|$365.91 × 3.717		$1360		
2.0|$213.50 × 3.717			$794	
2.5|$137.25 × 3.717				$510
PW(a)	$4115	$2280	$1794	$1835

Select the 2 hp pump and 1½ in. diameter pipe combination.

Example 6.17

A property management firm has an empty building that its engineering manager indicates can leased for a retail space, food market, or fitness gymnasium. The engineering manager submitted the following conversion estimates. The firm's MARR = 12% on all lease investments. In addition to the following alternatives, include the "Do Nothing" option in alternatives estimates.

Alternative	Initial conversion cost P_0	Monthly lease revenue A	Salvage value end of 5 years S
Retail space	$52,500	$5,400	$31,500
Food market	$137,500	$17,250	$82,500
Fitness gym	$99,750	$11,600	$42,750

Since property management firms are heavily dependent on monthly cash flows to stay viable, EUAW analysis will be used. The equivalence equation for each alternative is,

$EUAW(a) = -P_0(A/P, 1\%, 60) + A + S(A/F, 1\%, 60)$

Do nothing : $EUAW = -\$0(0.0222) + \$0 + \$0(0.0122) = \$0$

Retail space : $EUAW = (-\$52,500)(0.0222) + \$1000 + \$31,500(0.0122)$
$= \$218.80$

Food market: $EUAW = (-\$137,500)(0.0222) + \$2500 + \$82,500(0.0122)$
$= \$454.00$

Fitness gym:$EUAW = (-\$99,750)(0.0222) + \$1700 + \$42,750(0.0122)$
$= \$7.10$

Invest in converting the empty building to a food market and seek a marketer for a lease contract. Note that the fitness gymnasium EUAW is equivalent to the do-nothing option and should not be pursued.

6.9 Investment Worth Metrics in Spreadsheet Analyses

In chapter five, it was noted that spreadsheets have become the standard for acquiring, managing, and analyzing engineering managerial economic data. Many of the primary investment worth metric estimates have been illustrated in the examples in chapter five and this chapter. In this section, we will illustrate the use of spreadsheets of two common investment worth estimates: (1) bond pricing and (2) building an amortization schedule.

Bond Pricing. Governments and corporations issue bonds to raise funds for public projects and for expansion of productive capacity. Several types of bonds are sold in financial markets.

Treasury notes and bonds. The United States Treasury issues notes and bonds to pay for public projects and finance the national debt.

Treasury bills. These are short-term securities, 13–52 weeks, typically sold at a discount to par value (also known as face value). When the bill matures, its par value is paid to the holder. The difference between the purchase price and par value represents the interest earned.

Municipal bonds. State and local governments issue bonds to finance public projects that cannot be funded by taxes. There are two basic types of municipal bonds. *General obligation bonds* are backed by the credit worthiness of the issuing agent. *Revenue bonds* are repaid from a specific income source, typically usage fees, pledged in their contracts.

Corporate bonds. *Debentures* are backed by the credit worthiness of the corporation. *Asset backed bonds* are collateralized by a pool of assets, such as property or equipment, loans, leases, credit card debt, royalties, or receivables.

Zero-coupon bonds. Governments and corporations sell zero-coupon bonds at a deep discount. Interest is not paid. Rather the value of the bond increases to its par value when it matures.

Callable bonds. The issuer may recall and pay off the bond before its maturity date. Organizations issue callable bonds when they believe that interest rates will decline in the future. With callable bonds, organization issues the bonds at a current interest rate. If the interest rate does decline, the organization will pay off the outstanding bonds and issue new bonds at the lower rate. If the interest rate remains the same or increases, the issuer may allow the callable bonds to remain outstanding to full maturity. To protect the bondholders who expect the higher interest rate, recallable bonds are typically issued with a non-callable option for a specified period.

Floating-rate bonds. These bonds are issued with adjustable interest rates if the general interest rate increases. This prevents investors from being "locked" into bonds with unattractive lower interest rates.

Corporate bond prices are quoted as a percentage of the bond's face value or in 1/8's fraction of $10 point values. For the base 1/8th point, each fraction paid in annual interest is $1.25. For example, a bond quoted as 96¾ is selling for $967.50. Treasury bonds are quoted in 1/32nds, 31.25 cents, of a point.

Bonds are traded on the bond market just like stocks on the stock market exchanges. Depending on the bond's interest rate relative to the current market interest rate, a bond will may sell at below or above its face value. This differential in the selling price versus the face value is the measure of **yield to maturity** or the actual interest earned over its holding period. A bond's **current yield** is the annual interest earned as a percentage of its current market price.

Example 6.18

Consider purchasing a ten-year $1000 corporate bond with an annual coupon rate of $80 paid semiannually. The current market price is $992.50. Find the yield to maturity and current yield.

$$P = \$992.50 \text{ Face value} = \$1000, i_n = \$80, n = 10 \text{ years}$$
$$m = 2 \text{ subperiods/year}$$
$$i = \$80/2 = \$40 \text{ or } 4\% \text{ per semiannual period}$$
$$\text{Periods} = 2 \times 10 = 20 \text{ periods}$$

$$P = A(P/A, i, p) + F(P/F, i, p)$$
$$\$992.50 = \$40(P/A, i, 20) + \$1000(P/F, i, 20)$$

Since the bond price $992.50 < $1000 face value, the yield to maturity interest rate must be higher than the nominal semiannual interest rate of 4.0%. Try, 4.5%.

$$\$992.50 = \$40(P/A, 4.5\%, 20) + \$1000(P/F, 4.5\%, 20)$$
$$\$992.50 = \$40(13.008) + \$1000(0.4146)$$
$$\$992.50 \neq \$934.92$$

Linear interpolation:

$$\left\{ a \left\{ \begin{array}{ll} 4\% & \$1,000 \\ i & \$992.50 \\ 4.5\% & \$934.92 \end{array} \right\} c \right\} d$$

$$\text{Ratio} = \frac{a}{b} = \frac{c}{d}$$

$$\frac{i - 4\%}{4.5\% - 4\%} = \frac{\$992.50 - \$1000}{-\$934.92 - \$1000}$$

$$i = 4\% + (4.5\% - 4\%) \left(\frac{\$992.50 - \$1000}{\$934.92 - \$1000} \right)$$

The semiannual yield to maturity = 4.0576% and the annual yield to maturity = 8.1152%. The semiannual current yield = $40/$992.50 = 0.0403 or 4.03%, and the annual current yield = 8.06%.

We can set up a MS Excel® spreadsheet with the input data and use its Goal Seek function to estimate the semiannual yield to maturity.

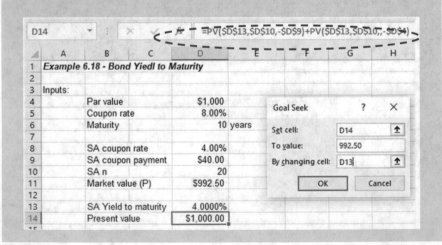

	A	B	C	D	E
1	**Example 6.18 - Bond Yiedl to Maturity**				
2					
3	Inputs:				
4		Par value		$1,000	
5		Coupon rate		8.00%	
6		Maturity		10 years	
7					
8		SA coupon rate		4.00%	
9		SA coupon payment		$40.00	
10		SA n		20	
11		Market value (P)		$992.50	
12					
13		SA Yield to maturity		4.0555%	
14		Present value		$992.50	
15					

Loan Amortization

Another application of spreadsheets to economic analysis is use in setting up an amortization schedule for the repayment of a loan with a constant A period payment. Spreadsheets assist in:

- Establishing the loan repayment amortization schedule.
- Showing the progression of principal and interest payments.
- Determining the balance due on a loan at any time in its life.
- Estimating the number of payments remaining on the loan at any time in its life.

Establishing the amortization schedule begins with calculating the annual payment $A = P(A/P, i_{mn}, mn)$, where i = interest rate per subperiods (quarterly, monthly, weekly, etc.), m = subperiods per year, and n = number of years that the loan is outstanding. The interest paid in each period is Interest($) = Pi, where P is the remaining principal owed at the beginning of the period and i is the per period interest rate. The amortization schedule is then estimated in tabular format as:

Period	Beginning balance	Payment (A)	Interest Paid	Principal paid	Ending balance

Example 6.19

Edward purchased a house one year ago and is making monthly payments (A). The closing price on the house was $300,000, financed for 30 years, and at an annual interest rate of 6.0%. Estimate the monthly payment and set up the first year's amortization schedule. (1) What is his monthly payment (excluding insurance and taxes)? (2) How much interest and principal did he pay on the 12th payment, and how much does he owe on the house (ending balance) after he makes the 12th payment? (3) If he decides to pay an extra 10% of his payment each month for the remainder of the loan, when will the loan be repaid?

$$P = \$300,000.00 \quad i_n = 6.0\% \quad n = 12 \times 30 = 360 \text{ months}$$
$$i = 6.0\%/12 = 0.5\%$$

Initial Balance	$300,000.00		(1) Payment (A)= $1,798.65			
Interest r.	6.00%					
Interest i.	0.50%					
Periods n	360					
	Month	Beg Bal	Payment	Interest	Principal	End Bal
	0					$300,000.00
	1	$300,000.00	$1,798.65	$1,500.00	$298.65	$299,701.35
	2	$299,701.35	$1,798.65	$1,498.51	$300.14	$299,401.20
	3	$299,401.20	$1,798.65	$1,497.01	$301.65	$299,099.56
	4	$299,099.56	$1,798.65	$1,495.50	$303.15	$298,796.40
	5	$298,796.40	$1,798.65	$1,493.98	$304.67	$298,491.73
	6	$298,491.73	$1,798.65	$1,492.46	$306.19	$298,185.54
	7	$298,185.54	$1,798.65	$1,490.93	$307.72	$297,877.82
	8	$297,877.82	$1,798.65	$1,489.39	$309.26	$297,568.56
	9	$297,568.56	$1,798.65	$1,487.84	$310.81	$297,257.75
	10	$297,257.75	$1,798.65	$1,486.29	$312.36	$296,945.38
	11	$296,945.38	$1,798.65	$1,484.73	$313.92	$296,631.46
	12	$296,631.46	$1,798.65	$1,483.16	$315.49	$296,315.96
(2)						
Interest(12)	$1,483.16					
Principal(12)	$315.49					
Balance(12)	$296,315.96					
(3)						
Periods w +10%	277.0					

He shortens his loan from 348 remaining payments to 277 remaining payments. He saves 71 months worth of payments.

6.10 Spreadsheets for Alternatives with Multiple Rate-of-Returns

In examples thus far, cash flows were considered to have an initial negative investment cash flow and a sequence of net positive benefit cash flows. This single transition in cash flow from negative to positive guaranteed a single unique rate-of-return solution. There are other cash flows with multiple negative-to-positive and positive-to-negative transitions for which there will be multiple rate-of-return solutions. The most common examples are high-cost upgrade or expansions during an alternative's useful life or a high salvage cost in the final year of an alternative's life. In general, there will be a rate-of-return solution for each minus-to-plus and plus-to-minus cash flow transitions.

To estimate the correct rate-of-return for an alternative that has multiple cash flow sign changes, we will use the internal rate of return definition.

The **internal rate of return** is the interest rate at which the net present worth or equivalent annual worth equals $0.

We can use spreadsheet financial functions to identify all multiple rate-of-return solutions.

1. Determine the net annual cash flows for the alternative under consideration.
2. Set up a range of interest rates (e.g., −70 to +100%).
3. Estimate the net present value or equivalent uniform annual worth of the net annual cash flows for each interest rate.
4. Perform linear interpolation for each sign change to determine the rate-of-return.
5. If there is only one positive root, it is the valid rate of return.
6. If there is only one negative root, it is the valid rate of return.
7. If there is a negative and positive root, only the positive root is considered as the valid rate of return.
8. When multiple roots are found, there will be a negative root that can be considered as invalid and at least one positive root that can be considered as the valid rate of return.

Example 6.20 illustrates multiple roots for an alternative with a high salvage value at the end of its useful life. Example 6.21 illustrates multiple roots for the upgrade or expansion during an alternative's life.

Example 6.20

Estimate the rate of return for an alternative with an initial cost of $300,000, annual net benefit cash flows of $80,000 for years 1 through 9, and a salvage cost of $120,000 in year 10.

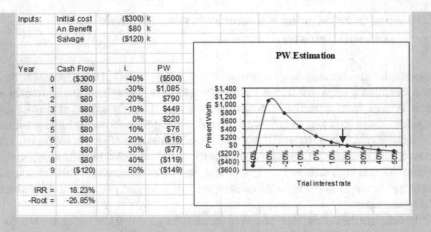

Inputs:	Initial cost	($300) k
	An Benefit	$80 k
	Salvage	($120) k

Year	Cash Flow	i.	PW
0	($300)	-40%	($500)
1	$80	-30%	$1,085
2	$80	-20%	$790
3	$80	-10%	$449
4	$80	0%	$220
5	$80	10%	$76
6	$80	20%	($16)
7	$80	30%	($77)
8	$80	40%	($119)
9	($120)	50%	($149)

| IRR = | 18.23% |
| -Root = | -26.85% |

Valid IRR = 17.90%.

Example 6.21

Estimate the rate of return for an alternative with an initial cost of $200,000, annual net benefit cash flows of $40,000 for years 1 through 5, an expansion cost of $200,000 in year 5 with net benefit cash flows of $80,000 for years 6–10. The alternative has a salvage cost of $100,000 at the end of year 10.

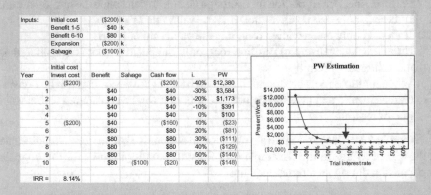

Inputs:	Initial cost	($200) k
	Benefit 1-5	$40 k
	Benefit 6-10	$80 k
	Expansion	($200) k
	Salvage	($100) k

Year	Initial cost Invest cost	Benefit	Salvage	Cash flow	i.	PW
0	($200)			($200)	-40%	$12,380
1		$40		$40	-30%	$3,584
2		$40		$40	-20%	$1,173
3		$40		$40	-10%	$391
4		$40		$40	0%	$100
5	($200)	$40		($160)	10%	($23)
6		$80		$80	20%	($81)
7		$80		$80	30%	($111)
8		$80		$80	40%	($129)
9		$80		$80	50%	($140)
10		$80	($100)	($20)	60%	($148)

| IRR = | 8.14% |

Valid IRR = 7.4%.

6.11 Summary

Definition: A *budget period* is the management equivalent of an accounting period (i.e., month, quarter, semi-annual, year, or multi-year tied to the product life cycle).

- The *zero-period budget* delineates the short-term budget and sets controls for current fiscal year revenues and expenses.
- A *life-cycle budget* is a medium-term budget greater than one year that is an estimate of all costs to design, develop, install productive assets, market, produce revenues, and retire a product. A life-cycle budget is tied directly to an organization's product life cycle and may range from two to ten or fifteen years.
- A *business-cycle budget* is a long-term budget across multiple product life-cycle budgets that predicts the investments needed in facilities and equipment to support a chain of products.

Undiscounted Payback Period:

Definition: *Undiscounted payback period* is number of time periods required to recover the initial cost cash flow(s) of an investment from the net cash flows (profit or other benefits) produced by that investment for an interest rate of zero.

Criterion: Select the alternative with the minimum payback period.

If CF_t = cash flow (cost or benefit) in period t, the payback period is defined as the smallest value of n that satisfies

$$\sum_n CF_t = 0$$

Discounted Payback Period:

Definition: *Discounted payback period* is the number of time periods required to recover the initial cost cash flow(s) of an investment from the net cash flows (profit or other benefits) produced by that investment at the MARR interest rate.

Criterion: Select the alternative with the minimum discounted payback period.

If CF_t = cash flow (cost or benefit) in period t, the discounted payback period is defined as the smallest value of n that satisfies

$$\sum_n CF_t (1+i)^{-t} = 0$$

Net Present Worth:

The **net present value** or **net present worth** of a project, NPV or NPW, is defined as the difference between discounted revenues and expenses at the MARR interest rate.

$$\text{NPW} = -C_0 + \sum_{n=1}^{N} \frac{r_n - c_n}{(1+i)^n} = -C_0 + \sum_{n=1}^{N} \frac{F_n}{(1+i)^n}$$

Criterion: Select the alternative that maximizes net present worth.

Equivalent Uniform Annual Worth:
The **equivalent uniform annual worth**, EUAW, of an alternative is defined as its annualized Net Present Worth.

$$\text{EUAW} = \text{NPW} \left[\frac{i(1\_+i)^N}{(1+i)^N - 1} \right] = \text{NPW}(A/P, i, N)$$

Criterion: Select the alternative that maximizes equivalent uniform annual worth.

Future Worth:
The **future value** or **future worth** of a project, FV or FW, is defined as the difference between compounded revenues and expenses at the MARR interest rate at the end of an alternative's life.

$$\text{FW} = C_0(F/P, i, N) + \sum_{n=1}^{N} (r_n - c_n)(F/P, i, N - n)$$

Criterion: Select the alternative that maximizes future worth.

Internal Rate of Return:
The **internal rate of return** is the interest rate at which the equivalent benefits are equal to the equivalent costs.

Criterion: Select the alternative that maximizes IRR $\geq$ MARR.

Differing combinations of alternatives and project lives include:

- Useful life of each alternative equals each other and the analysis period.
- Alternatives have useful lives different from each other and from the analysis period.
  - *Least common multiple life analysis.*
  - *Project life analysis.*

- Special case, EUAW of a continuing requirement.
- The analysis period is infinite, $n = \infty$.

Types of bonds:

- Treasury notes and bonds.
- Treasury bills.
- Municipal bonds.
- Corporate bonds.
- Zero-coupon bonds.
- Callable bonds.
- Floating-rate bonds.

Loan amortization:

$$A = P(A/P, i_{mn}, mn)$$

where $i =$ interest rate per subperiods (quarterly, monthly, weekly, etc.), $m =$ subperiods per year, and $n =$ number of years that the loan is outstanding. The amortization schedule is then estimated as:

Period	Beginning balance	Payment (A)	Interest paid	Principal paid	Ending balance

6.12 Key Terms

Amortization schedule
Analysis period
Business-cycle budget period
Capitalized cost
Common service period
Continuing requirement
Coupon rate
Discounted payback period
Equivalent uniform annual worth criterion
Equivalent uniform annual benefit
Equivalent uniform annual cost
Face value
Future worth
Infinite analysis period
Internal rate of return
Least common multiple life analysis
Life cycle budget period

Net present worth criterion
Par value
Project life analysis
Undiscounted payback period
Zero-budget period

Problems

1. Consider four mutually exclusive alternatives for a one-year short-term product contract in a job shop. For short-term products, the firm requires 4.0% MARR.

Cash flow	Product A	Product B	Product C	Product D
Initial cost	$17,500	$15,000	$4750	$21,000
Uniform monthly net income	$1900	$1400	$450	$2400

   a. Estimate the undiscounted payback period for each product.
   b. Estimate the discounted payback period for each product.

2. Specialty Plastics, Inc., is considering purchasing a new injection molding machine to improve control of shrinkage of molded parts. The improvement in shrinkage control will reduce losses by $95,000 per year. The injection molding machine can be purchased and installed for $340,000 today. It will have a useful life of seven years and will have a salvage value of $17,000. The new injection molding machine will not increase operating expenses but will require an additional maintenance expenses of $15,300 per year. Specialty Plastics requires a 12% rate of return on investments in new technology. Do you recommend purchasing the new injection molding machine? What is the EUAW and rate of return on the investment?

3. A new engineering graduate, who just turned 21, started her new engineering job. Her salary is $68,000 per year, which after income taxes, health insurance, and social security will yield take-home pay of $60,000 per year. She elects to have $500 per month deducted into an Individual Retirement Account (IRA) for retirement. Historically, the IRA has paid 3.0% annual nominal interest rate but compounded monthly. (a) If she works to the required retirement age of 67, how much will be in the IRA when she retires? (b) Currently, social security pays $3011 per month, which after taxes for someone in the 25% tax bracket nets $2260. If she desires to maintain her $54,000/12 = $4500 per month income during retirement, how many months will her IRA account last?

4.   For the following alternatives, which is preferred if the firms's MARR = 7.0%?
     Total revenue will be the same for both alternatives.

Parameter	A	B
First cost	$72,500	$52,500
Annual operating expenses	$5000	$6000
Annual maintenance exp's	$1800	$400
Overhaul (year 5)	$15,000	NA
Salvage value	$2500	$7500
Useful life (years)	8	4

5.   Reconsider problem 4 if alternative A's market value is 80% of its initial cost
     at the end of the two years after initial purchase, alternative B's market value is
     40% of its initial cost, and the project life is ten years.

6.   A maintenance manager is considering the budget for a machine on the produc-
     tion line. Over the next five years, the budget is expected to be $19,000 per year.
     During years six to ten, the budget is expected to increase by $6000 per year due
     to machine wear. Additionally, $9000 is budgeted in year four and year eight for
     overhaul of the machine. If the relevant MARR = 8.0%, what is the equivalent
     uniform annual cost for maintaining this machine?

7.   Assuming the alternative A, B, and C will be needed for continued use and
     MARR = 12.0%, which alternative should be selected based on annual cash
     flow analysis?

	A	B	C
Initial cost	$15,000	$22,500	$30,000
Annual benefit	$1500	$2650	$6325
Life	∞	30	8

8.   The state highway department is considering the construction of a new bridge.
     Based on the following data, which type of construction is preferred. The
     relevant interest rate is 5.0%.

	Steel	Concrete
Initial cost	$5,000,000	$7,000,000
Annual maintenance	$350,000	$250,000
Resurfacing	$3,500,000	$4,500,000
Resurfacing interval	10 years	15 years

9. A process engineer for a French fry manufacturing line is considering investment
    in one of three automatic slicing machines. The MARR for this investment is
    6.0%. Which alternative should be selected?

	A	B	C
Initial cost	$88,400	$107,100	$113,900
Increased revenue	$64,600	$52,700	$62,900
Annual O&M expense	$25,500	$15,300	$20,400
Salvage value	$22,100	$32,300	$37,400
Life	6	6	12

Chapter 7
Depreciation Effects on Investment Worth

Abstract We first introduced the concept of depreciation in Chapter two with the statement, "assets are valued at their acquisition cost and adjusted for depreciation and improvement costs." Likewise in chapter two, under the balance sheet category fixed assets, depreciation was subtracted from plant and asset initial cost to arrive at net fixed assets valuation. On the income statement, depreciation was subtracted as a part of operating expenses to calculate earnings before interest and taxes. However, the methods to estimate annual depreciation expenses were not discussed. This chapter discusses historical depreciation methods, the Modified Accelerated Cost Recovery System depreciation required under US tax code, unit-of-production depreciation, and depletion depreciation.

7.1 Depreciation Fundamentals

Fixed assets such as buildings, equipment, computer networks, and office furniture are acquired to directly or indirectly support production operations that generate future cash flows from the sale of products and services. Unlike direct labor, direct materials, and indirect materials consumed and expensed in the current budget accounting period, fixed assets exist across multiple periods. Thus, a different accounting method is needed to expense fixed assets initial costs across multiple budget accounting periods. Depreciation is the method used to expense fixed assets' initial costs.

> **Definition**: *Depreciation* is a decrease in fixed asset value; that is, market value is due to the use or value to the owner due to the aging.
> **Definition** (financial): Depreciation is the systematic allocation of the cost of a fixed asset over its depreciable life related to deterioration, or consumption of its useful life.

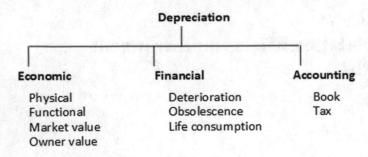

Fig. 7.1 Depreciation categories

As shown in Fig. 7.1, economic depreciation can be recognized as,

Economic Depreciation = Initial value (purchase price) − Market Value

From the accounting perspective, depreciation is ***not a cash flow*** in the period it is expensed. However, depreciation affects the accounting journal value of assets, and, as a budgeting period expense, depreciation affects the after-tax cash flow of an alternative or project. Financial depreciation can be recognized as

Financial Depreciation = Initial value (purchase price) − Consumed Life value

Figure 7.1 also shows that depreciation must also be considered from an accounting perspective.

Expenses: subtracted from business revenues during each accounting budget period.

- Labor.
- Utilities.
- Materials.
- Insurance, etc.
- Depreciation—allocation of initial equipment cost plus installation costs.

At the end of each accounting budget period,

$$\text{Book Value (balance sheet)} = \text{Initial value (purchase price)}$$
$$- \text{Sum(depreciation expenses)}$$

$$\text{Net Income} = \text{EBTn} - \text{Taxes} = \text{EBTn} - \text{EBTn} \times \text{Tax Rate}$$
$$\text{Net Income} = [\text{Rn} - (\text{On} + \text{Mn} + \text{Dn} + \text{In})]$$
$$- [\text{Rn} - (\text{On} + \text{Mn} + \text{Dn} + \text{In})]\,\text{TR}$$
$$\text{Net Income} = [\text{Rn} - (\text{On} + \text{Mn} + \text{Dn} + \text{In})]\,(1 - \text{TR})$$

From a cash flow perspective, depreciation must be added back to net income to estimate the after-tax net cash flow.

$$\text{Net Cash Flow} = [R_n - (O_n + M_n + D_n + I_n)](1 - TR) + D_n$$
$$\text{Net Cash Flow} = [R_n - (O_n + M_n + I_n)](1 - TR) - D_n(1 - TR) + D_n$$
$$\text{Net Cash Flow} = [R_n - (O_n + M_n + I_n)] - D_n + (D_n \times TR) + D_n$$
$$\text{Net Cash Flow} = [R_n - (O_n + M_n + I_n)] + (D_n \times TR)$$

7.1.1 US Tax Code Depreciation Fundamentals

Applicable US Department of the Treasury Internal Revenue Publications for this depreciation discussion:

- 179, Property Expense.
- 197, Intangibles.
- 551, Basis of Assets.
- 946, How to Depreciate Property.

Note: For personal business property purchased and used outside of the USA, consult the codes of the country in which the property is used. Tax codes of foreign governments must be observed for foreign assets.

Note: For US-headquartered companies, US IRS code still applies to depreciation of property used in foreign countries. Hence, separate journals, ledgers, balance sheets, income statements, statements of retrained earnings, and statements of cash position will have to be maintained in the USA and each foreign country.

IRS Publication 551, Basis of Assets

Tangible Asset Cost Basis:

- Initial asset cost plus.
- All fees and charges allowed under the Uniform Capitalization Rules with
- Allocated expenses for the purchase of multiple assets plus.
- Allowable increases or decreases in the cost basis.

necessary to place the asset in service fit for use.

Placed-in-service date is the day on which the asset begins to provide returns to the business by performing its intended function or by producing its intended output.

A placed-in-service date for each asset or class of assets must be recorded in the fixed assets journal. For historical depreciation methods, the recorded placed-in-service date specifies the remaining portion of depreciation that can be taken in the first and last years of service.

Depreciable life or **recovery period**: The period over which an asset is depreciated.

- Depreciation is a **non-cash cost**: money does not change hands during the recovery period.
- Depreciation is used to **allocate an asset's loss of value** over time.
- Depreciation is **deducted from revenue** and reduces the taxable income of a business over time.
- Depreciation **affects cash flow** on an **after-tax basis**.

An asset is depreciable if:

- The asset is used for business purposes in the production of income.
- The asset has a useful life that can be determined, and the useful life is longer than one year. Otherwise, the asset is expensed.
- The asset decays, gets used up, wears out, becomes obsolete or loses value from natural causes.

Types of Assets:
Tangible: can be seen, touched, and felt.

- **Real**: land, buildings, and things growing on, or attached to the land.
- **Personal**: equipment, furnishings, vehicles, office machinery, etc., are not defined as real property.

Intangible: has value but cannot be seen or touched; examples include patents, copyrights, and trademarks.

Note: Buildings and equipment are depreciable; land is not. Land is entered into the fixed asset journal and shown on the balance sheet at its purchase price plus costs and fees necessary to complete the purchase (real estate brokers, escrow fees, legal fees). Regardless of its current market value, land is maintained on the balance sheet at its initial purchase price until it is sold. Land value can be changed only when it is sold, and a new market value is established by the objective sales transaction.

7.1.2　Depreciation Estimation Fundamentals

Figure 7.2 shows the relationship between allocated depreciation expenses and asset book value. The horizontal axis is in time units of an asset's useful life (generally years). The vertical axis is the balance sheet book value of the asset at the end of each accounting budget period showing the effect on book value of each $d(t)$ depreciation expense. In general, an asset's book value declines linearly or at a decelerating rate as each depreciation expense is realized. The equation for the calculation of an asset's book value at each $d(t)$ is,

$$\text{Book value} = \text{Asset cost} - \text{Sum(Depreciation Charges)}$$

Fig. 7.2 Book value versus time relationship

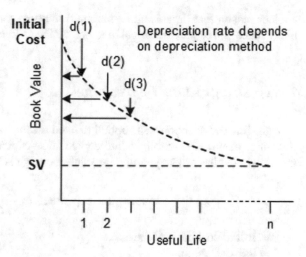

$$BV_j = \text{Asset cost} - \sum_1^t d_j \qquad (7.1)$$

7.2 Historical Depreciation Methods

Prior to 1981, four methods were used to estimate period depreciation allowances:

- Straight line.
- Sum-of-years'-digits.
- Declining balance.
- Declining balance switching to straight line.

1981–1986, Accelerated Cost Recovery System (ACRS).

- Asset assigned to category of property class life.
- Estimate salvage value.
- Shorter recovery periods than historical methods.

1986–present, Modified Accelerated Cost Recovery System (MACRS).

- Expanded number of property classes.
- Annual depreciation percentages modified to include half-year convention for first and last year.

Depreciation estimates are important for estimating after-tax cash flow effects. We will find that US IRS tax code depreciation methods are based on variations of historical methods. Many organizations continue to use historical depreciation methods for financial reporting to stockholders and outside stakeholders (hence, another reason that organizations maintain multiple journals and ledgers). Finally, many states still require historical methods for the estimation of state taxable income.

7.2.1 Straight-Line Depreciation

Straight-line depreciation was applied to fixed assets that were consumed uniformly over their useful lives. Straight-line depreciation expenses are an equal fraction of the initial cost basis. The period depreciation charge is

$$\text{Depreciation expense } d_t = \frac{B - S}{N} \tag{7.2}$$

B = Initial cost basis of the asset.
S = Salvage value.
N = Depreciable life in years.
For the historical methods, if an asset is placed in service during its first year, the first-year depreciation charge is just for the period of service of the remainder of the year.

$$d_1 = \frac{B - S}{N} P_{\text{remaining-year}}$$

Example 7.1

A fixed asset has a cost basis of \$8600 and has a useful life of 4 years, and a salvage value of \$200. It is expected to be used uniformly each day. It is placed in service at the end of the 5th month of the organization's fiscal year. What is the annual depreciation expense, and what is the first-year depreciation expense?

$$d_t = \frac{\$8600 - \$200}{4} = \$2100/\text{year}$$

$$P_{\text{remaining-year}} = 7/12$$

$$d_1 = \$2100(7/12) = \$1225$$

The remaining \$2100 − \$1225 = \$875 depreciation charge will be taken for the first five months of year 4 use.

7.2.2 Sum-of-Years-Digits Depreciation

An asset may provide useful output that declines over time due to wear or techno-logical obsolescence. For this asset type, we should apply an **accelerated depreci-ation method** to recognize the higher consumption of useful life early in the asset's life cycle. One such accelerated depreciation method is the **Sum-of-Years-Digits** (SYOD) depreciation. SYOD estimates higher depreciation expenses than straight-line depreciation in an asset's early life and smaller depreciation expenses at the end of its depreciable life. SYOD is calculated as

$$\text{Depreciation expense } d_t = \frac{N-t+1}{\sum_0^{N-1} N-j}(B-S) = \frac{N-t+1}{N(N+1)/2}(B-S) \quad (7.3)$$

B = Initial cost of asset.
S = Salvage value.
N = Depreciable life in years.
t = Depreciation year of asset life.
An algorithmic approach to estimating the multiplier is

$$\text{Multiplier} = \frac{\text{Years in reverse order}}{\text{Sum(Years digits)}}$$

As an example, for an asset with a five-year useful life, the multiplier for each year is

Year	SYOD multiplier
1	5/15
2	4/15
3	3/15
4	2/15
5	1/15

Example 7.2

Compute the SYOD depreciation schedule for an asset with the following expense and life information: $B = \$10,000$, $S = \$1000$, $N = 5$.

$$d_t = \frac{N-t+1}{15}(\$10,000 - \$1000)$$

Year	Multiplier	$d(t)$	End BV
0			$10,000
1	5/15	$3000	$7000
2	4/15	$2400	$4600
3	3/15	$1800	$2800
4	2/15	$1200	$1600
5	1/15	$600	$1000

7.2.3 Declining Balance Depreciation

The second accelerated depreciation method, declining balance allocates a fixed fraction of the asset's remaining book value as the annual depreciation expense.

$$\text{Depreciation expense } d_t = \frac{C}{N} \text{Book Value}_{t-1}$$

where $C = $ constant multiplier to be selected. The most widely applied values of C are 150% and 200% of straight-line depreciation. When $C = 1.5$, d_t is termed **150% declining balance**. When $C = 2$, or the 200% rate, d_t is termed **double-declining balance** (DDB) and is estimated as

$$\text{DDB } d_t = \frac{2}{N}(\text{Book Value}_{t-1}) = \frac{2}{N}\left(B - \sum_{t-1} d_j\right) \qquad (7.4)$$

$B = $ Initial cost of asset.
$N = $ Depreciable life in years.
$t = $ Current year of life.
Note: Eq. (7.4) does not include the salvage value lower limit. Since total depreciation charges cannot reduce asset value below its salvage value, the final depreciation charge may be only the proportion of the final d_t that reduces the asset worth to its salvage value.

Example 7.3
Rework Example 7.2 using double-declining balance.

Year	$(2/N)$ BV	$d(t)$	End BV
0			$10,000
1	(2/5) $10,000	$4000	$6000
2	(2/5) $6000	$2400	$3600
3	(2/5) $3600	$1440	$2160
4	(2/5) $2160	$864	$1296
5	(2/5) $1166	Not $518; $296	$1000

7.3 Modified Accelerated Cost Recovery System (MACRS)

The Economic Recovery Tax Act of 1981 established the Accelerated Cost Recovery System (ACRS). ACRS changed from straight-line depreciation to depreciating assets to shorter depreciation schedules based on cost recovery. Accelerated depreciation increased the depreciation expenses that organizations were able to claim early in an asset's life. The law's proponents believed that accelerated asset depreciation would, in turn, accelerate economic growth.

Congress modified ACRS in the United States Tax Reform Act of 1986, naming the new system as the **Modified Accelerated Cost Recovery System** (MACRS). MACRS introduced the following tax advantages over ACRS.

- "Property class lives" are less than "actual useful lives."
- Salvage values are required to be zero.
- Tables of annual percentages simplify computations.

MACRS depreciation schedules are set forth in IRS Publication 946 How to Depreciate Property, Section 4 Figuring Depreciation under MACRS. (Note: This discussion presents only the fundamentals of MACRS depreciation. Actual application is much more complex. Always consult the organization's tax accountant when estimating MACRS depreciation on final project estimates.)

MACRS uses the general depreciation system (GDS), which is just declining balance with switch to straight-line depreciation. The alternative depreciation system (ADS) provides longer recovery periods and uses straight-line depreciation. Under US law, ADS must be used for:

- Tangible assets used outside the USA.
- Any tax-exempt asset financed by tax-exempt bonds.
- Farming assets.

Depreciation is estimated using general method (IRS FORM 4562):

1. Determine asset cost basis.

2. Determine asset property class and recovery period.
3. Use the asset's placed-in-service date to locate MACRS depreciation percentages in Tables 7.1, 7.2 and 7.3.

Once the asset class and recovery period are determined IRS 946 Tables B-1 or B-2, consult the appropriate table in Appendix A, MACRS Percentage Table Guide for the annual percentage depreciation charges. Table 7.3 reproduces IRS 946 Chart 1 guidance for personal business property.

Table 7.4 reproduces IRS 946 Chart 2 guidance for residential rental and nonresidential real property.

In this text, we will restrict personal business property depreciation percentages to IRS 946, Table A-1 shown in Table 7.5. Likewise, we will restrict real property depreciation percentages to IRS 946, Table A-7a shown in Table 7.6.

Table 7.1 IRS 946 Table B-1: specific depreciable assets used in all business activities

Asset class	Description	Class life (in years)	Recovery periods (in years)	
			GDS (MACRS)	ADS
00.11	Office furniture, fixtures, and equipment	10	7	10
00.12	Information systems including computers	6	5	5
00.13	Data handling equipment: except computers	6	5	6
00.21	Airplanes (airframes and engines), except commercial or freight and all helicopters	6	5	6
00.22	Automobiles and taxies	3	5	5
00.23	Buses	9	5	9
00.241	Light general purpose trucks	4	5	5
00.242	Heavy general purpose trucks	6	5	6
00.25	Railroad cars and locomotives not owned by railroad transportation companies	15	7	15
00.26	Tractor units for use over-the-road	4	3	4
00.27	Trailers and trailer-mounted containers	6	5	6
00.28	Vessels, barges, tugs and similar water transportation	18	10	18
00.3	Land improvements	20	15	20
00.4	Industrial steam and electric generation and distribution systems	22	15	22

Table 7.2   IRS 946 Table B-2: depreciable assets used in the following activities

| Asset class | Description | Class life (years) | Recovery periods (in years) | |
			GDS (MACRS)	ADS
01.1	Agriculture	10	7	10
10.0	Mining	10	7	10
13.0	Offshore drilling	7.5	5	7.5
15.0	Construction	6	5	6
20.1	Manufacture food products	17	10	17
20.5	Manufacture food products—special handling devices	4	3	4
21.0	Manufacture tobacco products	15	7	15
22.1	Manufacture of textiles—knitted goods	7.5	5	7.5
24.4	Manufacture of wood product & furniture	10	7	10
26.1	Manufacture pulp and paper	13	7	10
27.0	Printing and publishing	11	7	11
28.0	Manufacture chemicals	9.5	5	9.5
30.1	Manufacture of rubber products	14	7	14
30.2	Manufacture finished plastic products	11	7	11
31.0	Manufacture leather products	11	7	11
32.1	Manufacture glass products	14	7	14
32.2	Manufacture of cement	20	15	20
32.3	Manufacture of stone/clay products	15	7	15
33.2	Manufacture nonferrous metals	14	7	14
33.3	Manufacture foundry products	14	7	14
33.4	Manufacture of primary steel mill products	15	7	15
34.0	Manufacture of fabricated metal products	12	7	12
35.0	Manufacture machinery and mechanical products	10	7	10
36.0	Manufacture of electronic components	6	5	6
37.11	Manufacture of motor vehicles	12	7	12

(continued)

Table 7.2 (continued)

Asset class	Description	Class life (years)	Recovery periods (in years)	
			GDS (MACRS)	ADS
37.2	Manufacture of aerospace products	10	7	10
37.31	Ship & boat building machinery & equipment	12	7	12
37.41 – 2	Manufacture locomotives (railroad cars)	11.5	7	11.5
39.0	Manufacture athletic, jewelry, other goods	12	7	12
41.0, 42.0	Motor transport—passengers & freight	8	5	8
45.0	Air transport	12	7	12
46.0	Pipeline Transportation	22	15	22
48.12	Telephone central office equipment	18	10	18
48.14	Telephone distribution	24	15	24
48.2	Radio and television broadcasting	6	5	6
48.31	Electric power and communications	19	10	19
49.11	Electric utility & power production—classes 49.11 to 49.5	50	20	50
50	Municipal water treatment	24	15	24
51	Municipal sewer	50	20	50
57.0	Distributed trades and services—wholesale, retail, personal and professional services	9	5	9
57.1	Distributed trades and services—billboard, service station, and petroleum marketing	20	15	20
79.0	Recreation	10	7	10
80.0	Theme and amusement park	12.5	7	12.5
	Technological & research equipment		5	

Table 7.3 IRS 946 Chart 1: personal business property

MACRS system	Depreciation method	Recovery period	Convention	Class	Mon/Qtr placed in service	Tables
GDS	200%	GDS/3, 5, 7, 10 Nonfarm	Half-Year	3, 5, 7, 10	Any	A-1
GDS	200%	GDS/3, 5, 7, 10 Nonfarm	Mid-Quarter	3, 5, 7, 10	1st Qtr 2nd Qtr 3rd Qtr 4th Qtr	A-2 A-3 A-4 A-5
GDS	150%	GDS/3, 5, 7, 10	Half-Year	3, 5, 7, 10	Any	A-14
GDS	150%	GDS/3, 5, 7, 10	Mid-Quarter	3, 5, 7, 10	1st Qtr 2nd Qtr 3rd Qtr 4th Qtr	A-15 A-16 A-17 A-18
GDS	150%	GDS/15, 20	Half-Year	15 & 20	Any	A-1
GDS	150%	GDS/15, 20	Mid-Quarter	15 & 20	1st Qtr 2nd Qtr 3rd Qtr 4th Qtr	A-2 A-3 A-4 A-5
GDS ADS	SL	GDS ADS	Half-Year	Any	Any	A-8
GDS ADS	SL	GDS ADS	Mid-Quarter	Any	1st Qtr 2nd Qtr 3rd Qtr 4th Qtr	A-9 A-10 A-11 A-12
ADS	150%	ADS	Half-Year	Any	Any	A-14
ADS	150%	ADS	Mid Quarter	Any	1st Qtr 2nd Qtr 3rd Qtr 4th Qtr	A-15 A-16 A-17 A-18

Table 7.4 IRS 946 Chart 2: Residential rental and nonresidential real property

MACRS system	Depreciation method	Recovery period	Convention	Class	Mon/Qtr placed in Service	Table
GDS	SL	GDS/27.5	Mid-Month	Residential Rental	Any	A-6
GDS	SL	GDS/31.5	Mid-Month	Nonresidential Real	Any	A-7 A-7a
ADS	SL	GDS/3, 5, 7, 10	Half-Year	3, 5, 7, 10	Any	A-13
	SL	GDS/3, 5, 7, 10	Mid-Quarter	3, 5, 7, 10	Any	A-13a

Table 7.5 IRS 946 Table A-1. 3-, 5-, 7-, 10-, 15-, and 20-year property; half-year convention

Recovery year	Applicable percentage for GDS property class					
	3-year property	5-year property	7-year property	10-year property	15-year property	20-year property
1	33.33	20.00	14.29	10.00	5.00	3.750
2	44.45	32.00	24.49	18.00	9.50	7.219
3	14.81	19.20	17.49	14.40	8.55	6.677
4	7.41	11.52	12.49	11.52	7.70	6.177
5		11.52	8.93	9.22	6.93	5.713
6		5.76	8.92	7.37	6.23	5.285
7			8.93	6.55	5.90	4.888
8			4.46	6.55	5.90	4.522
9				6.56	5.91	4.462
10				6.55	5.90	4.461
11				3.28	5.91	4.462
12					5.90	4.461
13					5.91	4.462
14					5.90	4.461
15					5.91	4.462
16					2.95	4.461
17						4.462
18						4.461
19						4.462
20						4.461
21						2.231

Table 7.6 IRS 946 Table A-7a nonresidential real property mid-month convention-39 years

Recovery year	Percentage—month placed in and removed from service					
	1	2	3	4	5	6
1	2.461	2.247	2.033	1.819	1.605	1.391
2–39 (mid)	2.564	2.564	2.564	2.564	2.564	2.564
40 (last)	0.107	0.321	0.535	0.749	0.963	1.177
Recovery year	7	8	9	10	11	12
1	1.177	0.963	0.749	0.535	0.321	0.107
2–39 (mid)	2.564	2.564	2.564	2.564	2.564	2.564
40 (last)	1.391	1.605	1.819	2.033	2.247	2.461

Once the correct depreciation percentages are determined, the year-to-year depreciation expense for the GDS asset is estimated as:

$$d_t = B \times r_t \tag{7.5}$$

d_t = depreciation deduction in year t.
B = cost basis (initial ready-for-use cost).
r_t = MACRS percentage rate in year t.

Example 7.4

A company installs a small research laboratory with \$230,000 of research equipment with an estimated salvage value of \$30,000 at the end of its 7-year technological life. Calculate the annual MACRS depreciation expenses and annual book values.

From IRS 946 Table B-2, research equipment is GDS 5-year personal business property. From IRS 946 Table A-1, the GDS 5-year depreciation percentages are shown in the following table.

Year t	Cost basis	MACRS r_t (%)	d_t	Cum d_t	Book value
0	\$230,000				\$230,000
1		20.00	\$46,000	\$46,000	\$184,000
2		32.00	\$73,600	\$119,600	\$110,400
3		19.20	\$44,160	\$163,760	\$66,240
4		11.52*	\$26,496	\$190,256	\$39,744
5		11.52	\$26,496	\$216,752	\$13,248
6		5.76	\$13,248	\$230,000	\$0

The year of switch to straight-line depreciation is indicated by an "*"

Note in Table 7.5 that each GDS property class life is depreciated over $n + 1$ years. This is due to the IRS 946 MCRS *half-year convention*. The half-year convention requires that *one-half the depreciation charge be taken in the first and last years*. This rule assumes that the asset is placed in service on average at mid-year (mid-quarter for the mid-quarter convention) and disposed of on average at mid-year in the final year of service. Table 7.7 gives estimation of the MACRS 5-year property percentages with the half-year convention. To simplify the discussion, we will assume a fixed asset with \$1.00 initial cost basis. For a 5-year life, the DDB multiplier is $2/5 = 0.40$. For year 1, the depreciation expense is $0.4(\$1.00) = \$0.40$, but by the MACRS half-year convention take only $0.5(\$0.40) = \$0.20$ depreciation expense. The straight-line depreciation for the remaining GDS life is $1/5.5 = 0.1818$ with depreciation

Table 7.7 Estimation of MACRS GDS 5-year property rates

Year	Beg BV	DDB	SLN	MACRS	End BV
1	1.0000	× 4/2 = 0.20	0.1818	0.20	0.80
2	0.8000	× 4/2 = 0.32	0.1777	0.32	0.48
3	0.4800	× 4/2 = 0.192	0.1371	0.192	0.288
4	0.2880	× 4/2 = 0.1152	0.1152*	0.1152	0.1728
5	0.1728	× 4/2 = 0.06912	0.1152	0.1152	0.0576
6	0.0576	× 4/2 = 0.02304	0.0576	0.0576	0.0000

The year of switch to straight-line depreciation is indicated by an "*"

expense of $0.1818(\$1.00) = \$0.1818$. Since MACRS depreciation $\$0.20 > \$0.1818$, use the MACRS depreciation. The book value at the end of the 1st year is $\$1.00 - \$0.20 = \$0.80$. For year 2, the MACRS depreciation expense is $0.4(\$0.80) = \$0.32$. The straight-line depreciation for the remaining GDS life is $1/4.5 = 0.2222$ with depreciation expense of $0.2222(\$0.80) = \$0.1717$. Since MACRS depreciation $\$0.32 > \$0.1717$, use the MACRS depreciation. The book value at the end of the 2nd year is $\$0.80 - \$0.32 = \$0.48$. For year 3, MACRS depreciation is $0.4(\$0.48) = \$0.1920$, and the straight-line depreciation is $(1/3.5)(\$0.48) = \$0.1371$. Again, use MACRS depreciation with end-of-year book value of $\$0.48 - \$0.192 = \$0.288$. For year 4, MACRS depreciation is $0.4(\$0.288) = \$0.1152$, and the straight-line depreciation is $(1/2.5)(\$0.288) = \$0.1152$. Since MACRS depreciation $\$0.1152 = \$0.1152$ straight-line depreciation, switch to straight-line depreciation for the remaining asset life. The book value at the end of year 4 is $\$0.288 - \$0.1152 = \$0.1728$. For year 5, the straight-line depreciation is $(1/1.5)(\$0.1728) = \$0.1152$ and the ending book value is $\$0.1728 - \$0.1152 = \$0.0576$. The year 6 depreciation is $\$0.0576$ and the ending book value is $\$0.00$. Double-declining balance with switch to straight-line depreciation is used to estimate the MACRS depreciation percentages for each GDS property class in Table 7.5.

Example 7.5

MAN property management group purchased land for $7,500,000 and constructed a building for lease to a logistics and warehousing firm. The building cost $120,000,000 to construct and was completed and leased to the logistics and warehousing firm in the 4th month of MAN's current fiscal year. The lease is for 7 1/2 years with option for the logistics and warehousing firm to purchase the building. If the logistics and warehousing firm does not exercise the option to purchase the building, MAN will sell the land and building to any other interest party. Using the percentages in Table 7.6, estimate the annual MACRS depreciation expenses and book value for MAN's 7 1/2 years of ownership.

As previously stated, land is not depreciable. The building cost basis is $120,000,000.

Percentage – Month Placed in and Removed from Service						
Recovery Year	1	2	3	4	5	6
1	2.461	2.247	2.033	1.819	1.605	1.391
2-39 (mid)	2.564	2.564	2.564	2.564	2.564	2.564
40 (last)	0.107	0.321	0.535	0.749	0.963	1.177
Percentage – Month Placed in and Removed from Service						
Recovery Year	7	8	9	10	11	12
1	1.177	0.963	0.749	0.535	0.321	0.107
2-39 (mid)	2.564	2.564	2.564	2.564	2.564	2.564
40 (last)	1.391	1.605	1.819	2.033	2.247	2.461

Since the building was placed in service in the 4th month of the current fiscal year, the first-year depreciation is 1.819%. All intermediate years' depreciation charges are at 2.564%. The 7 $1/2$-year contract will terminate in the 10th month of MAN's fiscal year. The final year's depreciation is 2.033%. The depreciation schedule is shown in the following table.

Year	$r(t)$	$d(t)$	$\Sigma d(t)$	End BV
0				$120,000,000
1	0.01819	$2,182,800	$2,182,800	$117,817,200
2	0.02534	$3,076,800	$5,259,600	$114,740,400
3	0.02534	$3,076,800	$8,336,400	$111,663,600
4	0.02534	$3,076,800	$11,413,200	$108,586,800
5	0.02534	$3,076,800	$14,490,000	$105,510,000
6	0.02534	$3,076,800	$17,566,800	$102,433,200
7	0.02534	$3,076,800	$20,643,600	$99,356,400
8	0.02033	$2,439,600	$23,083,200	$96,916,800

7.4 Depreciation at Asset Disposal

In disposing of fixed assets depreciated by MACRS, the engineering manager must consider its value at the time of disposal. If the asset is disposed of at the end of its or after its MACRS GDS property life, the asset has a book value of zero, and taxes must be paid on any gains made on the salvage value of the asset. If the asset is disposed of during its MACRS GDS property life, the manager must compare its market value and book value.

- **Depreciation recapture** (ordinary gains): If the asset is sold (market value) for more than its book value, "... any gain on the disposition generally is recaptured (included in income) as *ordinary income* up to the amount of the depreciation previously allowed or allowable for the property." (IRS 946, section 4) Taxes must be paid on the difference between its market value received and its book value.
- **Depreciation loss** (ordinary loss): If the asset's market value (sales price) is less than its book value, a tax credit will be realized on ordinary income.
- **Capital gains:** Capital gain must be realized when the asset is sold (market value) for more than its initial cost basis. The excess over the original cost basis is the capital gain. If the asset is held for less than one year, short-term capital gain tax rates apply. The federal capital gains tax rate on assets held for less than one year corresponds to ordinary corporate income tax rate of 21%. If the asset is held for more than one year, long-term capital gains tax rates apply as set forth in the following table.

Capital gain	Fed. tax rate (%)
$0 to $53,600	0
$53,601 to $469,050	15
$469,051 +	20

Regardless of whether it is a depreciation recapture or loss, the MACRS rule for disposal in any period during the GDS property life is *to take one/half of the allowable depreciation in the year of disposal*. Additional considerations in accounting for depreciation recapture or loss or capital gains can be found in:

- IRS section 1231 Property—defines rules for gains or losses for different asset classes.
- IRS Publication 946, *How to Depreciate Property*. General Asset Accounts (GAA)—defines rules for gains or losses for disposition of separate assets grouped into a GAA.

Example 7.6

Estimate the depreciation recapture or loss for a fixed asset disposed of in year 2 and 3 of its GDS 3-year property life with an initial cost basis of $100,000 and market values of $84, 000 at the end of year 1, $42,000 at the end of year 2, $21,000 at the end of year 3, and $10,000 salvage value every year thereafter.

Year	$r(t)$	$d(t)$	Book value	Market value	Recapture/loss
1	0.3333	$33,333	$66,667		
2	0.4445 × 0.5	$22,225	$44,442	$42,000	(−$2442) Loss
1	0.3333	$33,333	$66,667		
2	0.4445	$44,450	$22,217		
3	0.1481 × 0.5	$7405	$14,812	$21,000	$6188 Recapture

7.5 Unit-of-Production Depreciation

Unit-of-production depreciation method for fixed assets whose life consumption is proportional to its use in the extraction of natural resources (minerals, gas, oil, timber, etc.). The unit-of-production depreciation for these fixed assets is,

$$d_t = \frac{\text{Unit Production Year}(i)}{\text{Total Number Recoverable Units}} \times (B - S) \qquad (7.5)$$

Example 7.7

Extraction and processing equipment with an initial cost basis of $1,800,000 and salvage value $300,000 is used in the extraction and processing of limestone into gravel. It is estimated that the limestone volume will take 5 years to extract. Then the pit will be shut down and reclaimed in accordance with the US Office of Surface Mining Reclamation and Enforcement regulations. Estimate the annual unit-of-production depreciation expenses on the extraction and processing equipment using the unit-of-production method.

The first-year depreciation expense is:

$$d_t = \frac{105,000}{700,000} \times (\$1,800,000 - \$300,000) = \$225,000$$

Year	m³ sold	Proportion	d_t
1	105,000	0.15	$225,000
2	140,000	0.20	$300,000
3	175,000	0.25	$375,000
4	140,000	0.20	$300,000
5	140,000	0.20	$300,000
Total	700,000		

7.6 Depletion

Depreciation of natural resource, such as mineral, oil, gas, timber, etc., is estimated by one of two depletion methods. The rules for estimating depletion expenses are set forth in IRS Publication 535, *Business Expenses*, section 9 Depletion. Two methods are set forth for estimating depletion: **cost depletion** and **percentage depletion**. The selected method is used for estimating of annual depletion expenses and the book value of the remaining natural resource.

7.6.1 Cost Depletion Method

Cost depletion expenses are determined by the adjusted basis of the natural resource divided by the number of recoverable units of the natural resource multiplied number of units sold during the given year. The adjusted basis represents the depletion allowance and is re-estimated annually for tax reporting purposes. Re-estimating the number of recoverable units of the natural resource is an engineering problem. Cost depletion expense for the resource is

$$d_t = \frac{\text{Adjusted Basis}}{\text{Total Number Recoverable Units}} \times B \tag{7.6}$$

7.6.2 *Percentage Depletion Method*

Annual percentage depletion is computed based on annual income rather than the adjusted cost basis of the resource. Under the percentage depletion method, the annual depletion expense is estimated as

$$d_t = \frac{\text{Value Production Year}(i)}{\text{Value Total Number Recoverable Units}} \times B \qquad (7.7)$$

As in the cost method, the value of the total number of recoverable units must be re-estimated annually. Note that since the percentage depletion is estimated based on income rather than the cost basis of the property, the total depletion on asset may exceed the cost of the property. To avoid this, the annual allowance under the percentage method may not be more than 50% of the taxable income from the property.

Both depletion methods are complex requiring ownership allocation, annual re-estimates of the number of recoverable units, and carry back/forward depletion re-estimates. Bottom line, leave depletion estimates to tax accountants.

7.7 **Spreadsheets and Depreciation**

Spreadsheets have functions for estimating depreciationTable 7.8.

The VDB() function is used for MACRS double-declining balance with switch to straight-line depreciation. It includes the specification of starting and ending periods and default switch = TRUE to straight-line depreciation. For the case of only declining balance only, set switch = FALSE. The factor = 2 for double-declining balance. For 150% declining balance, set factor = 1.5.

Table 7.8 Spreadsheets and depreciation

Method	Personal property (non-real estate)
Straight-line	SLN (cost, salvage, life)
Declining balance	DDB (cost, salvage, life, period, factor)
Sum-of-years' digits	SYD (cost, salvage, life, period)
MACRS	VDB (cost, salvage, GDSlife, start_period, end_period, factor, no_switch)

Example 7.8

Use the VDB() function in a spreadsheet to estimate the depreciation schedule for the $230,000 research equipment of Example 7.4. Its estimated salvage value was $30,000 at the end of its 7-year technological life. Recall that its MACRS property class was 5-year.

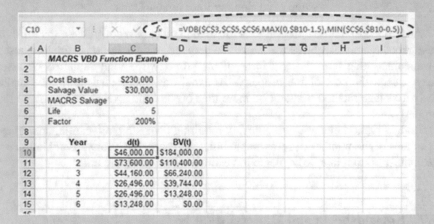

Note that the start-period $= \text{MAX}(0,\$B10\text{-}1.5)$. For $\$B10 = 1$, start-period $= \text{MAX}(0, -0.5) = 0$ and end-period $= \text{MIN}(5, 1\text{-}0.5) = 0.5$ to account for the MACRS half-year convention for year 1. For year 2, $\$B11 = 2$ and start-period $= \text{MAX}(0, 2\text{-}1.5) = 0.5$ and end-period $= \text{MIN}(5, 2\text{-}0.5) = 1.5$ for one full year depreciation expense. The same one-year depreciation expense occurs for years 3 to 5. In year 6, $\$B15 = 6$, start-period $= \text{MAX}(0, 6\text{-}1.5) = 4.5$, and end-period $= \text{MIN}(5, 6\text{-}0.5) = 5$ or 0.5 again accounting for the MACRS half-year convention in the final year.

7.8   Summary

Definition: *Depreciation* is a decrease in fixed asset value; that is, market value is due to the use or value to the owner due to aging.

Definition (financial): Depreciation is the systematic allocation of the cost of a fixed asset over its depreciable life related to deterioration, or consumption of its useful life.

IRS Publication 551, Basis of Assets
Tangible Asset Cost Basis:

- Initial asset cost.
- All fees and charges allowed under the Uniform Capitalization Rules.
- Allocated expenses for the purchase of multiple assets.
- Allowable increases or decreases in the cost basis.

necessary to place the asset in service fit for use.

Placed-in-service date is the day on which the asset begins to provide returns to the business by performing its intended function or by producing its intended output.

Depreciable life or **recovery period**: the period over which an asset is depreciated.

- Depreciation is a **non-cash cost**: money does not change hands during the recovery period.
- Depreciation is used to **allocate an asset's loss of value** over time.
- Depreciation is **deducted from revenue** and reduces the taxable income of a business over time.

Depreciation **affects cash flow** on an **after-tax basis**.

Types of Assets:

Tangible: can be seen, touched, and felt.

- **Real**: land, buildings, and things growing on, or attached to the land.
- **Personal**: equipment, furnishings, vehicles, office machinery, etc., are not defined as real property.

Intangible: have value but cannot be seen or touched; examples include patents, copyrights, and trademarks.

$$\text{Book value} = \text{Asset cost} - \text{Sum(Depreciation Charges)}$$

Straight-line Depreciation

$$\text{Depreciation expense } d_t = \frac{B - S}{N}$$

Sum-of-Years-Digits Depreciation

$$\text{Depreciation expense } d_t = \frac{N - t + 1}{\sum_0^{N-1} N - j}(B - S) = \frac{N - t + 1}{N(N + 1)/2}(B - S)$$

Declining Balance Depreciation

$$\text{Depreciation expense } d_t = \frac{C}{N}\text{Book Value}_{t-1}$$

Modified Accelerated Cost Recovery System

$$d_t = \text{B} \times r_t$$

d_t = depreciation deduction in year t.
B = cost basis (initial ready-for-use cost).
r_t = MACRS percentage rate in year t.
Depreciation at Asset Disposal

- **Depreciation recapture** (ordinary gains): If the asset is sold (market value) for more than its book value, "... any gain on the disposition generally is recaptured (included in income) as ***ordinary income*** up to the amount of the depreciation previously allowed or allowable for the property." (IRS 946, section 4) Taxes must be paid on the difference between its market value received and its book value.
- **Depreciation loss** (ordinary loss): If the asset's market value (sales price) is less than its book value, a tax credit will be realized on ordinary income.
- **Capital gains:** Capital gain must be realized when the asset is sold (market value) for more than its initial cost basis. The excess over the original cost basis is the capital gain.

Unit-of-Production Depreciation

$$d_t = \frac{\text{Unit Production Year}(i)}{\text{Total Number Recoverable Units}} \times (B - S)$$

Depletion—Cost Depletion Method

$$d_t = \frac{\text{Adjusted Basis}}{\text{Total Number Recoverable Units}} \times B$$

Depletion—Percentage Depletion Method

$$d_t = \frac{\text{Value Production Year}(i)}{\text{Value Total Number Recoverable Units}} \times B$$

7.9   Key Terms

Alternative depreciation system
Capital gain
Cost basis
Declining balance depreciation
Depletion
Depreciable life n
Depreciation
Depreciation loss
Depreciation recapture
Deterioration
Double-declining balance depreciation

General depreciation system
Loss on disposal
MACRS
Obsolescence
Percentage depletion
Personal property
Real property
Recovery period
Straight-line depreciation
Sum-of-years-digits depreciation
Unit-of-production depreciation

Problems

1. Develop straight-line, sum-of-years'-digits, double-declining balance, and MACRS GDS depreciation schedules for petroleum drilling equipment with an initial cost of $110,000 and salvage value of $10,000 at the end of its 10-year useful life.
2. Specialty Plastics, Inc., is considering purchasing a new injection molding machine to improve control of shrinkage of molded parts. The improvement in shrinkage control will reduce losses by $95,000 per year. The injection molding machine can be purchased and installed for $340,000 today. It will have a useful life of 7 years and will have a salvage value of $17,000. The new injection molding machine will not increase operating expenses but will require additional maintenance expenses of $15,300 per year. Specialty Plastics requires a 12% rate of return on investments in new technology. Estimate the annual taxable net income and pre-tax cash flow for this machine using (a) straight line, (b) sum-of-years'-digits, (c) double-declining balance, and (d) MACRS GDS depreciation.
3. A process engineer for a French fry manufacturing line is considering investment in one of three automatic slicing machines. The MARR for this investment is 6.0%. Estimate the annual taxable net income and pre-tax cash flow using MACRS GDS depreciation. Which alternative should be selected?

	A	B	C
Initial cost	$88,400	$107,100	$113,900
Increased revenue	$64,600	$52,700	$62,900
Annual O&M expense	$25,500	$15,300	$20,400
Salvage value	$22,100	$32,300	$37,400
Life	6	6	12

4.   Western Silver LLC purchased a silver mine with ore extraction and refinement
     equipment valued at $25,000,000 and a salvage value of $5,000,000 at the end
     of an 8-year project life. Develop the unit-of-production depreciation schedule
     for the equipment.

Year	Tons silver
1	5000
2	10,000
3	20,000
4	20,000
5	15,000
6	15,000
7	10,000
8	5000

Chapter 8
Tax Effects on Engineering Investments

Abstract Governments protect their citizens from outside interference, provide for the general welfare, and provide the parameters (laws and regulations) for everyday behavior for citizens as the basis of social order. Governments protect their citizens from outside influence by investing in and maintaining standing militaries. Governments provide for the general welfare by establishing policies promoting and investing in health, education, commerce, physical, and safety infrastructures. Governments provide social order by investing in governmental and social institutions and legal institutions. Governmental investments require money, and that money is generally raised in the form of taxes and fees. Governmental taxes can be into four major classes.

- Federal corporate and personal income taxes.
- State corporate and personal income taxes and sales taxes.
- Local property taxes.
- Federal, state, and local use taxes based on the type of use.

This discussion of the effect of taxes on engineering investments will be restricted to federal and state corporate income taxes. Corporations are taxed at the "entity" level. The corporation pays taxes on income earned. Partnerships, limited liability corporations, and proprietors are "flow through" entities. The business pays no taxes. Proprietors pay personal income taxes on gains or losses from the business. Managing partners pay personal income taxes on their respective portions of gains or losses generated from partnership business operations. The Internet portal for United States IRS tax code is http://www.irs.gov/Businesses/Corporations.

In general,

- Nearly all states and some localities levy taxes on corporate income.
- The rules for determining state taxes vary widely from state to state.
- Many states compute taxable income with reference to federal taxable income, with specific modifications.

T. S. Cotter, *Engineering Managerial Economic Decision and Risk Analysis*,
Topics in Safety, Risk, Reliability and Quality 39,
https://doi.org/10.1007/978-3-030-87767-5_8

- States do not allow a tax deduction for income taxes, whether federal or state.
- Most states deny tax exemption for interest income that is tax exempt at the federal level.

8.1   Classification of Taxable Income

There are two general types of business expenditures:

- Capital expenses:

  - Expenditures for depreciable assets. Generally, assets having a life in exceeding one year.
  - Expenditures for non-depreciable assets. Generally land, as land has no finite life.

- Operating expenses—materials, labor, overhead, rents, leases, interest, and equipment having a life of less than one year.

  From the income statement, taxable income for business firms is computed as:

$$\text{Taxable Income} = \text{Net revenue}$$
$$- \text{ Operating expenditures (except capital outlays)}$$
$$- \text{ Allocated fixed expenses}$$
$$- \text{ Depreciation and depletion charges}$$
$$- \text{ Allocated general and administrative expenses}$$
$$- \text{ Operating interest expenses.}$$

The federal corporate tax established in IRS Publication 542 Corporations. For tax years beginning after 2017, corporations, including qualified personal service corporations, figure their tax by multiplying taxable income by 21% (0.21).

State corporate income tax rates vary by state. Corporate income taxes are levied in 44 states. Although the other six states advertise no corporate income taxes, they effectively collect those taxes as commercial activity taxes or gross receipts taxes. In general,

- Forty-four states levy a corporate income tax. Rates range from 2.5% in North Carolina to 12% in Iowa.
- Six states—Alaska, Illinois, Iowa, Minnesota, New Jersey, and Pennsylvania—levy top marginal corporate income tax rates of 9% or higher.
- Ten states (Arizona, Colorado, Florida, Kentucky, Mississippi, Missouri, North Carolina, North Dakota, South Carolina, and Utah) have top rates at or below 5%.
- Nevada, Ohio, Texas, and Washington impose gross receipts taxes instead of corporate income taxes.

- South Dakota and Wyoming are the only states that do not levy a corporate income or gross receipts tax. South Dakota generates the bulk of its tax revenue by levying a general sales tax and select sales excise taxes. Wyoming relies on property tax collections. Over 60% of these revenues come from minerals production.

Table 8.1 summarizes corporate income taxes by state.

8.2 Economic Analysis Taking Taxes into Account

From Chap. 7 discussion, depreciation must be added back to net income to estimate the after-tax net cash flow.

$$\text{Net Cash Flow} = [R_n - (O_n + M_n + D_n + A_n + I_n)](1 - TR) + D_n \qquad (8.1)$$

From Chap. 2 discussion, the after-tax net cash flow may be calculated from the accrual income state as follows.

Net Revenue
Less:
    Operating Expenses
    Maintenance Expenses
    Depreciation
    Administrative Expenses
    Interest Expenses
    Taxable Income
    Less: State Income Tax
    Federal Taxable Income
    Less: Federal Income Tax
    Net Income

$$\text{Net Cash Flow} = \text{Net Income} + \text{Depreciation}$$

Alternatively, the after-tax net cash flow may be calculated directly from the cash flow income statement as follows.

Table 8.1 State income tax rates (tax year 2020)

State	Tax rate (%)	Tax brackets Low high	Number Brackets
Alabama	6.5	– Flat Rate–	1
Alaska	1.0–9.4	$1–$222 k	10
Arizona	4.9	– Flat Rate–	1
Arkansas	1.0–6.5	$1–$100 k	6
California	8.84	– Flat Rate–	1
Colorado	4.63	– Flat Rate–	1
Connecticut	7.5	– Flat Rate–	1
Delaware	8.7	– Flat Rate–	1
Florida	4.458	– Flat Rate–	1
Georgia	5.75	– Flat Rate–	1
Hawaii	4.4–6.4	$1–$100 k	3
Idaho	6.295	– Flat Rate–	1
Illinois	9.5	– Flat Rate–	1
Indiana	5.5	– Flat Rate–	1
Iowa	6.0–12.0	$1–$250 k	4
Kansas	4.0–7.0	$1–$50 k	2
Kentucky	5.0	– Flat Rate–	1
Louisiana	4.0–8.0	$1–$200 k	5
Maine	3.5–8.93	$1–$3.5 m	4
Maryland	8.25	– Flat Rate–	1
Massachusetts	8.0	– Flat Rate–	1
Michigan	6.0	– Flat Rate–	1
Minnesota	9.8	– Flat Rate–	1
Mississippi	3–5	$1–$10 k	3
Missouri	4.0	– Flat Rate–	1
Montana	6.75	– Flat Rate–	1
Nebraska	5.58–7.81	$1–$100 k	2
Nevada	–	No corp tax	
New Hampshire	7.7	– Flat Rate–	1
New Jersey	6.5–10.5	$1–$1 m	4
New Mexico	4.8–5.9	$1–$500 k	2
New York	6.5	– Flat Rate–	1
North Carolina	6.9	– Flat Rate–	1
North Dakota	1.41–4.31	$1–$50 k	3
Ohio	–	No corp tax	
Oklahoma	6.0	– Flat Rate–	1

(continued)

Table 8.1 (continued)

State	Tax rate (%)	Tax brackets Low high	Number Brackets
Oregon	6.6–7.6	$1–$1 m	2
Pennsylvania	9.99	– Flat Rate–	1
Rhode island	7.0	– Flat Rate–	1
South Carolina	5.0	– Flat Rate–	1
South Dakota		No corp tax	
Tenn	6.5	– Flat Rate–	1
Texas	–	Franchise Tax	
Utah	4.95%	– Flat Rate–	1
Vermont	6.0–8.5	$1–$25 k	3
Virginia	6.0	– Flat Rate–	1
Washington		No corp tax	
West Virginia	6.5	– Flat Rate–	1
Wisconsin	7.9	– Flat Rate–	1
Wyoming		No corp tax	
Washington DC	8.25	– Flat Rate–	1

Net Cash Income
Less:
    Cash Operating Expenses
    Cash Maintenance Expenses
    Cash Administrative Expenses
    <u>Cash Interest Expenses   </u>.
    Cash Taxable Income
    <u>Less: State Income Tax   </u>.
    Cash Federal Taxable Income
    <u>Less: Federal Income Tax </u>.
    Net Cash Flow Income

Example 8.1

A corporation was formed to produce kitchen utensils. The firm purchased land for $2,200,000, erected a factory for $9,000,000, and installed equipment for $6,500,000. The facility began operation in the fourth month of its fiscal year. For the first year, gross income was $5,400,000 and operating and maintenance expense were $1,200,000. The firm uses MACRS depreciation and operates in a state that imposes a flat 5.0% income tax. Estimate the taxable income, state and federal taxes, after-tax net income, and after-tax cash flow income.

IRS 946 Table B-2 does not specify kitchen utensils equipment. So, the firm must use Asset Class 39.0 7-year "other" property for equipment depreciation.

Equipment—GDS 7-year property: 1st Yr Depr = $6,500,000 × 0.1429 = $928,850.

Building 1st Yr Depr = $9,000,000 × 0.01819 = $163,710.

Total 1st yr MACRS depreciation = $928,850 + $163,710 = $1,092,560.

Gross Income	$5,400,000
Less:	
Expenses	($1,200,000)
Depreciation	($1,092,560)
Taxable Income	$3,107,440
State Tax @ 5.0%	($ 155,372)
Federal Taxable Income	$2,952,068
Federal Taxes @ 21%	($ 619,934)
Earnings After Taxes	$2,332,134
Add back depreciation	+$1,092,560
After-tax Cash Flow Income	$3,424,694

8.2.1 Combined Federal and State Income Taxes

As discussed in Chap. 2, organizations pay state taxes on taxable income, subtract state taxes to arrive at federal taxable income, and pay taxes on federal taxable income. This yields a combined tax rate of,

$$\text{Combined Tax Rate} = \text{TR}_S + \text{TR}_F(1 - \text{TR}_S) \tag{8.2}$$

where TR_S = state income tax rate and TR_F = federal income tax rate. For Example 8.1, the combined tax rate is

$$\text{Combined tax rate} = 0.05 + 0.21(1 - 0.05) = 0.2495$$

Checking, total taxes paid and earnings after tax for Example 8.1 are

$$\text{Combined taxes paid} = \text{Taxable income} \times \text{Combined tax rate}$$
$$= \$3,107,440 \times 0.2495 = \$775,306$$

$$\text{Earnings After Tax} = \$3,107,440 - \$775,305 = \$2,332,135$$

Example 8.2

A firm is considering the acquisition of manufacturing equipment which costs $150,000. The equipment has a useful life of 5 years. It is estimated that the firm will realize a gross income of $120,000 per year from the new equipment with operating expenses of $80,000 per year. The firm will use straight-line depreciation for the equipment. At the end of its 5-year life, the equipment will have a salvage value of $45,000. The firm has combined corporate income taxes of 26%. Estimate the after-tax rate of return.

Time	0	1	2	3	4	5
Revenue		$120,000	$120,000	$120,000	$120,000	$120,000
Expenses	(−$150,000)	(−$80,000)	(−$80,000)	(−$80,000)	(−$80,000)	(−$80,000)
Depr		(−$21,000)	(−$21,000)	(−$21,000)	(−$21,000)	(−$21,000)
Salvage						$45,000
Taxable Inc		$19,000	$19,000	$19,000	$19,000	$64,000
Tax @ 26%		(−$4,940)	(−$4,940)	(−$4,940)	(−$4,940)	(−$16,640)
Net income		$14,060	$14,060	$14,060	$14,060	$47,360
Cash flow	(−$150,000)	$35,060	$35,060	$35,060	$35,060	$68,360

Try $i = 10\%$:

$$\$0 = (-\$150,000) + \$35,060 \,(P/A, 10\%, 5) + \$33,300 \,(P/F, 10\%, 5)$$
$$\$0 = (-\$150,000) + \$35,060 \,(3.791) + \$33,300 \,(0.6209)$$
$$\$0 \neq \$3,588.43$$

Try $i = 12\%$:

$$\$0 = (-\$150,000) + \$35,060 \,(P/A, 12\%, 5) + \$33,300 \,(P/F, 12\%, 5)$$
$$\$0 = (-\$150,000) + \$35,060 \,(3.605) + \$33,300 \,(0.5674)$$
$$\$0 \neq (-\$4,714.28)$$

Linear interpolation: $\text{IRR} = 10\%$
$$+ (12\% - 10\%)\left[(\$0 - \$3588.43) / (-\$4714.28 - \$3588.43)\right]$$
$$= 10.86\%.$$

8.3 Capital Gains and Losses for Non-depreciated Assets

When a non-depreciated capital asset is sold or exchanged, its change in value during ownership must be recognized. If the market value selling price is greater than the original cost basis, a capital gain must be recognized in the year of disposal. Conversely, if the market value selling price is less than the original cost basis, a capital loss must be recognized in the year of disposal and may be carried back to offset prior taxes paid or forward to offset future taxes owed.

Tax treatment of corporate capital gains and losses

Corporation	
Capital gain	Taxed as ordinary income
Capital loss	Deduct capital losses only to extent of capital gains. Excess losses in current year can be carried back 2 years to offset prior taxes paid and, if not completely absorbed, is then carried forward for up to 20 years to offset future taxes owed

8.4 After-Tax Cash Flows with Spreadsheets

In practice, after-tax analysis of engineering operations and projects requires spreadsheets. Realistic cash flow analyzes with required multiple data input through interlinking spreadsheets with summary annual cash flow and economic criterion. Example 8.3 presents a more realistic analysis of an actual project summary spreadsheet.

> **Example 8.3**
> A medium-size manufacturing business is installing a computer network to automate data collection. In the past, data collection was by manual records, and the firm's industrial engineer estimates that it costs the business $32,000 per year in lost productivity for manual recording and entering the data into spreadsheets. It is estimated that the computer network will have a 5-year technological life with a salvage value of $20,000 at the end of year 5 at which time it will be upgraded. Estimate the after-tax rate of return.

	K13		⨯	f_x	=$C13+$D13+$I13					
⊿	A	B	C	D	E	F	G	H	I	J	K

	A	B	C	D	E	F	G	H	I	J	K
1		**Example 8.3 - Delta Tax Effect with Depreciation Recapture**									
2											
3		Cost	($120,000)								
4		Annual Benefit	$32,000								
5		Tech. Life		5 years							
6		MACRS GDS life		5 years							
7		Salvage Value	$20,000								
8		Tax Rate	24%								
9											
10			Untaxed	Taxed			Recaptured	Taxable	Tax @		
11	Year		Initial Cost	BT CF	MACRS	Salvage Value	Depr.	Income	24%	AT Income	AT CF
12		0	($120,000)								($120,000.00)
13		1		$32,000	$24,000.00			$8,000.00	($1,920.00)	$6,080.00	$30,080.00
14		2		$32,000	$38,400.00			($6,400.00)	$1,536.00	($4,864.00)	$33,536.00
15		3		$32,000	$23,040.00			$8,960.00	($2,150.40)	$6,809.60	$29,849.60
16		4		$32,000	$13,824.00			$18,176.00	($4,362.24)	$13,813.76	$27,637.76
17		5		$32,000	$6,912.00	$20,000	$6,176.00	$31,264.00	($7,503.36)	$23,760.64	$44,496.64
18											
19				Cum Depr	$106,176.00				IRR =		11.24%

IRS Publication 946, Table B-1 lists computers as asset class 00.12 with a GDS 5-year property life. By the MACRS half-year convention, take only ½ of the final year of ownership's allowable depreciation amount.

$$D(5) = \$120,000 \times 0.1152 \times 0.5 = \$6912- \text{ cell E17}$$

Depreciation recapture is estimated as,

$$\text{Recapture} = \$20,000 - (\$120,000 - \$106,176) = \$6176$$

Include depreciation recapture in the year 5 taxable income

$$\text{Taxable Income} = \$32,000 + (-\$6912) + \$6176 = \$31,264$$

After-tax cash flow in year 1,

$$CF(1) = \$32,000 - \$1920 = \$30,080$$

or

$$CF(1) = \text{After} - \text{tax Income} + \text{Depreciation}$$
$$CF(1) = \$6,080 + \$24,000 = \$30,080$$

After-tax cash flow in year 5,

$$CF(5) = \$32,000 + \$20,000 - \$7504 = \$44,496$$

As shown in the spreadsheet cell K19, the internal rate of return is 11.24%.

8.5 Summary

Governmental taxes can be into four major classes.

- Federal corporate and personal income taxes.
- State corporate and personal income taxes and sales taxes.
- Local property taxes.
- Federal, state, and local use taxes based on the type of use.

Taxable income for business firms is computed as:

Taxable Income = Net revenue

                     − Operating expenditures (except capital outlays)

                     − Allocated fixed expenses

                     − Depreciation and depletion charges

                     − Allocated general and administrative expenses

                     − Operating interest expenses.

Economic analysis taking taxes into account:

$$\text{Net Cash Flow} = [R_n - (O_n + M_n + D_n + I_n)](1 - TR) + D_n$$

Net Revenue
<u>Less: Op Expenses, Depreciation, Admin Expenses, Interest</u>
Taxable Income
<u>Less: State Tax</u>
Federal Taxable Income
<u>Less: Federal Taxes</u>
Net Income

$$\text{Net Cash Flow} = \text{Net Income} + \text{Depreciation}$$

Alternatively,

Net Cash Income
Less: Cash Op Expenses, Admin Expenses, Interest
Cash Taxable Income
Less: State Tax
Federal Taxable Income
Less: Federal Taxes
Net Cash Flow Income

Combined federal and state income taxes,

$$\text{Combined tax rate} = Sr + Fr(1 - Sr)$$

8.6 Key Terms

After-tax cash flow
Capital expense
Capital gain or loss
Combined tax rate
Depreciated asset
Federal tax rate
Non-depreciated asset
Operating expense
State tax rate
Taxable income.

Problems

1. A corporation operates in a state that imposes 9.6% corporate income tax. This fiscal year the corporation had taxable income of $725,000. What is the state income tax and federal income tax the corporation must pay? What is the total taxes paid? What is the corporation's after-tax net income (EAT)?
2. A fabricated metal products manufacturer is purchasing special tools for a new contract. The tools will cost $110,000 and have a market value of $7000 at the end of the 6-year contract. The tools will be depreciated under MACRS. The corporation's combined tax rate is 27.0%. The before-tax cash flows are as follows.

Year	Before-tax cash flow
1	$33,000
2	$33,000
3	$37,500
4	$44,000
5	$15,000
6	$11,000

What is the manufacturer's rate of return on this contract?

3. New Wave Drones, Inc., is building a new manufacturing facility on land it already owns for production of its new artificially intelligent drone model. The building will cost $4,125,000, and the manufacturing equipment will cost $2,325,000 with installation cost of $202,500. Operating expenses are expected to be $1,500,000 the first year and maintenance expenses are expected to be $450,000 the first year. Both are expected to increase at 6% per year. New Wave Drones depreciates with manufacturing equipment using MACRS and is in a state with a 5% corporate income tax rate. The new manufacturing facility is scheduled to open for production in the third month of this fiscal year and product the new AI drone through the end of the fifth fiscal year at which time the facility will be converted to production of the second-generation AI drone. The market value of the building is expected to increase at the 2.5% inflation rate, and the equipment market value is estimated at $230,250 at the end of year five. The new AI drone is expected to have a 5-year useful life with net sales as follows.

Year	Net sales
1	$3,150,000
2	$4,800,000
3	$5,700,000
4	$6,750,000
5	$5,550,000

For this investment, New Wave Drones requires MARR = 15.0%. What is the annual net income and cash flow for this investment? By the NPW criterion, is the new AI drone economically viable?

4. Specialty Plastics, Inc., is considering purchasing a new injection molding machine to improve control of shrinkage of molded parts. The improvement in shrinkage control will reduce losses by $95,000 per year. The injection molding machine can be purchased and installed for $340,000 today. It will have a useful life of 7 years and will have a salvage value of $17,000. The new

injection molding machine will not increase operating expenses but will require an additional maintenance expenses of $15,300 per year. Specialty Plastics uses MARCS depreciation and operates in a state with a 9.0% corporate income tax rate. Estimate the annual after-tax net income and after-tax cash flow for this machine. At the required MARR = 12.0%, should Specialty Plastics invest in the new injection molding machine?

Chapter 9
Inflation Effects on Engineering Investments

Abstract Historically, the economies of all countries exhibit year-to-year price inflation. As USA examples, in 1970 the cost of:

- A gallon of gasoline was $0.36. In mid-2020, the same gallon of gasoline costs $2.52. Note, in mid-2015, a gallon of gasoline cost about $3.50, but the inflation-adjusted cost was $2.27. The difference was due to an international shortage causing demand to exceed supply. By mid-2016, the price had dropped to $2.14 per gallon with an inflation-adjusted price of $2.29. The collapse in gasoline price was a part of the collapse in worldwide petroleum prices due to (1) an increase in shale oil production in the USA between 2014 and 2016, (2) a growing global crude petroleum supply that was greater than that demanded by global economic growth resulting in a supply glut, and (3) OPEC nations unable to prop up petroleum prices with production cuts.
- A cup of coffee was $0.25. In mid-2020, the same cup of coffee cost about $2.62 with an inflation-adjusted price of $1.70. The average higher actual cost is due to the increased demand for specialty coffees.
- A gallon of milk was $1.15. In mid-2020, the same gallon of milk cost $3.45 with an inflation-adjusted price of $7.88. The price difference is due to the USDA subsidy purchases of dairy product through food purchase programs. Under the current The Dairy Margin Coverage program (DMC) of 2018, USDA purchases cover the price difference between the price farmers can get for a gallon of milk based on demand and the price they would have to charge to cover their cost of dairy feed.
- A dozen eggs was $0.62. In mid-2020, the same dozen eggs costs $1.54 with an inflation-adjusted price of $4.25. The price difference is due to USDA subsidies

T. S. Cotter, *Engineering Managerial Economic Decision and Risk Analysis*,
Topics in Safety, Risk, Reliability and Quality 39,
https://doi.org/10.1007/978-3-030-87767-5_9

for other farm products. Removing USDA subsidies would lead to price increases in food from 1 to 67%.

- An automobile was $3,542. In mid-2020, the cost of an average mid-size family automobile is $36,718 with an inflation-adjusted price of $24,279. The most important factors that have influenced automobile prices have been environmental standards, global market conditions, tax levels for public and private users, consumers' purchasing power, and within industry competition.

9.1 The Meaning and Effect of Inflation on Investment Worth

As can be observed in the above examples, inflation is a significant component of price increase with other factors moderating the actual price from the inflation-adjusted price. Since all engineering managerial economic decisions involve future cash flows, it is important that the effects of inflation on future asset prices, revenues, and expenses be incorporated into economic analyses.

- Inflation causes a loss in purchasing power over time for a fixed amount of money.
- Inflation tends to cause goods and services to cost more over time in terms of the same fixed amount of money.
- Because of inflation, a fixed amount of money in a given period will not have equivalent purchasing power in prior or later periods.
- Inflation is pervasive. Many industrialized countries like to see inflation maintained at about 3% per year to match expected economic growth.
- All engineering managerial economic analyses need to include inflation effects.

  Economists generally recognize four causal sources of inflation.

- *Money supply increases*: money available to consumers in the general economy increases faster than the value of the goods available.
- *Exchange rates*: prices differentials reflect the comparative value of currencies in different countries.
- *Cost-push inflation*: producers raise prices to cover costs.
- *Demand-pull inflation*: consumers bid up prices by attempting to buy more than is available.

  Any or all these causal sources may drive inflation in any given period.

9.2 Incorporating Inflation in Engineering Managerial Economic Estimates

To incorporate inflation into annual cash flows, we apply the following definitions.

Inflation rate (f): The proportion year-over-year annual increase in the monetary units (number of dollars) needed to purchase the same unit of goods or services.
Real interest rate (i_R): The "real" interest rate earned on paid on a unit of money excluding inflation. The real interest rate is also referred to as the *inflation-free interest rate*.
Market interest rate (i_M): The interest rate earned or paid on a unit of money on open market including the effect of year-over-year inflation.

Using these definitions, we can estimate the future amount of $1.00 one year from now as.

$$F = \$1(1 + i_R)(1 + f)$$
$$F / \$1 = 1 + f + i_R + (i_R)(f)$$
$$(F / \$1) - 1 = i_R + f + (i_R)(f)$$

$$i_m = i_R + f + (i_R)(f) \tag{9.1}$$

Example 9.1
Given the real interest rate and inflation rate, estimate the market interest rate.

$$i_R = 0.030 \quad f = 2.4\% i_M = ?$$

$$i_M = i_R + f + (i_R)(f)$$
$$i_M - 0.030 + 0.024 + (0.030)(0.024)$$
$$i_M = 0.0547 \text{ or } 5.47\%.$$

Corresponding to the real interest rate and the market interest rate, we need to define their dollar counterparts.

- **Actual dollars** ($A\$$): cash money—the kind you carry in your pocket. Sometimes called inflated dollars (currency) because its purchasing power includes inflation effects. Use the market interest rate (i_M) to estimate future actual dollars on an investment.
- **Real dollars** ($R\$$): constant purchasing power dollars (currency) expressed as a base year. (e.g., 1984 CPI or 1992 PPI-based dollars or a project's year 0 dollars). These are inflation-free dollars. Use the real interest rate (i_R) to estimate future real dollars on an investment.

Given these definitions, the following relationships exist among i_R, $R\$, f, i_M$, and $A\$$.

- When using actual dollars ($A\$$), use the market interest rate (i_M).
- When using real dollars ($R\$$), use the real interest rate (i_R).

At any point in time $t \leq n$, the inflation rate f can be substituted for i_M or i_R as

$$R\$ = (1+f)^{-n} A\$ = (P/F, f, t)A \tag{9.2}$$

$$A\$ = (1+f)^n R\$ = (F/P, f, t)R \tag{9.3}$$

Figure 9.1 provides a roadmap for converting among actual dollars $A\$$ and real dollars $R\$$ at any point in a project's life.

- At base year time 0, $P(A\$) = P(R\$)$.
- Starting at $t = n$ with $F(A\$)$, computing $P(R\$)$ can be accomplished in one of two ways.

  - The first approach uses $F(A\$)$ and converts it to its equivalence at $t = 0$.

$$P(A\$) = F(A\$)(1+i_M)^{-n}$$

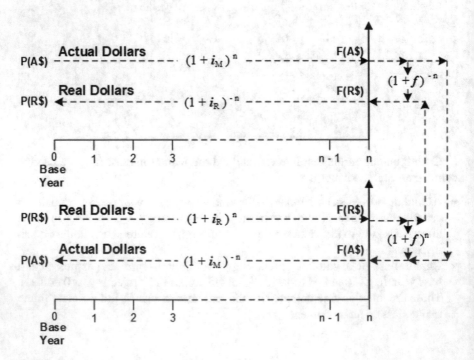

Fig. 9.1 Relationships among f, i_R, $R\$$, i_A, and $A\$$

- The second approach converts $F(A\$)$ to $F(R\$)$ and then finds the constant dollar amount at $t = 0$.

$$F(R\$) = F(A\$)(1 + f)^{-n}$$
$$P(R\$) = F(R\$)(1 + i_R)^{-n}$$

Since $P(A\$) = P(R\$)$, equating the results of the two approaches gives,

$$F(A\$)(1 + i_M)^{-n} = F(A\$)(1 + f)^{-n}(1 + i_R)^{-n}$$

$$i_M = (1 + f)(1 + i_R) - 1 \tag{9.4}$$

Solving for i_R yields,

$$i_R = [(1 + i_m)/(1 + f)] - 1 \tag{9.5}$$

Cross-multiplying and simplifying (9.4) yield Eq. (9.1).

$$i_M = 1 + i_R + f + (f)(i_R) - 1$$

$$i_M = i_R + f + (f)(i_R)$$

Example 9.2

A small private university's art center was completed 50 years ago at a total cost of $20.5 million. At that time, a wealthy alumnus donated $20 million to be used for future replacement. The university invested the $20 million in bonds that pay $i_M = 6.0\%$ annually, and the mean inflation has been 3.5% per year. Administration is now considering building a new replacement art center.

a. Given $i_M = 6.0\%$, what are the bonds worth today in actual dollars.
b. What are the bonds worth today in the real dollars at the time of the donation?
c. What is the replacement cost of a new art center?

a. $F(A\$) = \$20,000,000(1 + 0.06)^{50}$
$= \$20,000,000(F/P, 6.0\%, 50) = \$368,403,086$

b. $R\$ = F(A\$)(1 + 0.035)^{-50} = A\$(P/F, 3.5\%, 50)$
$= \$368,403,086(1 + 0.035)^{-50} = \$65,963,816$

c. $F(A\$) = \$20,000,000\,(1 + 0.035)^{50}$
    $= \$20,000,000(F/P, 3.5\%, 50) = \$111,689537$

We can check the R$ estimate by calculating the real interest rate and using it to estimate the future value of the $20 million donation.

$$i_R = [(1 + i_m)/(1 + f)] - 1$$

$i_R = [(1 + 0.06)/(1 + 0.035)\} - 1 = 0.0241546 \text{ or } 2.41546\%$
$F(R\$) = \$20,000,000\,(1 + 0.0241546)^{50} = \$65,953,816.$

9.3   Investment Analysis in Real Dollars Versus Actual Dollars

As shown in Fig. 9.1, estimates of net present worth, equivalent uniform annual worth, future worth, and rate of return will yield consistent investment decision if cash flows are in either real or actual dollars. Analysis can be conducted in either real or actual dollars and the criterion converted to the other cash flow equivalent. The analyst just needs to distinguish cash flows as being either:

- Real dollars (currency) expressed in terms of the **purchasing power of base year dollars** and discounted by the *real interest rate*.
- *Actual* dollars (currency) discounted by a *market interest rate* (real interest rate *combined* with the inflation rate).

Cash Flow Domain	Applicable Interest Rate
Real dollars R$	Real (inflation-free) rate i_R
Convert between domains with inflation rate f	
Actual dollars A$	Market rete i_M (includes inflation f)

Example 9.3

The Surface Computer Corporation is evaluating two new proposed CPU technologies proposed by two competing chip manufacturers. Both companies propose to deliver equivalent product performance at the end of a 4-year development period. From the proposed chip development costs in the following table, which company should surface choose if its market MARR = 20% and inflation is expected to be 4.5% mean rate over the development period?

MAD Chips. Development cost will be $15,750,000 first year increasing at 7.0% per year.

Itel Inc. development cost will be $17,000,000 per year in today's real dollars.

$$i_R = [(1 + i_m)/(1 + f)] - 1$$

$$i_R = [(1 + 0.20)/(1 + 0.045)] - 1 = 0.1483 \text{ or } 14.83\%$$

Estimate MAD Chips yearly geometric cost increase. Then convert these yearly costs to Real$.

Year	MAD Chips A$	Actual$	(P/F, 4.5%, n)	Real$
1	$15.750(1.07)^0 = $	$15.750	$(1.045)^{-1}$	$15.072
2	$15.750(1.07)^0 - $	$16.853	$(1.045)^{-2}$	$15.432
3	$15.750(1.07)^0 = $	$18.032	$(1.045)^{-3}$	$15.802
4	$15.750(1.07)^0 = $	$19.294	$(1.045)^{-4}$	$16.180

Covert Itel Inc.'s Real$ into Actual$.

Year	Itel Inc R$	(F/P, 4.5%, n)	Real$
1	$17.000	$(1.045)^1$	$17.765
2	$17.000	$(1.045)^2$	$18.564
3	$17.000	$(1.045)^3$	$19.400
4	$17.000	$(1.045)^4$	$20.273

Finally, estimate the present worth of the Real$ cash flows at i_R and the Actual$ cash flows at i_M.

1st Year	$15.750	$17.000	Inflation	MARR		
Change	7.0%		4.5%	20%		
	Costs as stated		Real dollar analysis		Actual dollar analysis	
	MAD Chips	Itel Inc.	MAD Chips	Itel Inc.	MAD Chips	Itel Inc.
	A$	R$				
1	$15.750	$17.000	$15.072	$17.000	$15.750	$17.765
2	$16.853	$17.000	$15.432	$17.000	$16.853	$18.564
3	$18.032	$17.000	$15.802	$17.000	$18.032	$19.400
4	$19.294	$17.000	$16.180	$17.000	$19.294	$20.273
		i_R & i_M	14.83%	14.83%	20.00%	20.00%
		PW	$44.57	$48.70	$44.57	$48.70

The same decision is made for either Real$ or Actual$ cash flows. Select MAD Chips to minimize the present worth of development cost.

Example 9.4

A firm is considering the acquisition of manufacturing equipment which costs $15,000. The equipment has a useful life of 5 years. It is estimated that the firm will realize a gross income of $12,000 R$ per year from the new equipment with operating expenses of $8,000 R$ per year. The firm will use straight-line depreciation for the equipment. At the end of its 5-year life, the equipment will have a salvage value of $4500 R$. The firm operates in a state with no corporate income taxes and is in the 21% federal tax bracket. The firm uses MARR = 11.0% market rate including expected inflation $f = 3.0\%$ over the 5-year analysis period. Using NPW, should the firm invest in the equipment?

First convert annual R$ revenues and expenses and the year 5 salvage value to A$.

Year	Revenue A$	Expenses A$
1	$12,000(1.03)^1 = $12,360$	$8,000(1.03)^1 = $8,240$
2	$12,000(1.03)^2 = $12,731$	$8,000(1.03)^2 = $8,487$
3	$12,000(1.03)^3 = $13,113$	$8,000(1.03)^3 = $8,742$
4	$12,000(1.03)^4 = $13,506$	$8,000(1.03)^4 = $9,004$
5	$12,000(1.03)^5 = $13,911$	$8,000(1.03)^5 = $9,274$

$$\text{Salvage Value}(5) = \$4500\,(1.03)^5 = \$5217$$

Year	0	1	2	3	4	5
Initial cost	($15,000)					
Gross revenue A$		$12,360	$12,731	$13,113	$13,506	$13,911
Ann expenses A$		($8,240)	($8,487)	($8,742)	($9,004)	($9,274)
Depreciation		($2,100)	($2,100)	($2,100)	($2,100)	($2,100)
Salvage A$						$5,217
Taxable Income		$2,020	$2,144	$2,271	$2,402	$7,754
Fed tax @ 21%		($424)	($450)	($477)	($504)	($1,628)
Net income		$1,596	$1,693	$1,794	$1,898	$6,126
A-T A$ cash Flow	($15,000)	$3,696	$3,793	$3,894	$3,998	$8,226
NPW A$	$1,770					
IRR A$	15.04%					
A-T R$ cash Flow	($15,000)	$3,588	$3,576	$3,564	$3,552	$7,095
IRR R$	11.69%					

Given a positive NPW, the firm should invest in the equipment.

$$i_R = [(1 + i_m)/(1 + f)] - 1$$

$$i_R = \{(1 + 0.11)/(1 + 0.03)\} - 1 = 0.0777 \text{ or } 7.77\%$$

Given that IRR $R\$ = 11.69\% > 7.77\% = i_R$, the firm should invest in the equipment.

9.4 Cash Flows That Inflate at Different Rates

In managing engineering organizations and projects, engineering managers often encounter commodities or utilities that inflate at different rates. For cash flows that inflate at different rates,

1. Use the respective individual inflation rates.
2. State the annual cash flows in actual dollars, and use the market interest rate to estimate NPW, EUAW, or IRR.

Example 9.5

An engineering manager is preparing her 5-year operating budget. Find the NPW estimate of the utilities for her manufacturing facility given the following data. Market rate MARR = 15.0%.

Last year's utility costs	5-year predicted average inflation rates
Electricity $185,000	Electricity costs increase at 2.5% per year
Water $8,000	Water costs increase at 4.3% per year
Natural gas $83,000	Natural gas costs increase at 2.8% per year

Since the engineering manager is using last year's utilities costs, those are considered as the base year 0 for restating them to future Actual$ in the spreadsheet below. The annual Actual$ for utilities are summed, and the NPV() function is used to estimate the present worth of the utilities summed Actual$ cash flows at the MARR = 15.0%. For each utility, the annual cost estimate is,

$$\text{Actual\$} = (\text{Cost Year}_0)(1 + f)^t$$

E10		×	f_x	=$E9*(1+$D10)						
⊿ A	B	C	D	E	F	G	H	I	J	K
1	Example 9.6: Inflation at Different Rates per Period									
2										
3	Inputs		MARR	15.0%						
4			Budget (years)	5						
5		Last year's cost		$185,000			$8,000		$ 83,000		
6										
7	Estimates		Utilities							
8		Year	Elect. Inf. Rate	Electricity	Water Inf Rate	Water	Gas Inf Rate	Natural gas	Total	NW
9		1	2.50%	$189,625	5.40%	$8,432	2.80%	$ 85,324	$283,381	$1,003,372
10		2	2.40%	$194,176	3.70%	$8,744	3.60%	$ 88,396	$291,316	
11		3	2.00%	$198,060	4.70%	$9,155	5.70%	$ 93,434	$300,649	
12		4	1.70%	$201,427	2.80%	$9,411	6.20%	$ 99,227	$310,065	
13		5	1.40%	$204,247	4.00%	$9,788	12.10%	$ 111,234	$325,268	

9.5 Different Inflation Rates Per Period

The typical case is that general inflation changes from year to year. Figure 9.2 shows that the US CPI inflation rate has varied from a low of −0.36% in 2009 to a high of 13.5% in 1980 with a geometric mean of 3.91%. Likewise, inflation rates for commodities and utilities vary from year to year. Figure 9.3 shows that the US PPI inflation rate varied from a low of −8.8% in 2009 to a high of 9.8% in 2008 with a geometric mean of 2.0%. For the period 1990 to 2019, there was only moderate correlation of 0.68 between the CPI and PPI inflation rates. For cash flows that inflate at different rates,

- Apply the inflation rates in the years in which they are expected to occur.
- State the annual cash flows in actual dollars, and use the market interest rate to estimate NPW, EUAW, or IRR.

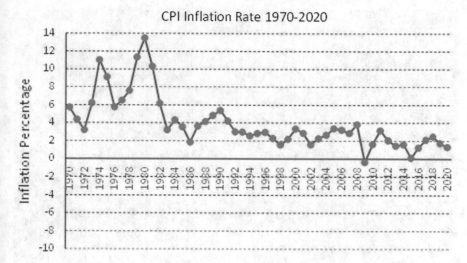

Fig. 9.2 US CPI Inflation Rate 1970 to 2020

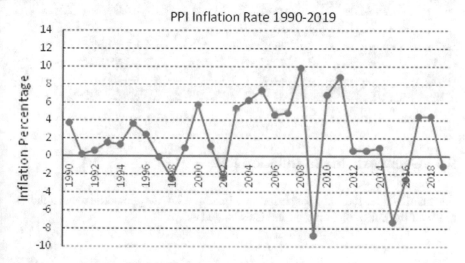

Fig. 9.3 US PPI Inflation Rate 1990 to 2019

Example 9.6

The engineering manager in Example 9.5 has found that the utilities inflation predictions are not constant over the 5-year budget period. New utilities inflation rates are reported in the following table. Re-estimate the NPW of utilities expenses over the 5-year budget period. Market rate MARR = 15.0%.

Now, the annual utility cost estimate is,

$$\text{Actual\$} = (\text{Cost Year}_{t-1})(1 + f_t)^t$$

Last year's utility costs	5-year predicted inflation rate per year
Electricity $185,000	Electricity costs increase: 2.5, 2.4, 2.0, 1.7,
Water $8000	1.4% Water costs increase: 5.4, 3.7, 4.7, 2.8, 4.0%
Natural gas $83,000	Natural gas costs increase: 2.8, 3.6, 5.7, 6.212.1%

E10		×	f_x	=$E9*(1+$D10)						
A	B	C	D	E	F	G	H	I	J	K
1	Example 9.6: Inflation at Different Rates per Period									
2										
3	Inputs		MARR	15.0%						
4			Budget (years)	5						
5		Last year's cost		$185,000			$8,000		$ 83,000		
6										
7	Estimates		Utilities							
8		Year	Elect. Inf. Rate	Electricity	Water Inf Rate	Water	Gas Inf Rate	Natural gas	Total	NW
9		1	2.50%	$189,625	5.40%	$8,432	2.80%	$ 85,324	$283,381	$1,003,372
10		2	2.40%	$194,176	3.70%	$8,744	3.60%	$ 88,396	$291,316	
11		3	2.00%	$198,060	4.70%	$9,155	5.70%	$ 93,434	$300,649	
12		4	1.70%	$201,427	2.80%	$9,411	6.20%	$ 99,227	$310,065	
13		5	1.40%	$204,247	4.00%	$9,788	12.10%	$ 111,234	$325,268	

9.6 Geometric Mean Inflation Rate

For inflation rates that vary per period as in Example 9.5, the geometric mean inflation rate over the entire planning period is estimated as

$$f_G = \left(\prod_{t=1}^{n} (1 + f_t) \right)^{1/n} - 1 \qquad (9.6)$$

The same formula can be applied to any sequence of rate increase or decrease to estimate the mean rate of change. For differing interest rates per period for a sequence of periods, the geometric mean interest rate of change is

$$i_G = \left(\prod_{t=1}^{n} (1 + i_t) \right)^{1/n} - 1 \qquad (9.7)$$

Example 9.7

Estimate the geometric mean inflation rates for the utilities of Example 9.5.
 5-year predicted inflation rate per year:
 Electricity costs increase: 2.5, 2.4, 2.0, 1.7, 1.4%

$$f_G = [(1 + 0.025)(1 + 0.024)(1 + 0.020)(1 + 0.017)(1 + 0.014)]^{1/5} - 1$$
$$= 0.01999 \text{ or } 2.0\%$$

Water costs increase: 5.4, 3.7, 4.7, 2.8, 4.0%

$$f_G = [(1 + 0.054)(1 + 0.037)(1 + 0.047)(1 + 0.028)(1 + 0.040)]^{1/5} - 1$$
$$= 0.0412 \text{ or } 4.12\%$$

Natural gas costs increase: 2.8, 3.6, 5.7, 6.2, 12.1%

$$f_G = [(1 + 0.028)(1 + 0.036)(1 + 0.057)(1 + 0.062)(1 + 0.121)]^{1/5} - 1$$
$$= 0.0603 \text{ or } 6.03\%$$

9.7 Price and Cost Change with Indexes

Price indexes are another way of measuring the relative price changes of goods and services over a period. In the USA (as in many countries), price indexes record the relative price changes of goods and services in the national economy.

- Indexes can be for a specific commodity or utility or a composite for a bundle of commodities or utilities.
- Indexes can be used to measure price changes for individual items like labor and materials of the producer price index (PPI) or general costs like consumer products of the consumer price index (CPI).

The term "relative price change" means that price changes are relative to a selected base year for which the price is set equal to 100. The formula for estimating a price index is

$$I = \big[((\text{price}(n) - \text{price(base)})/\text{price(base)}) \times 100\big] + 100 \qquad (9.8)$$

Simplifying,

$$PI = [((price(n) / price(base)) - (price(base)/ price(base))) \times 100] + 100$$
$$PI = [((price(n) / price(base)) -1) \times 100] + 100$$
$$PI = (price(n) / price(base)) \times 100-100 + 100$$

$$PI = (price(n)/price(base)) \times 100 \qquad (9.9)$$

where price(n) = unit price in year n and price(base) = unit price in the selected base year. To estimate year-to-year percentage price changes,

$$API(n) = ((Index(n) - Index(n - t))/Index(n - t)) \times 100 \qquad (9.10)$$

where Index(n) = Index in year n and Index$(n - t)$ = Index in the year of interest. Estimate the average rate of price increase as the geometric mean,

$$PI(avg) = (Index(n)/Index(base))^{1/n} - 1 \qquad (9.11)$$

where Index(n) = Index in year n and Index(base) = Index in the base year.

Composite cost indexes measure the historical prices of *groups* or *bundles* of assets or commodities. The US Bureau of Labor Statistics tracks the *consumer price index* (CPI) and the *producer price index* (PPI). The CPI is a measure of the average change over time in the prices paid by urban consumers for a market basket of consumer goods and services. Currently, the reference base for most CPI indexes is 1982–1984 = 100. Additionally, expenditure weights are updated every two years to keep the CPI current with changing consumer preferences. Table 9.1 presents the CCI for 1980 to 2019.

The producer price index (PPI) program measures the average change over time in the selling prices received by domestic producers for their output. Some sectors

Table 9.1 CPI 1980 to 2019

Year	CPI	Year	CPI	Year	CPI	Year	CPI
1980	82.4	1990	130.7	2000	172.2	2010	218.1
1981	90.9	1991	136.2	2001	177.1	2011	224.9
1982	96.5	1992	140.3	2002	179.9	2012	229.6
1983	99.6	1993	144.5	2003	184.0	2013	233.0
1984	103.9	1994	148.2	2004	188.9	2014	236.7
1985	107.6	1995	152.4	2005	195.3	2015	237.0
1986	109.6	1996	156.9	2006	201.6	2016	240.0
1987	113.6	1997	160.5	2007	207.3	2017	245.1
1988	118.3	1998	163.0	2008	215.3	2018	251.1
1989	124.0	1999	166.6	2009	214.5	2019	255.7

covered include agriculture, construction, manufacturing, and mining. The prices included in the PPI are from the first commercial transaction for many products and some services. Currently, the reference base for the PPI index is 1982 = 100. The Bureau of Labor Statistics publishes thousands of product price indexes broken into three large categories.

- Industry Level—The industry-based classification measures the cost of production at the industry level. It tracks the changes in prices received for an industry's output outside the sector itself by calculating industry net output. BLS product price index includes over 535 industry-specific listings.
- Commodity—This classification ignores the industry of production and combines goods and services by similarity and product make-up. More than 3.700 indexes cover products and about 800 cover services. The indices are arranged by end-use, product, and service.
- Commodity-Based Final Demand-Intermediate Demand (FD-ID)—The FD-ID system groups commodity indexes for goods, services, and construction into subproduct classes, which account for the specific buyer of the products. The end-user or buyer is termed as either the final demand (FD) or the intermediate demand (ID) user. This classification considers the physical assembly and processing required for these goods. The Bureau of Labor Statistics publishes over 600 FD-ID targeted indexes. Some indices are adjusted for seasonality.

Example 9.8

Using the CPI indexes in Table 9.1, estimate the cost of a product in 2019 that cost $1,000 in 2009.

$$CPI(2009) = 214.5 \quad CPI(2019) = 255.7$$

Applying Formula (9.10),

$$API(n) = ((Index(n) - Index(n - t))/ Index(n - t)) \times 100$$
$$API(2019) = ((255.7 - 214.5)/214.5) \times 100 = 19.21\%$$

$$Price(2019) = Price(2009)(1 + API)$$
$$Price(2019) = \$1000(1 + 0.1921) = \$1192.$$

9.8 Inflation Effect on After-Tax Calculations

In practice, managerial engineering economic analyses require cash flows over multiple budget periods from multiple inputs that may inflate (or deflate) at different rates. To produce accurate cash flow and economic criterion estimates, economic analyses must account for inflation (or deflation). In general, the tax effects of inflation will reduce the rate of return on an investment or the cost of interest on a purchase.

- Depreciation is taken in year 0 (asset's purchase year) constant dollars (currency).
- Thus, inflation results in increased taxable income and lower after-tax rate of return.

As an example of the lower after-tax rate of return, consider the inflation-adjusted after-tax Actual\$ rate of return versus the Real\$ rate of return of Example 9.4. Using the definition of internal rate of return, the inflation-adjusted Actual\$ after-tax rate of return was,
After-Tax IRR:

$$\$0 = -\$15,000 + \$3696(P/F, i, 1) + \$3793(P/F, i, 2)$$
$$+ \$3894(P/F, i, 3) + \$3998(P/F, i, 4) + \$8226(P/F, i, 5)$$

$$\text{Actual\$ IRR} = 15.04\%$$

The Real\$ after-tax rate of return was,
After-Tax IRR:

$$\$0 = -\$15,000 + \$3696(1.03) - 1(P/F, i, 1) + \$3793(1.03) - 2(P/F, i, 2)$$
$$+ \$3894(1.03) - 3(P/F, i, 3) + \$3998(1.03) - 4(P/F, i, 4)$$
$$+ \$8226(1.03) - 5(P/F, i, 5)$$

$$\text{Real\$ IRR} = 11.69\%$$

Example 9.9 directly illustrates the effects of inflation on after-tax rate of return.

Example 9.9

A \$21,000,000 investment in an automatic inventory system with no salvage value is expected to save \$5,250,000 real dollars per year in labor and fixed costs for 6 years. The automated inventory system is MACRS asset class 00.12 information systems with depreciation GDS 5-year property life. The company

operates in a state that yields a 30% combined income tax rate. Estimate the
rate of return for (1) no inflation and (2) inflation at 7.0% per year.

Inputs

Initial investment	$21,000	x 1,000
Annual benefit	$5,250	x 1,000
Salvage value	$0	
Life (years)	6	
Tax rate	30%	
Inflation	7%	

No Inflation

Year	Before-tax Cash Flow	MACRS Rate	Depr.	Taxable Income	Taxes	After-tax Cash Flow
0	($21,000)					($21,000)
1	$5,250	20.00%	$4,200	$1,050	($315)	$4,935
2	$5,250	32.00%	$6,720	($1,470)	$441	$5,691
3	$5,250	19.20%	$4,032	$1,218	($365)	$4,885
4	$5,250	11.52%	$2,419	$2,831	($849)	$4,401
5	$5,250	11.52%	$2,419	$2,831	($849)	$4,401
6	$5,250	5.76%	$1,210	$4,040	($1,212)	$4,038
IRR	13.0%					9.79%

Inflation at 7%

Year	Before-tax Cash Flow	MACRS Rate	Depr.	Taxable Income	Taxes	After-tax Actual$ Cash Flow	After-tax Real$ Cash Flow
0	($21,000)					($21,000)	($21,000)
1	$5,618	20.00%	$4,200	$1,418	($425)	$5,192	$4,853
2	$6,011	32.00%	$6,720	($709)	$213	$6,224	$5,436
3	$6,431	19.20%	$4,032	$2,399	($720)	$5,712	$4,662
4	$6,882	11.52%	$2,419	$4,462	($1,339)	$5,543	$4,229
5	$7,363	11.52%	$2,419	$4,944	($1,483)	$5,880	$4,192
6	$7,879	5.76%	$1,210	$6,669	($2,001)	$5,878	$3,917
IRR	20.9%					16.1%	8.47%

With 7% inflation, the after-tax return on investment is reduced to 8.47%
from 9.79% with real dollar analysis under no inflation.

9.9   Summary

Inflation causes a loss in purchasing power over time for a fixed amount of money.

Inflation tends to cause goods and services to cost more over time in terms of the same fixed amount of money.

Defining f = inflation rate, i_R = real interest rate, and i_M = market inflation rate, we can estimate i_M as

$$i_m = i_R + f + (i_R)(f)$$

Actual dollars ($A\$$): cash money also called inflated dollars (currency).

Real dollars ($R\$$): constant purchasing power dollars (currency) expressed as a base year.

Estimate the real interest rate i_R from the market rate i_M using the formula,

$$i_R = [(1 + i_m)/(1 + f)] - 1$$

Cash flow domain	Applicable interest rate
Real dollars $R\$$	Real (inflation-free) rate i_R
Convert between domains with inflation rate f	
Actual dollars $A\$$	Market rate i_M (includes inflation f)

Cash Flows that Inflate at Different Rates:

- Use the respective individual inflation rates.
- State the annual cash flows in *actual dollars,* and use the *market interest rate* to estimate NPW, EUAW, or IRR.

Different Inflation Rates per Period:

- Apply the inflation rates in the years in which they are expected to occur.
- State the annual cash flows in actual dollars, and use the market interest rate to estimate NPW, EUAW, or IRR

Geometric Mean Inflation Rate

$$f_G = \left(\prod_{t=1}^{n} (1 + f_t) \right)^{1/n} - 1$$

Geometric Mean Interest Rate:

$$i_G = \left(\prod_{t=1}^{n} (1 + i_t) \right)^{1/n} - 1$$

Price Indices

$$PI = \left[((\text{price}(n) - \text{price}(\text{base}))/\text{price}(\text{base})) \times 100 \right] + 100$$

$$PI = (\text{price}(n)/\text{price}(\text{base})) \times 100$$

$$API(n) = ((\text{Index}(n) - \text{Index}(n - t))/\text{Index}(n - t)) \times 100$$

$$PI(\text{avg}) = (\text{Index}(n)/\text{Index}(\text{base}))^{1/n} - 1$$

9.10 Key Terms

Actual dollars
Base year
Composite cost index
Cost-push inflation
Deflation
Demand-pull inflation
Exchange rate
Inflation rate
Market interest rate
Money supply
Price index
Purchasing power
Real dollars
Real interest rate.

Problems

1. Inflation is expected to remain relatively constant at about 3.5% per year for the next ten years. How much money will be required in 10 years to purchase will an item that costs $100 today?
2. If inflation is 2.5% per year and a bank is loaning money at 6.0% annually, what is the real interest the bank earns on it loans.
3. An investor purchased a 6% tax-free municipal bond at a face value of $1000. The bond pays $60 per year for 10 years. At maturity, the bond returns the original $1000. (a) What is the real rate of return on the bond? (2) If inflation is 2.0% annually, what is the market rate of return on the bond? (c) At the real rate of return, what is the present value of the bond? (d) At the market rate of return, what is the present value of the bond?

4. An economist predicts that prices will increase a total of 55% by the end of next 8 years. Further, she predicts that prices will increase a total of 25% over the subsequent 12 years. What is the annual inflation rate for the 20-year period?

5. An engineering manager examines the records for the average price of an electronic component over the last five years. Calculate the year-to-year basis of price increase. What is the manager's estimate of the component's inflation rate for next year?

Year	Price	Annual f
5 years ago	$170.00	
4 years ago	$171.70	
3 years ago	$174.90	
2 years ago	$180.00	
1 year ago	$182.00	
Last year	$185.20	

6. If $30,000 is deposited in a 6% savings account and inflation is 3%, what is the actual and real dollar value of the account at the end of 20 years? If the time value of money is 4%, what is the present worth of the account at the start of investing?

7. Radiation monitoring equipment must be purchased to monitor low-level X-ray leakage. Two alternative units are available that can fulfill the requirements. The X-Detect unit costs $12,600 and has a 4-year useful life. The Monitor-X unit costs $18,000 and has an 8-year useful life. Equipment costs are inflating at 9% annually. Based on an 8-year project period and MARR $= 15\%$, which monitoring equipment should be purchased?

8. An engineering manager is evaluating the options to overhaul a piece of equipment. The machine will be needed for the next 10 years. MARR $= 8\%$ for this type of investment. A contractor has suggested three alternatives.

    (1) Compete overhaul for $9600 that will yield 10 years of operation.
    (2) Major overhaul for $7200 that will provide 6 years of service and a minor overhaul for $4000 in year 6 to extend the equipment life 4 more years.
    (3) A minor overhaul now, in year 4, and in year 8. Each minor overhaul will cost $4000.

    Inflation is 4% annually. Which alternative should the engineering manager select?

9. Texas Chemicals is considering a construction project to expand its MTBE production over the next 3 years. Current construction costs and inflation rates are given in the following table. Texas Chemicals requires MARR $= 20\%$ for this type of project. General price inflation is expected to be steady at 4.0% annually for the 3-year construction project. What is the present worth of the 3-year construction project and its future worth at the completion?

10. Consider two mutually exclusive investments with cash flows stated in year 0 dollars ($\times$ $1 K). Both alternatives have a three-year life with salvage cost = salvage value. The annual inflation rate is 3.0%. The company operates in a state with 8.9% corporate income tax. These investments qualify for straight-line depreciation. The organization's after-tax MARR is 8.0%. Use rate of return analysis to determine which alternative is preferable.

Year	Invest 1	Invest 2
0	($420)	($300)
1	$225	$175
2	$225	$175
3	$225	$175

Chapter 10
Incremental Analysis

Abstract To this point it has been assumed that an engineering manager selects from one of multiple economically acceptable investment alternatives. In economic terms, the investment alternatives are assumed to be **mutually exclusive**. Under net present worth and equivalent uniform annual worth at a stated MARR interest rate, the mutually exclusive project with the highest NPW or EUAW is always preferred. This is not the case for internal rate of investment analysis. *Under the IRR criterion, the investment with the highest IRR may not be the preferred alternative.* To avoid this problem, we estimate the internal rate of return on the difference in cash flows of pairwise alternatives. The criterion for alternative selection is now maximizing *delta-IRR*.

10.1 Introduction

To understand this concept, we first need to define incremental analysis.

Definition: *Incremental analysis* is the examination of cash flow differences between alternatives to determine if the difference in the increased cost of the higher initial cost alternative is justified by the difference in increased benefits.

$$\text{Incremental Cash Flow} = \text{Cash Flows(Higher initial cost alt.)}$$
$$- \text{Cash Flows (Lower initial cost alt.)}$$

If the rate of return on the difference in the cash flows is greater than the MARR interest rate for a project type, then the discounted difference in benefit cash flows of the higher cost alternative exceeds the initial cost differential, and the higher cost alternative is preferred. This problem is illustrated in Example 10.1.

Example 10.1

Consider two simple alternatives each with an initial cost and a year one benefit cash flow. Alternative 1's initial cost is $100, and its year-1 benefit cash flow is $150. Alternative 2's initial cost is $200, and its year-1 benefit cash flow may range from $250 to $300. The MARR is 10%. Over the benefit cash flow range of Alternative 2, which alternative is preferred by the present worth criterion, the IRR criterion, and the delta-IRR criterion?

The following spreadsheet outputs illustrate the relationship between the present worth criterion, individual IRR criterion, and the delta-IRR criterion.

Alternative 2 Benefit		$250	
MARR		10%	
Analysis:			
	Alternative 1	Alternative 2	Delta
Initial cost	($100)	($200)	($100)
Benefit	$150	$250	$100
PW(10%)	$36.36	$27.27	
IRR	50%	25%	0%

Case 1: Alternative 2 benefit = $250

By IRR, Alternative 1 is preferred with IRR = 50% versus Alternative 2 IRR(2) = 25%. Delta-IRR = 0% because the initial delta cost (-$100) is exactly offset by the delta benefit $100. Delta-IRR = 0% < 10% = MARR and Alternative 1 is preferred. This is consistent with maximizing net present worth, because NPW(1) = $36.36 > $27.27 = NPW(2).

Alternative 2 Benefit		$260	
MARR		10%	
Analysis:			
	Alternative 1	Alternative 2	Delta
Initial cost	($100)	($200)	($100)
Benefit	$150	$260	$110
PW(10%)	$36.36	$36.36	
IRR	50%	30%	10%

Case 2: Alternative 2 benefit = $260

By IRR, Alternative 1 is preferred with IRR = 50% versus Alternative 2 IRR(2) = 30%. Delta-IRR = 10% because the delta benefit of $110 is greater than the initial cost (-$100). Delta-IRR = 10% = 10% = MARR and neither alternative is preferred. This is consistent with maximizing net present worth, because NPW(1) = $36.36 = $36.36 = NPW(2).

Alternative 2 Benefit		$280	
MARR		10%	
Analysis:			
	Alternative 1	Alternative 2	Delta
Initial cost	($100)	($200)	($100)
Benefit	$150	$280	$130
PW(10%)	$36.36	$54.55	
IRR	50%	40%	30%

Case 3: Alternative 2 benefit = $280

By IRR, Alternative 1 is preferred with IRR = 50% versus Alternative 2 IRR(2) = 40%. Delta-IRR = 30% because the delta benefit of $130 is greater than the initial cost (-$100). Delta-IRR – 30% > 10% – MARR and Alternative 2 is preferred. This is consistent with maximizing net present worth, because NPW(1) – $36.36 < $54.55 – NPW(2).

Alternative 2 Benefit		$300	
MARR		10%	
Analysis:			
	Alternative 1	Alternative 2	Delta
Initial cost	($100)	($200)	($100)
Benefit	$150	$300	$150
PW(10%)	$36.36	$72.73	
IRR	50%	50%	50%

Case 4: Alternative 4 benefit = $300

By IRR, neither alternative is preferred with IRR(1) = 50% = IRR(2). Delta-IRR = 50% because the delta benefit of $150 is greater than the initial cost (-$100). Delta-IRR = 50% > 10% = MARR and Alternative 2 is preferred. This is consistent with maximizing net present worth, because NPW(1) = $36.36 < $72.73 = NPW(2).

From Example 10.1, we see that maximizing delta-IRR is consistent with maximizing the net present worth of the pairwise alternatives.

The question arises, "Why just not use NPW or EUAW for project selection?" Two scenarios arise in application that inhibit direct use of NPW or EUAW in selecting among alternative investments.

- The relevant MARR may not be known. This case typically arises in small to medium size new operations. These organizations are typically still managed by the entrepreneur–owner, and industrial engineering or cost accounting departments are usually not established with sufficient talent and data to estimate MARRs for various investment risk categories.
- The relevant MARR may be known, but the rate of return criterion is preferable with proposing project alternative to investors. The difference between delta-IRR and the MARR indicates the relative gain in return for the same risk.

In the case where the MARR is not known, the *graphical delta-IRR sensitivity analysis* provides the capability to perform "what-if" delta-IRR analyses to determine the range over which an alternative is preferred relative to the other alternatives under consideration. In the case where the MARR is known, the *challenger–defender delta-IRR analysis* provides a direct relative return versus risk analysis.

10.2   Graphical Incremental Rate of Return Sensitivity Analysis

Sensitivity analysis has been applied for decades in management science to examine the effects of change due to uncertainty on an optimal (maximizing, minimizing, and delta from target) solution. Sensitivity analysis allows an engineering manager to determine the range over which an optimal solution holds.

Definition: *Sensitivity analysis* is a determination of the amount of variation ($\pm$) in an estimate necessary to change the decision to select a particular alternative. The point at which the decision changes from one alternative to another is the *point of indifference*.

In managerial economic analysis, graphical differences in alternatives NPW or EUAW represent the delta-IRR between alternatives over a given range of interest rates. At each point of indifference, the delta-IRR between alternatives goes to zero resulting in an *IRR critical decision point*.

The general steps in performing a graphical incremental rate of return sensitivity analysis are as follows:

1.   Identify all acceptable alternatives that fulfill similar system outcomes—difficult in "real-world" situations.
2.   Set up a table of interest rate ranges, and for each interest rate, compute the NPW or EUAW.
3.   Construct a graph of NPW or EAUW versus interest rates for all alternatives.
4.   Determine which alternative provides maximum NPW or EAUW over differing ranges of interest rates.
5.   Determine the points of indifference. They are the IRR critical decision points of the alternatives under consideration.
6.   List the NPW or EAUW maximizing interest rate ranges for each alternative.

Incremental rate of return sensitivity analysis may be performed for a range of interest rates where the period cash flows are held constant or vary. Both analyses will be illustrated in the next two examples.

Example 10.2

An organization must select one of three mutually exclusive alternatives. The decision-maker needs to select the most cost-effective machine, but he is uncertain as to what MARR to use. Perform a graphical DIRR sensitivity analysis to help the decision-maker make the correct economic decision.

Parameter	Alt. 1	Alt. 2	Alt. 3
Initial cost	($300)	($425)	($600)
UAB	$95	$110	$120
Life in years	6	8	10

Assume "continuing requirement" for EUAW one-life analysis. The spreadsheet analysis is provided below.

Inputs:			
	Alt1	Alt2	Alt3
Initial cost	($300)	($425)	($600)
UAB	$95	$110	$120
Life	6	8	10

Output:			
Interest Rate	EUAW(1)	EUAW(2)	EUAW(3)
0%	$45.00	$56.88	$60.00
2%	$41.44	$51.98	$53.20
4%	$37.77	$46.88	$46.03
6%	$33.99	$41.56	$38.48
8%	$30.11	$36.04	$30.58
10%	$26.12	$30.34	$22.35
12%	$22.03	$24.45	$13.81
14%	$17.85	$18.38	$4.97
16%	$13.58	$12.15	($4.14)
18%	$9.23	$5.77	($13.51)
20%	$4.79	($0.76)	($23.11)
22%	$0.27	($7.43)	($32.94)
24%	($4.32)	($14.22)	($42.96)
26%	($8.99)	($21.14)	($53.17)

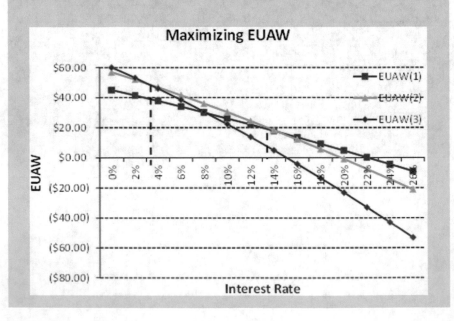

The IRR critical decision points are 3.20 and 14.55%. From 0 to 3.20%, prefer Alternative 3 to maximize EUAW. Between 3.20% and 14.55%, prefer Alternative 2 to maximize EUAW. Above 14.55%, prefer Alternative 1 to maximize EUAW.

Example 10.3

Three alternatives are available for investment. There is uncertainty about the arithmetic gradient of alternatives A and C, the interest rate at which the company can finance the projects, and the life of the project. Nominal gradients are $G(A) = -\$100$ and $G(C) = \$100$. Nominal interest rate is 15%, and nominal life is 10 years. Given the following equivalent uniform annual cash flow (EUA_CF) equations, create graphs for sensitivity analyses to gradient, interest rate, and project life. The respective cash flow equations are as follows:

$$\text{EAU\_CF}(A) = -\$1000\,(A/P, i, n) + \$1,000 - G(A/G, i, n)$$
$$\text{EAU\_CF}(B) = -\$5000\,(A/P, i, n) + \$1,300$$
$$\text{EAU\_CFi} = -\$5000\,(A/P, i, n) + \$1,000 + G(A/G, i, n)$$

Gradient Sensitivity Analysis:

MARR	15%	Gradient	$100			
Life	10	(A/G,i,n)	3.383			

	Alternative			Gradient	A	B	C
	A	B	C	$0	$800.75	$502.99	$3.74
Initial Cost	($1,000)	($4,000)	($5,000)	$50	$631.59	$502.99	$172.90
Eq Benefit	$1,000	$1,300	$1,000	$100	$462.43	$502.99	$342.06
Gradient	($100)	$0	$100	$150	$293.27	$502.99	$511.22
				$200	$124.11	$502.99	$680.38

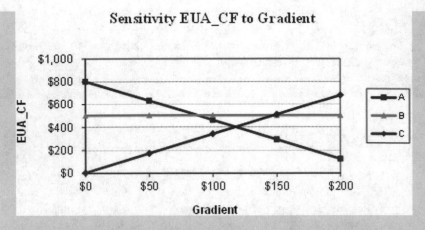

To maximize EUAW, prefer alternative *A* over the gradient range of $0 to
$88. From gradient $89 to $147, prefer alternative *B*. Greater than or equal to
$148 gradient, prefer alternative *C*.

Graphical delta-IRR Sensitivity Analysis:

MARR	15%	Gradient	$100
Life	10	(A/G,i,n)	3.383
		Alternative	
	A	B	C
Initial Cost	($1,000)	($4,000)	($5,000)
Eq Benefit	$1,000	$1,300	$1,000
Gradient	($100)	$0	$100

Interest	A	B	C
0%	$450.00	$899.98	$949.96
3%	$457.12	$831.08	$839.50
6%	$461.93	$756.53	$722.86
9%	$464.40	$676.72	$600.68
12%	$464.55	$592.06	$473.54
15%	$462.43	$502.99	$342.06
18%	$458.12	$409.94	$206.79
21%	$451.74	$313.34	$68.27
24%	$443.41	$213.59	($73.02)
27%	$433.25	$111.08	($216.64)
30%	$421.41	$6.15	($362.20)

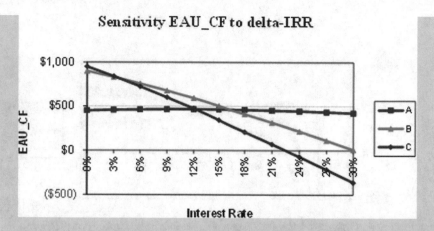

Sensitivity EAU_CF to delta-IRR

To maximize EUAW, prefer alternative C over an interest rate range of 0–3.6%. Prefer alternative *B* over and interest rate range of 3.7–16.4%. Greater than or equal to 16.5%, prefer alternative *A*.

Life Sensitivity Analysis:

				Life	A	B	C
MARR	15%	Gradient	$100	0	-1072280	-4,292,014	-5365692
Life	10	(A/G,j,n)	3.383	2	338.3721	-1,160	-2029.07
				6	526.0441	243	-111.465
				8	499.0171	409	163.8824
		Alternative		10	462.4284	503	342.0593
	A	B	C	12	424.6988	562	468.4166
Initial Cost	($1,000)	($4,000)	($5,000)	14	389.0707	601	562.7983
Eq Benefit	$1,000	$1,300	$1,000	16	356.8277	628	635.4862
Gradient	($100)	$0	$100	18	328.3825	647	692.4998
				20	303.7248	661	737.7064

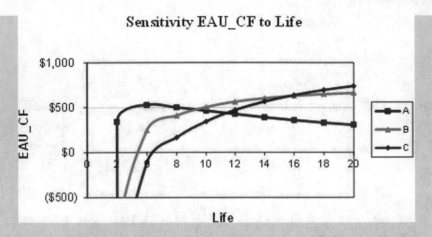

To maximize EUAW, prefer alternative *A* and a project life less than or equal to 9 years. Prefer alternative *B* and a project life of 9–16 years. For a project life greater than or equal to 16 years, prefer alternative *C*.

10.3   Challenger–Defender Incremental Rate of Return Analysis

In the case where the relevant MARR is known but it is desirable to examine the relative return versus risk analysis, we apply the challenger–defender incremental rate of return analysis. The steps to perform a challenger–defender delta-IRR analysis are as follows:

1. Identify all acceptable alternatives that fulfill similar system outcomes—difficult in "real-world" situations.
2. Compute rate of return for each alternative. Keep alternatives with IRR ≥ MARR.
3. Rank remaining alternatives by ascending order of initial investment.
4. Make a pairwise analysis of the contender and present selection. For investment:
5. If ΔIRR ≥ MARR, select the contender.
6. If ΔIRR < MARR, keep the present selection.
7. Sequentially compare preferred alternative from step 4 with next alternative in the list from step 3.
8. Continue until all pairwise comparisons have been made.

Example 10.4

Perform challenger–defender delta-IRR analysis for the three alternatives in Example 10.2 for MARR = 10%.

Inputs:			
MARR	10%		
	Alt1	Alt2	Alt3
Initial cost	($300)	($425)	($600)
UAB	$95	$110	$120
Life	6	8	10

To equalize differences in alternative lives (6 vs. 8 vs. 10 years), use the least common multiple life of 120 years.

Incremental Analysis						
Comparison	Do Nothing			Alt1		
Year	Initial Cost	UAB	Cash Flow	Initial Cost	UAB	Cash Flow
0	$0		$0	($300)		($300)
1		$0	$0		$95	$95
2		$0	$0		$95	$95
3		$0	$0		$95	$95
4		$0	$0		$95	$95
5		$0	$0		$95	$95
6		$0	$0	($300)	$95	($205)
7		$0	$0		$95	$95
8		$0	$0		$95	$95
9		$0	$0		$95	$95
10		$0	$0		$95	$95
11		$0	$0		$95	$95
12		$0	$0	($300)	$95	($205)
110		$0	$0		$95	$95
111		$0	$0		$95	$95
112		$0	$0		$95	$95
113		$0	$0		$95	$95
114		$0	$0	($300)	$95	($205)
115		$0	$0		$95	$95
116		$0	$0		$95	$95
117		$0	$0		$95	$95
118		$0	$0		$95	$95
119		$0	$0		$95	$95
120		$0	$0		$95	$95

Alt2			Alt3		
Initial Cost	UAB	Cash Flow	Initial Cost	UAB	Cash Flow
($425)		($425)	($600)		($600)
	$110	$110		$120	$120
	$110	$110		$120	$120
	$110	$110		$120	$120
	$110	$110		$120	$120
	$110	$110		$120	$120
	$110	$110		$120	$120
($425)	$110	($315)		$120	$120
	$110	$110		$120	$120
	$110	$110	($600)	$120	($480)
	$110	$110		$120	$120
	$110	$110		$120	$120
	$110	$110	($600)	$120	($480)
	$110	$110		$120	$120
($425)	$110	($315)		$120	$120
	$110	$110		$120	$120
	$110	$110		$120	$120
	$110	$110		$120	$120
	$110	$110		$120	$120
	$110	$110		$120	$120
	$110	$110		$120	$120
	$110	$110		$120	$120

Estimate each alternative's common multiple life rate of return.

RoR	22.12%	19.77%	15.10%	all > MARR

Perform pairwise delta-IRR analysis ordering from smallest to largest initial cost. MARR = 10%.

Alt1 - DoNothing	DeltaIRR	Alt2 - Alt1	DeltaIRR	Alt3 - Alt2	DeltaIRR
($300)	22.12%	($125)	14.55%	($175)	3.20%
$95	Select Alt 1	$15	Select Alt 2	$10	Keep Alt2
$95		$15		$10	
$95		$15		$10	
$95		$15		$10	
$95		$15		$10	
($205)		$315		$10	
$95		$15		$10	
$95		($410)		$435	
$95		$15		$10	
$95		$15		($590)	
$95		$15		$10	
($205)		$315		$10	

Note that the selection of Alternative 2 by delta-IRR at MARR = 10% is consistent with maximizing EUAW (or NPW) in Example 10.2. Note also that the relative difference in risk (14.55% − 3.20% = 11.35% loss) for keeping Alternative 2 versus selecting in Alternative 3 is greater than that of selecting Alternative 2 over keeping Alternative 1 (22.12% − 14.55% = 7.57% loss).

10.4 Observations on Incremental Rate of Return Analysis

Some final observations compare incremental rate of return analysis versus net present worth and equivalent annual worth analyses.

1. Unless the MARR is known, neither NPW or EUAW is possible.
2. NPW and EUAW often require less computation than RoR.
3. In some situations, RoR is easier to explain (i.e., comparing a single IRR to MARR). In other situations, NPW or EUAW may be easier to explain or more relevant.
4. Follow the established business policy for investment analysis. Often one or two methods are specified by standard operating procedures.

10.5 Summary

Definition: *Incremental analysis* is the examination of cash flow differences between alternatives to determine if the difference in the increased cost of the higher initial cost alternative is justified by the difference in increased benefits.

Incremental Cash Flow = Cash Flows(Higher initial cost alt.)

— Cash Flows (Lower initial cost alt.)

Definition: *Sensitivity analysis* is a determination of the amount of variation (±) in an estimate necessary to change the decision to select a particular alternative. The point at which the decision changes from one alternative to another is the *point of indifference*.

The general steps in performing a graphical incremental rate of return sensitivity analysis are as follows:

1. Identify all acceptable alternatives that fulfill similar system outcomes—difficult in "real-world" situations.
2. Set up a table of interest rate ranges, and for each interest rate, compute the NPW or EUAW.
3. Construct a graph of NPW or EAUW versus interest rates for all alternatives.
4. Determine which alternative provides maximum NPW or EAUW over differing ranges of interest rates.
5. Determine the points of indifference. They are the IRR critical decision points of the alternatives under consideration.
6. List the NPW or EAUW maximizing interest rate ranges for each alternative.

The steps to perform a challenger–defender delta-IRR analysis are as follows:

1. Identify all acceptable alternatives that fulfill similar system outcomes—difficult in "real-world" situations.
2. Compute rate of return for each alternative. Keep alternatives with IRR ≥ MARR.
3. Rank remaining alternatives by ascending order of initial investment.
4. Make a pairwise analysis of the contender and present selection. For investment:
5. If ΔIRR ≥ MARR, select the contender.
6. If ΔIRR < MARR, keep the present selection.
7. Sequentially compare preferred alternative from step 4 with next alternative in the list from step 3.
8. Continue until all pairwise comparisons have been made.

10.6   Key Terms

Challenger–defender delta-IRR analysis
Graphical delta-IRR sensitivity analysis
IRR critical decision point
Mutually exclusive
Point of indifference
Sensitivity analysis.

Problems

1. Consider alternatives do nothing, *A*, and *B*. Perform challenger–defender incremental analysis, and set up a choice table incremental analysis. Which alternative should be selected for MARR = 12.0%?

Year	Alt A	Alt B
0	(−$22,500)	(−$19,000)
'1–5	$6700	$5700

2. An organization is considering two alternatives, each with no salvage value. Perform challenger–defender incremental analysis, and set up a choice table incremental analysis. Which alternative should be selected for MARR = 10.0%?

Year	Alt X	Alt Y
0	(−$22,500)	(−$19,000)
UAB	$6700	$5700
Useful life (years)	4	8

3. Pakossis Construction Company is considering the purchase of trenchless auger boring equipment to reduce subcontracting costs for water and gas line installation. Perform challenger–defender incremental analysis, and set up a choice table incremental analysis. Which alternative should be selected for MARR = 8.0%?

Cash flow	Auger tiger	McAuger
Initial cost	(−$100,500)	($169,000)
Annual savings	$77,000	$33,000
Annual O&M Expenses	(−$45,000)	(−$22,000)
Salvage value	$15,000	$35,000
Useful life	5	10

4. Consider three mutually exclusive alternatives. Perform challenger–defender incremental analysis, and set up a choice table incremental analysis. Which alternative should be selected for MARR = 12.0%?

Year	Alt X	Alt Y	Alt Z
0	(−$22,000)	(−$24,000)	(−$22,500)
1	$5500	$4000	$3500
2	$5500	$7000	$6500
3	$10,000	$11,000	$10,000
4	$10,000	$14,000	$14,000

Chapter 11
Replacement Analysis

Abstract All assets have a finite life and require replacement. The engineering managerial economics question is when is the optimum time to replace an asset given the physics and economics of asset use?

11.1 The Replacement Problem

There are four primary factors that must be considered in answering the asset replacement question.

- *Obsolescence*: Asset's technology is surpassed by newer technology making the asset's benefit/cost unacceptable relative to assets that possess the newer technology.
- *Depletion*: The gradual loss of an asset's market value due to its consumption.
- *Deterioration*: Loss in an asset's value due to use or aging.
- *Economic Service Life*: The remaining useful economic life that results in the minimum annual equivalent cost of ownership.

In addition to these factors, the engineering manager must consider other aspects in planning for and logistically replacing an existing asset.

- If a unit fails, must it be removed permanently from service, or repaired?
- Are standby units available if the system fails?
- Do components or units fail independently of the failure of other components?
- Is there a budget constraint?
- If the unit can be repaired after failure, is there a constraint on the capacity of the repair facility?
- Is only one replacement allowed over the planning horizon? Are subsequent replacements allowed at any time during the study period?
- Is there more than one replacement unit (price and quality combination) available at a given point in time?
- Do future replacement units differ over time? Are technological improvements considered?

© The Author(s), under exclusive license to Springer Nature Switzerland AG 2022
T. S. Cotter, *Engineering Managerial Economic Decision and Risk Analysis*,
Topics in Safety, Risk, Reliability and Quality 39,
https://doi.org/10.1007/978-3-030-87767-5_11

- Is preventative maintenance included in the model?
- Are periodic operating and maintenance costs constant or variable over time?
- Is the planning horizon finite or infinite?
- Are consequences other than economic effects considered? (i.e., environmental, sustainability, socio-technical issues)
- Are income tax consequences considered?
- Is "inflation" considered?
- Does replacement occur simultaneously with retirement, or are there nonzero lead times?
- Are cash flow estimates deterministic or stochastic?

In managerial economic analyses the central issue is determining an asset's *economic service life*. Fixed asset costs can be partitioned into three categories. *Capital costs* are the sum of one-time fixed expenses for the design of productive assets; fees for permits and regulatory compliance; acquisition and installation of land, facilities, and equipment; and commissioning legal fees. Asset annual *operating expenses* include annual non-direct labor and non-direct materials such as the consumption of utilities and supplies, rents, and insurance to support productive activities. Asset annual *maintenance expenses* include annual expenses for maintenance personnel, replacement parts, supplies, and preventive restorative labor. Fixed asset costs must be in comparable units for the defender and challenger assets under consideration.

> *Defender*—the existing asset currently in service.
>
> *Challenger*—the best alternative asset selected from incremental analysis available to replace the defender asset.

11.2 Economic Service Life of an Asset

To determine an asset's economic service life, we apply EUAW of a continuing requirement analysis of Chap. 6.

Definition: The *economic service life* of an asset is the number of time periods remaining service n_{ESL} that maximizes its EUAW of ownership or, correspondingly, minimizes it EUAC of service.

The challenger's cash flow economic life EUAW is

$$
\begin{aligned}
\mathrm{Max}(\mathrm{EUAW}_{\mathrm{ESL}}(\mathrm{challenger})) = {} & [(-P_0) \\
& + \Sigma_p[(\mathrm{DR}_p - O_p - M_p - I_p)(P/F, i, n_{\mathrm{ESL}}) \\
& + \mathrm{MV}p(P/F, i, n_{\mathrm{ESL}})](1 - \mathrm{TR}_p)(A/P, i, n_{\mathrm{ESL}})]
\end{aligned} \tag{11.1}
$$

where P_0 = initial installed capital cost, ΔR_p = difference in revenue or other benefit cash flows over the defender revenues or benefit cash flows in year p, O_p challenger operating expenses in year p, M_p = challenger maintenance expenses in year p, I_p = interest expense associated with the challenger in period p, TR_p = tax rate in period p, and MV_p = challenger market value or salvage value in period p, and p = 1, 2, ..., n_{ESL}, n_{ESL} + 1, ..., n_{MAX} (maximum remaining useful life in years of either the defender or challenger). In this case, $\Delta R_p > \$0$ is the result of efficiency improvements from technological advances in the challenger. Where $\Delta R_p = \$0$, no technological efficiency improvements, Eq. (11.1) becomes

$$\text{Max}(\text{EUAW}_{\text{ESL}}(\text{challenger})) = [(-P_0)$$
$$+ \Sigma_p[(-O_p - M_p - I_p)(P/F, i, n_{ESL})$$
$$+ \text{MV}p\,(P/F, i, n_{ESL})]\,(1 - TR_p)\,(A/P, i, n_{ESL}) \tag{11.2}$$

The defender's EUAW_{ESL} or EUAC_{ESL} cash flow equations are the same except that we set $P_0 = \text{MV}_0$ the defender's market value in period 0, which is the opportunity cost that could have been received if we had sold the defender asset in period 0 of the analysis period. For the defender, Eqs. (11.1) and (11.2) become

$$\text{Max}(\text{EUAW}_{\text{ESL}}(\text{defencer})) = [(-\text{MV}_0)$$
$$+ \Sigma_p[(\text{DR}_p - O_p - M_p - I_p)(P/F, i, \ n_{ESL})$$
$$|\ \text{MV}p(P/F, i, n_{\text{ESL}})]\,(1 \ \text{TR}_p)\,(\Lambda/P, i, n_{\text{ESL}}) \tag{11.3}$$

$$\text{Max}(\text{EUAW}_{\text{ESL}}(\text{challenger})) = [(-\text{MV}_0)$$
$$+ \Sigma_p[(-O_p - M_p - I_p)(P/F, i, n_{\text{ESL}})$$
$$+ \text{MV}p(P/F, i, n_{\text{ESL}})]\,(1 - \text{TR}_p)(A/P, i, n_{\text{ESL}}) \tag{11.4}$$

The defender asset will be retained while its economic life EUAW is greater than the challenger's economic life EUAW.

- If the defender's economic service life EUAW < challenger's economic service life EUAW, replace the defender with the challenger.
- Finite Horizon Project Life: If the defender's economic life EUAW > challenger's economic life EUAW, keep the defender. Select the combination of remaining years ownership = finite project life n that maximizes the joint EUAW (defender(p), challenger($n - p$)).
- Infinite Horizon Project Life: If the defender's economic life EUAW > challenger's economic life EUAW, keep the defender. For each subsequent year, estimate the defender's EUAW (ESL + p), where p = 1, 2, ..., n. Select the combination remaining years ownership of the defender where EUAW(ESL + p) > EUAW(challenger) that maximizes the joint EUAW(defender(ESL + p), challenger(ESL)).

11.3 Tax Laws Affecting Replacement Analysis

Equations (11.1)–(11.4) specify that economic service life is estimated on an after-tax basis. Hence, it is important to consider tax laws that may affect determination of an asset's economic service life. Prior to January 1, 2018, exchange US IRS Section 1031 distinguished between retirement and replacement of an asset.

Retirement—the asset is salvaged, and the service it rendered is discontinued. **Gain or loss is realized**.

Replacement—an asset is removed from service, and another asset is acquired and placed in its place to continue service. Termed *like-for-like exchange*. **Gain or loss is not realized.**

Section 1031 defined rules for recognizing gains or losses for different classes of personal, real, and intangible property. If an asset was retired during its useful life, a capital gain or loss equal to the difference between its market value and book value was realized. If the asset was retired at the end of its useful life, a capital gain or loss equal to the difference between its market value and salvage value was realized. Conversely, if an asset was replaced, gain or loss is not realized. If a defender asset is replaced during its useful life,

$$\text{Initial book value of the new asset} = \text{purchase price}$$
$$- \text{market value (old)} + \text{book value (old)}$$

If the asset was replaced at the end of its useful life,

$$\text{Initial book value of the new asset} = \text{purchase price}$$
$$- \text{salvage value (old)} + \text{book value(old)}.$$

Under the Tax Cuts and Jobs Act, Section 1031 now applies only to exchanges of real property and not to exchanges of personal or intangible property. An exchange of real property held primarily for sale still does not qualify as a like-kind exchange. Effective January 1, 2018, exchanges of machinery, equipment, vehicles, artwork, collectibles, patents, and other intellectual property and intangible business assets generally do not qualify for non-recognition of gain or loss as like-kind exchanges. Real properties generally are of like-kind, regardless of whether they are improved or unimproved. However, real property in the USA is not like-kind to real property outside the USA.

Example 11.1

Retirement example: Wood Products, Inc., purchased one acre of land adjacent to its property five years ago for $100,000, drilled a natural gas well for $5000, and installed the well head for $32,500. The well head has been depreciated as MACRS seven-year property. The facility is transitioning from gas-fired production to wind turbine electricity-based production and no longer needs the land and well head and has sold it to a neighboring production operation for $130,000. Current well head market value is $9000. The facility operates in a state with an 8.0% corporate tax. Estimate the after-tax gain (loss) from the sale of the land and well head.

Land Cost	$100,000	Combined tax rate =		27.32%
Well	$15,000			
Well head	$32,500	MACRS Depreciation		
GDS property class	7 years	Year	d(t)	Book Value
Current life	5 years	1	$4,643	$27,857
Sales price	$130,000	2	$7,959	$19,898
Well head market value	$9,000	3	$5,685	$14,213
State tax rate	8.0%	4	$4,061	$10,152
Federal tax rate	21.0%	5	$1,450	$8,702

Income Effect	
Sales price	$130,000
Less:	
Land/well book value	($115,000)
Well head book value	($8,702)
Gain (-Loss)	$6,298
Net tax	($1,721)
After tax gain (-loss)	$4,578
Less Well head BV	$8,702
Land/Well	
Capital Gain (Loss)	($4,124)
Well head book value	
Market value	$9,000
Book Value	$8,702
Ordinary Gain (Loss)	($298)

Notes:

1. The well is attached to the land and is not depreciable.
2. In year 5, apply the *half-year convention* to the well head MACRS depreciation.
3. The sale of a business (facility and equipment) for a lump sum is governed by IRS Publication 544, *Sales and Other Dispositions of Assets*. The "residual method" must be used to allocate gain or loss to each individual asset. Consult your tax accountant.

Example 11.2

Replacement example: Reconsider now that Wood Products, Inc., in Example 11.1 trades the land, well, and well head for the to the neighboring production facility for land with a market value of $130,000 on which it plans to build the wind turbine electricity generating facility. The land on which the gas well is installed has appreciated in market value to $116,000. The well itself has no market value.

Land Cost	$100,000	Combined tax rate =		27.32%
Well	$15,000			
Well head	$32,500	MACRS Depreciation		
GDS property class	7 years	Year	d(t)	Book Value
Current life	5 years	1	$4,643	$27,857
Sales price (market value)	$130,000	2	$7,959	$19,898
Land market value (gas well)	$116,000	3	$5,685	$14,213
Well head market value	$9,000	4	$4,061	$10,152
State tax rate	8.0%	5	$1,450	$8,702
Federal tax rate	21.0%			
		Income Effect		
		New land market value		$130,000
		Land/well market value		($116,000)
		Land/well book value		$115,000
		Well head market value		($9,000)
		Well head book value		$8,702
		Net change initial book value		$128,702
		Well head		
		Market value		$9,000
		Book Value		$8,702
		Ordinary Gain (Loss)		($298)
		New land initial book value		$128,403

Note: Consult IRS Publication 544.

11.4   Planning Horizon

The economic service life depends on the follow observations about asset performance in applied operational and project cases.

- Scale-up or start-up inefficiencies occur in the first year of operation. These arise from correcting original design deficiencies and adapting asset design to operational or project constraints. These inefficiencies result in reduced benefit cash flows and increased engineering and operational expenses over those budgeted in the first year of useful life.

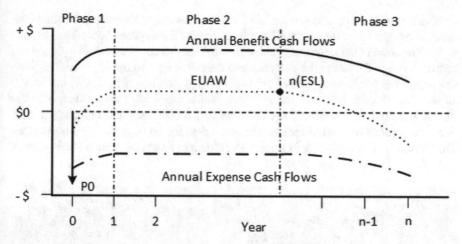

Fig. 11.1 Three phases of asset operational benefit/cost performance

- After scale-up or start-up inefficiencies are resolved or mitigated, there will be a sequence of operational periods where increased benefit cash flows or lower operational expenses than those of the previously implemented asset.
- Nearing the end of an asset's operational life, benefit cash flows decline due to wear and obsolescence inefficiencies, rising operational expenses from asset wear out, and rising maintenance expenses due to increasing asset failure.

These three general periods of any asset useful life are illustrated in Figure 11.1. Estimation of economic service life EUAW implies a continuing requirement. Where a continuing requirement exists (operational life is unknown or long-term > n_{ESL}), estimate the economic service one-life EUAW or EUAC for the defender and challenger and select the replacement period based on EUAW(total) or EAUC(total cost). Many asset investments, however, are for only a finite operations or project life. Application of Eqs. (11.1) to (11.4) must be modified for finite planning horizons. By planning horizon, the operational or project service period from the current defender and future challengers is unknown in the infinite horizon case. In the finite horizon case, the operational or service period has a definite duration. More formally,

Finite horizon: only a set number n of replacement cycles is needed to the end of operational or project life.

Infinite horizon: the asset fulfills a continuing requirement for a large, or infinitely large n → ∞ number of replacement cycles.

Replacement cycle: the number of times the asset is replaced. For example, replacement cycle 3 indicates that the defender asset has been replaced by three challenger assets.

Replacement time: the number of analysis periods p from project initiation to the placed-in-service date; includes the time needed to plan, budget, manufacture, install, and scale up or debug the replacement.

Estimation of economic service life for finite planning horizon assets requires the analyst to consider end-of-useful-life cash flows for the required operational life for the defender and challenger. Use the *project life* method of Sect. 6.6 to estimate the respective cash flows for the defender and the challenger to the required life given the appropriate following condition. Sequentially estimate the EUAW, or EUAC, of retaining the defender one more year and replacing it with the challenger to the end of required life, retaining the defender two more years and replacing it with the challenger to the end of require life, etc. Select the combination that maximizes EUAW(total). The following examples will illustrate finite and infinite horizon cases for $\Delta R_p = \$0$ and for $\Delta R_p > \$0$.

Example 11.3

Infinite planning horizon example:

Computer server generation A, used to control a manufacturing process, was purchased and installed four years ago for $27,500. It was estimated to have a useful life of ten years with a salvage value of $1000. Its current market value is $7000 with a market value of $6000 at the end of the analysis period one, declining at $1000/period thereafter. Its annual operating expenses for electricity is $650 per year. The facility engineering manager just re-negotiated the electricity contract down to $500 per year for years 6 to 10. Maintenance expenses were $10,000 in year 5, but the engineering manager just negotiated a service contract with the server manufacturer for $8000 per year in years 6 and 7 increasing by $500 per year in years 8–10 to cover expected end-of-life repairs. The organization operates in a state with a corporate income tax rate $= 8.9\%$.

The same server manufacturer now offers more technologically advanced server generation C for $26,500 with replacement time of two months, a ten-year useful life, and salvage value $0. Due to upgrades, server C will have estimated annual operating expenses starting at $600 per year for electricity. If the server is purchased with an extended warranty, the first three years maintenance will be covered by the server manufacturer and year 4 to 10 maintenance expense will be $7000 per year. Its market value will be $15,000 at the end of its first year of operation, will decline $2500/year in years 2–5, and will decline $1000/year in years 6–10.

All cash flows are in real dollars. The organization's MARR $= 15\%$ real interest rate. Perform one cycle replacement analysis, but assuming an infinite planning horizon since manufacturing control will be required into the future.

Inputs:	Server A	Server C			
Initial Cost	($27,500)	($26,500)		State tax rate	8.9%
Life (years)	10	10		Federal tax rate	21.0%
Market Value	$7,000			Combined tax rate	28.0%
MV Gradient	($1,000)	($1,500)			
Salvage Value	$1,000	$1,500			
Operating (electricity	($650)	($600)			
Annual Maintenance	($8,000)	($7,000)			
Main. Gradient 7-10	($500)	($500) years 2 - 5			
	$0	($700) years 8 - 10			
GDS property years	5	5			
MARR	15%				

Per IRS 946 Table B-1 (Table 7.1), computer server networks are asset class 00.12, information systems including computers with GDS property life of five years.

Year	Period	Defender Server A Initial Cost	Mkt Value	Opr(Elect)	Main Exp	MACRS	BkValue	MktValue	SlvgValue
0		($27,500)							
1						($5,500)	$22,000		
2						($8,800)	$13,200		
3						($5,280)	$7,920		
4	0		$7,000			($3,168)	$4,752		
5	1			($650)	($10,000)	($3,168)	$1,584	$6,000	
6	2			($500)	($8,000)	($1,584)	$0	$5,000	
7	3			($500)	($8,000)			$4,000	
8	4			($500)	($8,500)			$3,000	
9	5			($500)	($9,000)			$2,000	
10	6			($500)	($9,500)			$1,000	$1,000
	7								
	8								
	9								
	10								

Note the dual timelines, one for the original server A and the second for server C with year 0 set equal to year 4, the year in which the replacement decision is being considered.

Also note that the first four years, server A MACRS depreciation is grayed to indicate past estimates not relevant to the replacement decision but necessary for estimating book value.

Server A economic service life analysis. Server A $n_{ESL} = 2$ more years.

Year	Period	CashExp(1)	CashExp(2)	CashExp(3)	CashExp(4)	CashExp(5)	CashExp(6)	EUAW	
0									
1									
2									
3									
4	0								
5	1	($5,183)	($6,777)	($6,777)	($6,777)	($6,777)	($6,777)	($7,233)	
6	2		($2,075)	($5,673)	($5,673)	($5,673)	($5,673)	($6,570)	MAX
7	3			($3,239)	($6,117)	($6,117)	($6,117)	($7,306)	
8	4				($4,318)	($6,477)	($6,477)	($7,691)	
9	5					($5,398)	($6,837)	($7,934)	
10	6						($6,477)	($8,106)	

Server C economic service life analysis. Server C $n_{ESL} = 3$ years due to the warranty effect.

Year	Period	Initial Cost	Oper(Elect)	Main Exp	MACRS	Bk Value	Mkt Value	Salvage
0								
1								
2								
3								
4	0	($26,500)						
5	1			($600)	$0	($5,300)	$21,200	$15,000
6	2			($600)	$0	($8,480)	$12,720	$12,500
7	3			($600)	$0	($5,088)	$7,632	$10,000
8	4			($600)	($7,000)	($3,053)	$4,579	$7,500
9	5			($600)	($7,000)	($3,053)	$1,526	$5,000
10	6			($600)	($7,000)	($1,526)	$0	$4,000
	7			($600)	($7,000)			$3,000
	8			($600)	($7,000)			$2,000
	9			($600)	($7,000)			$1,000
	10			($600)	($7,000)			$0

Year	Period	NetCost(1)	NetCost(2)	NetCost(3)	NetCost(4)	NetCost(5)	NetCost(6)	NetCost(7)	NetCost(8)	NetCost(9)	NetCost(10)	EUAW
0												
1												
2												
3												
4	0											
5	1	($2,244)	$1,054	$1,054	$1,054	$1,054	$1,054	$1,054	$1,054	$1,054	$1,054	($17,719)
6	2		$3,650	$1,945	$1,945	$1,945	$1,945	$1,945	$1,945	$1,945	$1,945	($8,225)
7	3			$3,816	$994	$994	$994	$994	$994	$994	$994	($6,582) MAX
8	4				($1,841)	($4,614)	($4,614)	($4,614)	($4,614)	($4,614)	($4,614)	($7,084)
9	5					($1,443)	($4,614)	($4,614)	($4,614)	($4,614)	($4,614)	($7,258)
10	6						($1,064)	($5,042)	($5,042)	($5,042)	($5,042)	($7,167)
	7							($3,311)	($5,470)	($5,470)	($5,470)	($7,376)
	8								($4,030)	($5,470)	($5,470)	($7,419)
	9									($4,750)	($5,470)	($7,436)
	10										($5,470)	($7,436)

Decision:

Since the defender's two-year economic life EUAW $= (--\$6570) > (-\$6582)$ = EUAW challenger three-year economic life, keep the defender.

Example 11.4

Finite planning horizon example:

Machining Specialists, Inc., is considering replacing its CNC end mill with a newer unit, which has a higher productivity. The current CNC end mill was purchased five years ago to fulfill a ten-year contract to manufacture specialty aircraft parts for a new private jet for Boecraft Corporation. The current machine cost $55,000 and had an expected useful life of ten years with a salvage value of $5000. The current machine has a market value of $18,500. A new, more technologically advanced CNC end mill can be purchased and installed for $67,250 and is expected to increase throughput over the defender as shown in the following table. The new machine has an estimated useful life of 12 years. Machining Specialists requires a 12% MARR for machining equipment and operates in a state with 5.0% corporate income tax rate. Relevant operating revenues, operating expenses, maintenance expenses, and market values are presented in the following table. Perform finite planning horizon replacement analysis for these two machines.

	Current CNC machine			
Year	Revenue	Opr Exp	Main Exp	Market value
1	$12,000	($900)	($800)	$53,000
2	$14,000	($800)	($400)	$40,750
3	$14,000	($800)	($400)	$31,500
4	$14,000	($800)	($400)	$24,000
5 (0)	$14,000	($800)	($400)	$18,500
6 (1)	$14,000	($800)	($450)	$14,250
7 (2)	$14,000	($800)	($450)	$12,000
8 (3)	$13,000	($900)	($500)	$9750
9 (4)	$12,000	($1000)	($500)	$7400
10 (5)	$10,000	($1100)	($550)	$5000

	Challenger CNC machine			
Year	Revenue	Opr Exp	Main Exp	Market value
1				
2				
3				
4				
5 (0)				
6 (1)	$14,000	($1000)	($1200)	$63,750
7 (2)	$16,750	($1000)	($1000)	$49,000
8 (3)	$19,500	($1000)	($800)	$37,750
9 (4)	$19,500	($1000)	($600)	$29,000
10 (5)	$19,500	($1000)	($400)	$22,500

From IRS 946 Table B-2 (Table 7.2), CNC end mills are MACRS asset class 34.0 with GDS property life of seven years.

Inputs:	Current CNC	New CNC			
Initial Cost	($55,000)	($67,250)		State tax rate	5.0%
Useful Life	10	12	years	Federal tax rate	21.0%
MARR	12%	12%		Combined tax rate	25.0%
MACRS GDS property		7	years		

Defender CNC end mill economic service life analysis. Defender $n_{ESL} = 1$ more year.

Year	Current CNC Period	Initial Cost	Mkt Value	Revenue	Opr(Elect)	Main Exp	MACRS	BkValue	MktValue	SlvgValue
0		($55,000)								
1			$53,000	$12,000	($900)	($800)	($7,857)	$47,143		
2			$40,750	$14,000	($800)	($400)	($13,469)	$33,673		
3			$31,500	$14,000	($800)	($400)	($9,621)	$24,052		
4			$24,000	$14,000	($800)	($400)	($6,872)	$17,180		
5	0		$18,500	$14,000	($800)	($400)	($4,909)	$12,272		
6	1			$14,000	($800)	($450)	($4,909)	$7,363	$14,250	
7	2			$14,000	($800)	($450)	($4,909)	$2,454	$12,000	
8	3			$13,000	($900)	($500)	($2,454)	$0	$9,750	
9	4			$12,000	($1,000)	($500)			$7,400	
10	5			$10,000	($1,100)	($550)				$5,000

CashExp(1)	CashExp(2)	CashExp(3)	CashExp(4)	CashExp(5)	Defender EUAW	
$13,865	$10,794	$10,794	$10,794	$10,794	$7,395	Max
	$13,865	$10,794	$10,794	$10,794	$6,956	
		$11,160	$9,318	$9,318	$6,089	
			$7,880	$7,880	$5,296	
				$6,267	$4,931	

Challenger CNC end mill economic service life analysis. Challenger n_{ESL} = 3 years. With defender n_{ESL} = 1 year and challenger n_{ESL} = 3 years < 5 years required life, Condition 4 analysis is required.

Year	Period	Initial Cost	Revenue	Opr(Elect)	Main Exp	MACRS	BkValue	MktValue
0								
1								
2								
3								
4								
5	0	($67,250)						
6	1		$14,000	($1,000)	($1,200)	($9,607)	$57,643	$60,525
7	2		$16,750	($1,000)	($1,000)	($16,469)	$41,173	$57,500
8	3		$19,500	($1,000)	($800)	($11,764)	$29,410	$54,625
9	4		$19,500	($1,000)	($600)	($8,403)	$21,007	$51,900
10	5		$19,500	($1,000)	($400)	($6,002)	$15,005	$49,300

Year	Period	CashExp(1)	CashExp(2)	CashExp(3)	CashExp(4	CashExp(5)	EUAW	
0								
1								
2								
3								
4								
5	0							
6	1	$15,823	$11,253	$11,253	$11,253	$11,253	$1,028	
7	2		$31,558	$15,179	$15,179	$15,179	$8,162	
8	3			$38,090	$16,219	$16,219	$8,698	Max
9	4				$40,821	$15,530	$8,352	
10	5					$42,323	$7,851	

Replace the defender with the challenger.

Decision:

Replace the defender with the challenger. Keep the challenger three years to its economic life EUAW = $8698 and replace the challenger with itself and keep it two years to the end of the project life. This will maximize the remaining project life EUAW = $8519.

EUAW for Replacement of Defender Remaining 5 years of the Contract.

Ttl Periods	RepCF(2+3)	RepCF(3+2)
EUAW	$8,446.34	$8,518.79
1	$8,162	$8,698
2	$8,162	$8,698
3	$8,698	$8,698
4	$8,698	$8,162
5	$8,698	$8,162

11.5 Closing Comments on Replacement Analysis

This chapter has presented only an introductory discussion of the fundamentals of economic asset replacement analysis and just one example each of infinite and finite planning horizon replacement analysis. Some closing comments are in order.

- Application of the economic service life concept to physical assets is adapted from finance theory.

  - A physical asset may be in optimal physical condition but may not be economically useful. Conversely, a physical asset may not be in optimum physical condition, but the cash flows of its challenger make the defender economically useful.

– Changes in governmental regulations may render and asset obsolete even though it is in optimal physical condition and economically useful.
– Generally Accepted Accounting Principles (GAAP) require a reasonable estimate of the required time for asset use. Economic service life estimates for the same defender and challenger may vary based on daily, weekly, monthly, quarterly, or yearly usage.
– Economic service life may also vary depending on the depreciation schedule applied.
– Cash flow variations in the initial install cost of an asset, revenues, and operating expenses associated with the asset, market value schedule, and salvage value will affect the economic service life estimate.

- Given these variables, the generic cash flow Eqs. (11.1) to (11.4) must be modified to match the actual cash flows of the defender and challenger assets under consideration.
- In the infinite planning horizon case, the decision criteria "if the defender's economic life EUAW is greater than the challenger's economic life EUAW, retain the defender ...," did not consider the repurchase price of the defender asset. To normalize for defender cash flow repurchase, the defender and challenger n_{ESL} should be estimated, respectively, through one replacement cycle cash flow each terminating in disposal. In the finite planning horizon case, if either the defender or challenger useful life is less than the required life, repurchase and salvage value cash flows must be incorporated into n_{ESL} estimates.
- As illustrated in Example 11.4, a suboptimal EUAW(total) or EAUC(total cost) economic service life may be the best solution for a finite planning horizon required life operation. In this case, either or neither defender or challenger n_{ESL} may be included in the best EUAW(total) or EUAC(total cost) solution.

In summary, the combinatorics of cash flow, useful life, required life, and governmental regulations yield almost an infinite number of patterns. It is the engineering manager analyst's responsibility to accurately determine the defender and challenger cash flows of the situation to produce accurate n_{ESL} and EUAW(total) estimates. Not addressed in this chapter, sensitivity analyses may be required to identify ranges over which n_{ESL}, EUAW(total), and replacement decisions hold.

11.6 Summary

Defender—the existing asset currently in service.

Challenger—the best alternative asset selected from incremental analysis available to replace the defender asset.

Definition: The *economic service life* of an asset is the number of time periods remaining service n_{ESL} that maximizes its EUAW of ownership or, correspondingly, minimizes it EUAC of service.

The challenger's cash flow economic life EUAW is

$$\text{Max}(\text{EUAW}_{ESL}(\text{challenger})) = [(-P_0) + \Sigma_p[(\text{DR}_p$$
$$- O_p - M_p - I_p)(P/F, i, n_{ESL}) + \text{MV}p(P/F, i, n_{ESL})]$$
$$(1 - \text{TR}_p)(A/P, i, n_{ESL})$$

The defender's cash flow economic life EUAW is

$$\text{Max}(\text{EUAW}_{ESL}(\text{defencer})) = [(-\text{MV}_0) + \Sigma_p[(\text{DR}_p$$
$$- O_p - M_p - I_p)(P/F, i, n_{ESL})$$
$$+ \text{MV}p(P/F, i, n_{ESL})](1 - \text{TR}_p)(A/P, i, n_{ESL})$$

US IRS Section 1031:

Retirement—the asset is salvaged, and the service it rendered is discontinued. **Gain or loss is realized**.

Replacement—an asset is removed from service, and another asset is acquired and placed in its place to continue service. Termed *like-for-like exchange*. **Gain or loss is not realized.**
Planning horizon:

Finite horizon: only a set number n of replacement cycles is needed to the end of operational or project life.

Infinite horizon: the asset fulfills a continuing requirement for a large, or infinitely large $n \rightarrow \infty$ number of replacement cycles.

Replacement cycle: the number of times the asset is replaced. For example, replacement cycle 3 indicates that the defender asset has been replaced by three challenger assets.
Replacement time: the number of analysis periods p from project initiation to the

placed-in-service date; includes the time needed to plan, budget, manufacture, install, and scale up or debug the replacement.

11.7   Key Terms

Depletion
Deterioration
EUAC (total cost)
EUAW (total)
Economic service life n_{ESL}
Finite planning horizon
Infinite planning horizon
Obsolescence
Required life
Replacement
Replacement cycle
Replacement time
Retirement
Section 1031 property.

Problems

1. A mid-western company purchased a 5-acre tract of farmland three years ago with the intent to build a supply warehouse for its finished products, but the local market demand did not materialize the land cost $8000 per acre and has appreciated in value by 6.0% per year. Consider the gain/loss or new asset valuation for the following two scenarios.

    (a)   The land will be sold at its current market value.
    (b)   The land will be exchanged for a 5-acre tract in an east coast state where demand for the company's products is high. Due to population demand, land prices per acre in the east coast state are significantly higher, and the 5-acre tract market value is $67,000.

2. A special purpose machine for the manufacture of finished plastic products is to be purchased for $225,000. The machine is required for five-year contract with a new customer. Annual revenues, expenses, and market/salvage real dollar values are reported in the following table. The machine will be depreciated under MACRS and the organization operates with a combined 27% tax rate. If MARR = 10%, what is the machine's economic life?

Year	Revenues	O&M Exp's	Mkt value
1	$120,000	($37,500)	$192,000
2	$160,000	($48,000)	$121,500

(continued)

(continued)

Year	Revenues	O&M Exp's	Mkt value
3	$160,000	($79,500)	$78,000
4	$140,000	($97,000)	$25,500
5	$120,000	($117,000)	$0

3. An existing research and development test instrument that costs $16,000 two years ago has a market value of $12,000 today declining at (-$1500) per year (arithmetic gradient) until it reaches it six-year life salvage value of $2000 at the end of its six-year useful life. The asset was purchased to fulfill a 12-year development project. The test instrument expenses (−$4000) per year to operate. A new test instrument is under consideration as the replacement. The new instrument costs $18,000 but reduces annual test expenses to (−$3000) per year. The new asset will have a market value of $8500 at the end of the first year of use declining at (−$1500) per year (arithmetic gradient) to its salvage value of $4000 at the end of its four-year useful life. The organization operates in a state with a combined state and federal tax rate of 24% and uses MARR = 10% for this project. The selected asset is depreciated under MACRS. Perform replacement analysis and determine the combination of defender or challenger economic lives that maximizes the project EUAW.

4. Pro-Trailers, Inc., is considering the purchase of a computer numerical controlled welding robotic system to replace the automated welding system that required human setup and intervention. The existing automated welding system was purchased three years ago at an initial cost of $1,650,000. The current auto-mated welding system has a salvage value of $0 at the end of its ten-year useful life. Its remaining market values are given in the following table. The computer numerical controlled welding robotic system costs $1,250,000 due to advances in technology and has a salvage value of $125,000 at the end of its ten-year useful life. Its annual market values are given in the follow table. The computer numerical controlled welding robotic system will save $800,000 per year due to elimination of the human support labor. Both welding systems are depre-ciated as MACRS GDS seven-year property. Pro-Trailers operates in a state with a combined federal and state income tax of 30%, and Pro-Trailers requires MARR = 10% on this type of investment. Estimate the EUAW economic life of the defender and challenger and determine if the defender should be replaced by the challenger.

Year	Automated	Computer numerical controlled
1	–	$1,137,500
2	–	$1,025,000
3	$1,155,000	$912,500
4	$990,000	$800,000
5	$825,000	$687,500

(continued)

(continued)

Year	Automated	Computer numerical controlled
6	$660,000	$575,000
7	$495,000	$462,500
8	$330,000	$350,000
9	$165,000	$237,500
10	$0	$125,000

Part III
Managing Engineering Investments

Chapter 12
Determining the Appropriate MARR

Abstract The previous chapters assumed that a stated MARR for a particular asset investment was known. Depending on organizational maturity and accounting policies, MARR may or may not be known for every asset investment type. In small, start-up, entrepreneurial companies with only rudimentary accounting practices, the MARR will most likely be unknown. In more mature, medium-sized companies with maturing management and cost accounting practices, a general MARR may be available for all asset investments. In large corporations with mature management, cost accounting systems, and industrial engineering, a schedule of MARRs will be published by asset risk category. This chapter provides an introductory discussion of how MARR is determined from risk–return analysis. The discussion is a continuation of the finance cycle introduced in Chap. 1.

12.1 Risk–Return Fundamentals

In Chap. 2, we defined MARR as follows.

Definition: *MinimumAttractive Rate of Return (MARR)*—the risk adjusted, weighted, minimum acceptable rate of return, or hurdle rate that must be earned on a project or investment, given its opportunity cost of foregoing other projects or investments.

From Chap. 1, MARR may be measured as

$$\text{MARR} = \sum_i w_{Li} \times I_{Li} \times (1 - T_c) + \sum_j w_{Si} \times \text{RoE}_{Si} \qquad (12.1)$$

where $w_{Li} = L_i/(\Sigma_i\, L_i + \Sigma_j\, S_j)$, $w_{Sj} = S_j/(\Sigma_i\, L_i + \Sigma_j\, S_j)$, I_{Li} = interest rate charged on loan i, T_C = the organization's combined tax rate, and RoE_{Si} = return on equity interest rate for stock S_j.

Equation (12.1) recognizes that, ultimately, the sources of organizational financing, lenders, and stockholders set the MARR as a function of their risk assessment of organizational investment opportunities. This chapter reviews the sources and costs of financing organizational investments. The chapter then presents fundamental methods of determining MARR.

All cash flow estimates are inherently variable, and future estimates are often more variable than present estimates. Minor changes in any estimate(s) may alter the results of the economic analysis and the decision regarding the best investments. Chapter 10 introduced the concept of sensitivity analysis to determine the range over which a criterion may vary for the current decision to remain the best decision. The point at which the decision changes from one alternative to another was defined as the *point of indifference* or the *critical decision point*. It is the variability in future cash flow estimates that is the source of risk. There is an upper critical decision point at which an investment's risk of loss exceeds its rate of return (MARR).

Before we can establish risk–return fundamentals, we need a working definition of risk. In application, there are three broad categories of risk.

Risk is observed in those situations in which the potential outcomes can be described by well-known stochastic probability distributions.
Imprecision is observed in those situations in which the potential outcomes cannot be described by well-known stochastic probability distributions but can be estimated by human subjective probabilities.
Uncertainty is observed in those situations in which the potential outcomes cannot be described by well-known stochastic probability distributions and cannot be estimated by human subjective probabilities.

This chapter will deal with only the first category, *risk*. In engineering managerial economics,

Definition: *Risk* is the probability of earning a rate of return on an investment that varies from the investment's expected or predicted value.

Risk is a measure of variability.

Definition: *Variability* arises from random structural and residual error differences from expected or predicted values in the criterion under consideration.
Structural—bias difference between long-run average realized return on investment and the expected or predicted rate of return.
Random—random difference between realized rate of return and the long-run average structural rate of return.
Definition: A *criterion* is a principle, standard, or metric by which the desirability of a phenomenon can be evaluated.

In engineering managerial economics, the criterion is an investment's rate of return or a portfolio of investments weighted average rate of return relative to the expected or predicted rate of return $\geq$ MARR for that investment or portfolio of investments.

$$\text{Structural difference} = \text{Avg}[\text{RoR}(\text{Investment})] - E[\text{RoR}\}$$

$$\text{Random difference} = \text{RoR}[\text{Investment}] - \text{Avg}\{\text{RoR}(\text{Investment})\}$$

Economic risk is the probability that the RoR(investment) < MARR.

$$P(\text{RoR}(\text{investment}) < \text{MARR})$$
$$= \text{Structural difference} + \text{Random difference} < \text{MARR}$$

Figure 12.1 illustrates the structural and random variance components of RoR risk. The structural difference is the ± Avg[RoR(investment)—E[RoR] shown by the horizontal arrows. RoR(investment) is the ± variation of the realized RoR under the curve of its respective ± Avg[RoR]. The areas under the respective distributions are the P(RoR(investment) < MARR). The lower tail area for the dashed blue curve was the planned risk of RoR < MARR. If the + Avg[RoR] distribution is realized, P(RoR(investment) < MARR) is much less than planned. If the −Avg[RoR] distribution is realized, P(RoR(investment) < MARR) is much greater than planned.

Figure 12.2 points to the fundamental relationship between risk and return—the return on an investment should be proportional to the risk involved. R0 = the risk-free interest rate on Treasury bonds. The Real$ risk-free rate is calculated by subtracting the current inflation rate from the yield of the Treasury bond matching the investment duration. As variance increases below MARR, risk increases.

Generally, however, in corporations with mature management, cost accounting, and industrial engineering functions, RoR risk versus return will be decomposed into multiple MARR risk classes. As examples, installation of a pump required by EPA (low risk), installation of a fully automated production line for an existing product (medium risk), and construction of a new facility to manufacture an innovative new product (high risk) have increasing risk of failure to achieve the relevant MARR and correspondingly increasing consequences of the impacts of failure. Each risk category will have a relevant range of MARRs except the last. As some level, new investment opportunities incur excessive risk requiring such high interest rates and high common

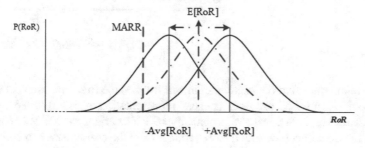

Fig. 12.1 RoR risk variance components

Fig. 12.2 Risk versus return
relationship

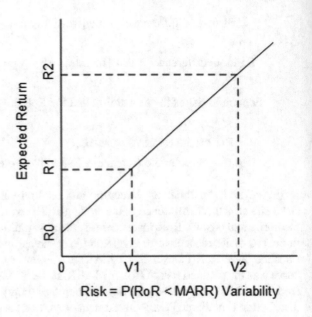

Fig. 12.3 General
risk–return categories

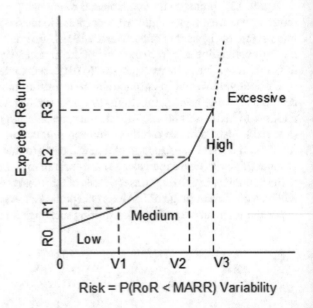

stock dividend rates that the organization will not be able to secure loans or be able
to sell stock. Figure 12.3 presents a general representation of risk–return categories.

Risk, however, is not just the probability of RoR variance < MARR. Risk also
involves the outcome, how much RoR variance, and the consequence to and impact
on the organization. Hence, risk has three primary dimensions.

- An outcome (difference from an expected or predicted RoR value).
- A probability of occurrence of that RoR < MARR.
- Consequence of the outcome (loss or gain given the outcome).

$$\text{Riak}_t = f\big((\text{Outcome}_t,\ \text{Probability}_t)\text{Consequence}_t\big) \qquad (12.1)$$

$$\big(\text{Outcome}_t,\ \text{Probability}_t\big) = f(\text{Input Events}_t,\ \text{Random Variation}_t)$$

$$\text{Impact}_t = \text{Probability}_t \times \text{Consequence}_t$$

where t = time which facilitates changes in outcomes, their associated probabilities and impacts, and the relationships among them. Risk also has a human perception dimension. Generally, there is an asymmetry between downside and upside risk. This is known as **risk aversion**.

Downside risk: In general, a project with a greater probability of earning low or negative returns is considered more risky than other projects with lesser probability of low or negative returns.

Upside risk: A project with a greater probability of higher-than-expected returns is, in general, considered less risky than other projects with lesser probability of similar high returns.

12.2 Sources of Financing

From Chap. 2, there are three sources of asset financing for an organization: borrowed money, money from the sale of stock, and retained earnings.

Borrowed money is debt financing. Debt financing includes both short-term borrowing and the sale of long-term bonds. Short-term and long-term loans refer to the time over which a loan is paid back. Short-term loans are generally unsecured and repaid within 30 days, a few months, or the current fiscal year. Short-term loans include

- Line-of-credit bank loans to maintain a minimum level of operating cash. A line of credit is a flexible loan from a local bank that makes available of a defined amount of money that an organization can access as needed and repay either immediately or over time. Interest is charged only on the amount of money borrowed over the time it borrowed.
- Cash advances from a local bank with repayment within a few weeks. Cash advances are often used for unplanned emergency purchases.
- Accounts payable—money owed to suppliers of materials or services, which is typically repaid with some percent discount if repaid in a short period or repaid in full in 30 days. Some organizations depend on rolling accounts payable as a source of short-term financing.

- Credit card purchases with repayment monthly. Virtually all organizations use credit card purchases as a form of short-term financing.

Long-term loan financing is either the sale of bonds or acquisition of long-term loans. Long-term financing is commonly used to support long-term initiatives, such as making acquisitions, opening a new production facility, financing common stock share repurchases, and preparing for rising interest rates.

- Bond financing represents a loan made by an investor. Organizations issue bonds directly to investors. A bond may be secured (backed by the value of assets it finances) or unsecured. A bond includes the terms of the loan, interest payments that will be made, and the time (maturity date) at which the bond principal will be repaid. The coupon interest payment is the return that bondholders earn for loaning their funds to the issuer.
- A long-term loan represents financing from commercial or investment banks. Long-term loans typically involve a contract for a set amount of money with equal repayments over multiple years until the loan is repaid. Long-term loans have a distinct advantage over equity financing (sale of stock) because loan interest payments are considered as operational expenses paid before taxes resulting in deductions equal to the tax rate organizations receive on interest payments.

Equity financing results from the sale of preferred on common stock to investors. Preferred stocks are stock shares issued without ownership and voting rights. Preferred stockholders have "preference" over common stockholders in payment of dividend (preferred dividends must be paid before issuing common stock dividends) and in receiving cash from assets in the event of organizational liquidation. Common stockholders are owners of the company and have voting rights on corporate issues, such as the board of directors and accepting takeover bids. Common stockholders may receive dividends out of net profit if the organization has cash available after long-term reinvestment.

Retained earnings is money (retained net profits) generated from the operation of the company. Retained earnings is reported in the equity section of the balance sheet but is invested in new productive assets that are listed on the assets side of the balance sheet. Retained net profits should be invested in new productive assets only if the rate of return on the assets is greater than or equal to the common stock dividend rate.

12.3 Cost of Financing and Investment Opportunities

As specified in Chap. 2, the debt ratio measures the proportion of total capital derived from borrowed funds.

$$\text{Debt ratio} = \frac{\text{Total debt}}{\text{Total assets}}$$

The debt ratio can range from zero to 100%. An organization's capital structure is not independent of the proportion of debt. Borrowing affects the riskiness of an organization's capital structure. The amount of risk is measured by the organization's risk premium.

Cost of Borrowed Money = prime interest rate + risk premium

The prime rate is the interest rate banks charge their very best corporate customers with the strongest credit ratings. Customers with weaker credit ratings are charged more than the prime rate. This additional interest rate is termed the risk premium. As an organization's debt ratio increases toward 100%, the organization is considered to be more risky and will pay an increasingly higher risk premium.

As the debt ratio decreases toward zero, the organization is considered less risky, but its cost of equity, dividend rate paid on preferred and common stock, increases its cost of borrowed money. This increase occurs because stockholders expect a higher return rate for holding stocks long term than the interest rate that they can receive by depositing their money in bank saving accounts, certificates of deposit, or money market accounts. The second source of equity financing is retained earnings. Reinvestment of retained earnings likewise increases the cost of borrowed money, because earnings will be retained only if they can be invested in assets and products that have rates of return higher than stock dividend rates. Otherwise, excess earnings should be returned to stockholders in the form of dividends. Retained earnings represents an opportunity cost to stockholders. Earnings are retained and reinvested only if they present the opportunity for higher future returns on stocks held.

Hence, each organization has a mix of debt and equity financing that will minimize its cost of borrowed money. The challenge for the organization's finance manager is to find the mixed financing, or range of mixed financing, that minimizes the organization's total cost of borrowed money.

12.3.1 Weighted Average Cost of Capital

The proportions of debt, equity, and retained earnings are combined into the *weighted average cost of capital* (WACC).

$$\text{WACC} = \sum i_{\text{Liabilities}} P_{\text{Liabilities}} (1 - T) + \sum i_{\text{Equity}} P_{\text{Equity}} + i_{\text{Earnings}} P_{\text{Earnings}}$$

$$(12.2)$$

where P = proportion in liabilities, equity, and retained earnings, respectively, i = interest rate on each respective debt, stock, or retained earnings, T = combined tax rate, and $P_{\text{Liabilities}} + P_{\text{Equity}} + P_{\text{Earnings}} = 1.00$. Recall that interest on debt is paid before taxes, so the after-tax cost of debt is

$$\text{After tax interest rate} = (\text{Before tax interest rate})(1 - T) \qquad (12.3)$$

If the firm uses a mix of borrowing and equity that maintains exactly the same capital and risk structure in financing the project or portfolio of projects under consideration, then

$$WACC \approx MARR$$

Example 12.1

A medium-sized chemical distributor has the following financing mix. The distributor operates in state with 10.0% corporate income tax rate. Estimate the distributor's after-tax weighted average cost of capital.

Type	Amount	Interest rate
Bank loan	$2,000,000	5.0%
Bonds	$28,000,000	8.0%
Common stock	$60,000,000	12.0%
Retained earnings	$30,000,000	15.0%
Total	$120,000,000	

Combined tax rate = $Sr + Fr(1 - Sr) = 0.10 + 0.21(1 - 0.10) = 0.289$ or 28.9%.

Type	Amount	Interest rate (%)	P
Bank loan	$2,000,000	5.0	0.01667
Bonds	$28,000,000	8.0	0.23333
Common stock	$60,000,000	12.0	0.50000
Retained earnings	$30,000,000	15.0	0.25000
Total	$120,000,000		1.00000

$WACC = [(0.05 \times 0.01667 + 0.08 \times 0.23333)(1 - 0.289) + 0.12 \times 0.50 + 0.15 \times 0.25$
$\quad WACC = 0.1114 \quad or \quad 11.14\%$

Also, we can estimate the WACC based on after-tax interest expense.

Type	Amount	Interest rate (%)	Annual interest	(1 − T)	After-tax interest
Bank loan	$2,000,000	5.0	$100,000	(1 − 0.289)	$71,100
Bonds	$28,000,000	8.0	$2,240,000	(1 − 0.289)	$1,592,640
Common stock	$60,000,000	12.0	$7,200,000	NA	$7,200,000
Retained earnings	$30,000,000	15.0	$4,500,000	NA	$4,500,000
Total	$120,000,000				$13,363,740

$$WACC = \$13,363,740/\$120,000,000 = 0.1114 \quad \text{or} \quad 11.14\%$$

12.3.2 Weighted Marginal Cost of Capital

The previous section presented the weighted average cost of capital as an estimate of an organization's MARR. However, as noted in the introduction to this section, the exact value of the WACC depends on how the amount of new capital affects the financing mix. Financial markets recognize that funding sources are limited. Accordingly, banks and investors will not make funds available beyond certain risk limits. If a firm tries to extend its financing beyond market-determined risk limits, it will encounter an increasing weighted marginal cost of capital, WMCC. A firm determines the marginal cost of capital from external financing for increasing risk levels by consulting an investment banker. The banker studies the organization's financial and risk structures and presents a report similar to the following table.

Range of new financing ($)	Weighted cost of capital (%)
$0 to $1,500,000	11.0
$1,500,001 to $3,000,000	11.5
$3,000,001 to $4,500,000	12.5
$4,500,001 to $5,500,000	13.5
≥$5,500,000	15.0+

The table indicates that the organization can obtain a mix of additional debt and stock financing up to $1,500,000 without affecting its current WACC = 11.0%. However, if the organization has investment opportunities beyond $1,500,000, it can obtain the additional financing only if it pays increasing interest rates in the schedule to compensate its debtors and stockholders for the corresponding increasing risk. For those alternatives with RoR > weighted cost of capital (%), estimate the weighted marginal cost of capital as

$$WMCC = \text{Re-estimate}(WACC + \text{acceptable alternatives})$$

If an organization uses a mix of borrowing and equity that increases the weighted marginal cost of capital due to increasing risk structure in financing the project or portfolio of projects under consideration, then

$$WMCC \approx MARR$$

Example 12.2

Suppose that the organization in Example 12.1 has MARR = 11.14% and the following proposed projects ranked by IRR. Use the above WCC/new-financing schedule to estimate the organizations weighted marginal cost of capital.

Project	IRR (%)	Project cost	Cum$
A	18.0	$1,500,000	$1,500,000
B	17.0	$1,250,000	$2,750,000
C	15.0	$1,500,000	$4,250,000
D	12.0	$1,750,000	$6,000,000
E	11.0	$1,000,000	$7,000,000
F	10.0	$ 500,000	$7,500,000
G	9.0	$ 500,000	$8,000,000

Plot the IRR (%) and WCC (%) for the cumulative project cost for each project investment. The organization can select only projects A, B, and C for a marginal cumulative investment of $4,250,000 at WMCC = 12.5%.

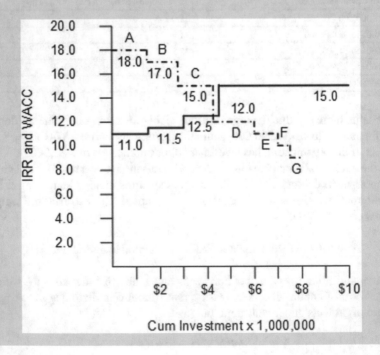

Analysis: Estimate the WNCC = Re-estimated(WACC) with 100% bond financing, current debt ratio mix of bonds and stock, and 100% stock financing for a state income tax rate of 10%. Compare the WNCC with the IRR = 15% opportunity cost of capital.

For scenario 1, use 100% bond financing at 12.50%.

State tax rate	10%					
Federal tax rate	21%			Combined tax rate		28.9%
Scenario 1: 100% Bond financing						
Type	Amount	Interest Rate	P	(1 - T)	
Bank Loan	$2,000,000	5.00%	0.0163	71.10%	0.000579226
Bonds	$28,000,000	8.00%	0.2281	71.10%	0.012974664
New Bonds	$4,250,000	12.50%	0.0346	71.10%	0.003077138
Common Stock	$60,000,000	12.00%	0.4888	100%	0.058655804
New Stock	$0	12.00%	0.0000	100%	0
Retained Earnings	$30,000,000	15.00%	0.2444	100%	0.036659878
Total	$122,750,000		1.0000	WNCC	0.1119
					11.19%

For scenario 2, use finance 25% with bonds and 75% with stock to maintain current debt ratio. Note: Stock financing must pay the current 12.0% stock dividend rate.

Scenario 2: Finance 25% with bonds and 75% with stock.					
Type	Amount	Interest Rate	P	(1 - T)	
Bank Loan	$2,000,000	5.00%	0.0163	71.10%	0.000579226
Bonds	$28,000,000	8.00%	0.2281	71.10%	0.012974664
New Bonds	$1,062,500	12.50%	0.0087	71.10%	0.000769285
Common Stock	$60,000,000	12.00%	0.4888	100%	0.058655804
New Stock	$3,187,500	12.00%	0.0260	100%	0.00311609
Retained Earnings	$30,000,000	15.00%	0.2444	100%	0.036659878
Total	$122,750,000		1.0000	WNCC	0.1128
					11.28%

For scenario 3, use 100% stock financing.

Scenario 3: 100% common stock financing					
Type	Amount	Interest Rate	P	(1-T)	
Bank Loan	$2,000,000	5.00%	0.0163	71.10%	0.000579226
Bonds	$28,000,000	8.00%	0.2281	71.10%	0.012974664
New Bonds	$0	12.50%	0.0000	71.10%	0
Common Stock	$60,000,000	12.00%	0.4888	100%	0.058655804
New Stock	$4,250,000	12.00%	0.0346	100%	0.004154786
Retained Earnings	$30,000,000	15.00%	0.2444	100%	0.036659878
Total	$122,750,000			1.01 WNCC	0.1130
					11.30%

All three financing mix scenarios are feasible relative to the retained earnings cost of 15%.

- 100% bond financing at 11.5% increases the WACC from 11.14% to WNCC = 11.19%.
- Debt ratio of 25% bond and 75% stock financing increases WACC from 11.14% to WNCC = 11.28%.
- 100% stock financing increases WACC from 11.14% to WNCC = 11.30%.
- 100% bond financing should be preferred if obtainable.

12.3.3 Opportunity Cost of Capital

WACC and WMCC assume that an organization has unlimited capital and resources to support new investments. In practice, organizational resources and capital are always constrained. It important to understand the effect on MARR due to the cost of capital under **capital rationing**. Capital rationing occurs when management places restrictions on the amounts of new investments or projects undertaken by an organization. Capital rationing is accomplished by imposing a higher cost of capital for investments under consideration or by setting a ceiling on specific portions of a budget. The objective of capital rationing is to ensure that *all selected projects have a higher ROR than the best rejected internal project*. **Opportunity cost** is the ROR of the best rejected *internal* project opportunity foregone due to capital rationing.

Opportunity Cost = Rate of Return on the best rejected project   (12.4)

Since new investments can be financed only from either retained earnings or new external financing, the opportunity cost of capital represents an upper bound on MARR.

Example 12.3

An organization is considering the independent investment projects in Example 12.2. The organization's current MARR $= 11.14\%$, and its stockholders require 12.0% stock dividend rate. Organizational management has set a budget of $3,000,000 for the selected new projects. Determine which projects should be selected under capital rationing by opportunity cost.

Budget	$3,000,000				Analysis:			
Available Projects								
Project	Cost	UAB	Useful Life	IRR(%)	Rank	CumProj$	Select	
A	$1,500,000	$333,750	10	18.0%	1	$1,500,000	Yes	
B	$1,250,000	$297,000	8	17.0%	2	$2,750,000	Yes	
C	$1,500,000	$334,350	8	15.0%	3	$4,250,000	No	Opportunity cost of capital
D	$1,750,000	$485,450	5	12.0%	4			
E	$1,000,000	$322,000	4	11.0%	5			
F	$500,000	$131,900	5	10.0%	6			
G	$500,000	$154,350	4	9.0%	7			

The opportunity cost of capital is the best rejected project C with IRR $=$ 15.0% $>$ 12.0% stock dividend rate of Example 12.1. Although financing is available, the organization does not have the capacity to take on project C.

12.4   Closing Comments on Determination of MARR

As noted in the introductory discussion, there is a mix of debt and equity financing that will minimize an organization's cost of borrowed money (debt and stock financing). A general ranking (as a guideline) from lowest to highest cost of financing for MARR estimates is

1. As a general guideline, the effect of the mix of debt and equity financing on MARR ranked from lowest to highest is:
2. Cost of debt financing (if debt interest $>$ current MARR, then increasing risk).
3. WACC or WMCC.
4. Cost of stock financing.
5. Opportunity cost of capital; this is the upper bound because the opportunity cost of capital for new projects should always be greater than the expected stock dividend rate.

In selecting a new MARR, add the cost of financing for all feasible new projects to the current financing mix and re-estimate the WACC. Select the financing mix that minimizes WACC ≈ MARR. Also, as noted in the introductory discussion of risk and illustrated by Fig. 12.2, well-managed organizations will further decompose RoR risk versus return into multiple MARR risk classes.

12.5 Summary

Definition: *Minimum Attractive Rate of Return (MARR)*—the risk adjusted, weighted, minimum acceptable rate of return, or hurdle rate that must be earned on a project or investment, given its opportunity cost of foregoing other projects or investments.

$$\text{MARR} = \sum_i w_{Li} \times I_{Li} \times (1 - T_c) + \sum_j w_{Si} \times \text{RoE}_{Si}$$

Risk is observed in those situations in which the potential outcomes can be described by well-known probability distributions.

Imprecision is observed in those situations in which the potential outcomes cannot be described by well-known probability distributions but can be estimated by subjective probabilities.

Uncertainty is observed in those situations in which the potential outcomes cannot be described by well-known probability distributions and cannot be estimated by subjective probabilities.

Definition: *Risk* is the probability of earning a return on an investment that varies from the investment's expected or predicted value.

Definition: *Variability* arises from random structural and residual error differences from expected or predicted values in the criterion under consideration.

> *Structural*—difference between long-run average realized return and expected or predicted return.
>
> *Random*—random difference between realized return and the long-run average return.

Definition: A *criterion* is a principle, standard, or metric by which the desirability of a phenomenon can be evaluated.

$$\text{Risk}_T = f\big((\text{outcome}_T, \ \text{probability}_T), \ \text{consequence}_T\big)$$

Downside risk: In general, a project with a greater probability of earning low or negative returns is considered more risky than other projects with lesser probability of low or negative returns.

Upside risk: A project with a greater probability of higher-than-expected returns is, in general, considered less risky than other projects with lesser probability of similar high returns.

$$WACC = \sum i_{Liabilities} P_{Liabilities} (1 - T) + \sum i_{Equity} P_{Equity} + i_{Earnings} P_{Earnings}$$

$$\text{After-tax interest rate} = (\text{Before-tax interest rate})(1 - \text{Tax Rate})$$

$$WMCC = \text{Re-estimate}(WACC + \text{acceptable alternatives})$$

$$\text{Opportunity Cost} = \text{Rate of Return on the best rejected project}$$

12.6 Key Terms

Cost of capital
Debt financing
Equity financing
Impact
Imprecision
Minimum Attractive Rate of Return
Opportunity Cost of Capital
Outcome
Random risk
Risk
Risk versus return
Structural risk
Uncertainty
Weighted Average Cost of Capital
Weighted Marginal Cost of Capital.

Problems

1. A firm's stockholders expect a 14% rate of return, and the firm has earned 16% on its retained earnings. There is $9,000,000 in common stock and $3,000,000 in retained earnings. The firm has $3,000,000 in loans at 6.0% interest rate and $10,000,000 in bonds that pay 5.0% per year. The firm has a combined income tax rate of 28.0%. What is the firms WACC?

2.   The firm in problem 1 has asset investment projects as shown in the first table and available financing in the second table. What is the weighted marginal cost of capital if the firm finances with 100% bonds, current percentage mix of liabilities and equities, and 100% stocks?

Project	First cost	Ann. benefit	Life (yrs)
1	$200,000	$43,000	15
2	$300,000	$59,750	10
3	$100,000	$33,400	5
4	$50,000	$12,500	10
5	$250,000	$75,000	5
6	$140,000	$28,000	15
7	$400,000	$125,000	5

Range of new financing	WCC (%)
$0 to $100,000	9.0
$100,001 to $150,000	10.0
$150,000 to $200,000	12.0
$200,001 to $300,000	15.0
$300,001 to $400,000	18.0
$400,001 to $500,000	22.0
>$500,001	NA

3.   What is the firm's opportunity cost of capital for the projects listed in the first table of problem 2? How will the available financing with respect to the opportunity cost of capital affect financing of the selected projects?

Chapter 13
Capital Budgeting Engineering Investments

Abstract An organization's viability depends on its ability to develop and grow to remain competitive. A well-managed organization expends a significant amount of its management's time and its resources to develop sound capital budgets to remain viable. Major capital investments in facilities, equipment, and supporting infrastructure are made and implemented through engineering projects. Capital budget projects fall in the medium-to-high risk–return categories.

- **Replacement: maintenance of existing business**—medium risk–return category. Typical projects include replacing worn or damaged equipment or infrastructure used to produce existing profitable products and services necessary to continue current operations. Usually, the profit level of existing products or services justifies continuing production, so the decision to move forward with a capital budget is automatic. The only decision to be made are (1) minimizing the investment to continue profitability, (2) timing for shutdown and repairs or upgrades, and (3) building safety inventory to maintain sales during shutdown.
- **Replacement for cost reduction**—medium risk–return category. This category includes costs to replace or upgrade functional equipment with remaining useful life. The expected outcome is a reduction in operating expenditures for direct labor, direct materials, indirect materials, or allocated fixed operating overhead. These capital budgeting decisions are more discretionary and require detailed capital budget analysis.
- **Expansion of existing products or markets**—medium–high-risk–return category. This category includes costs to increase output or expand the distribution of existing products or services, These capital budgeting decisions require explicit market analysis to determine the level of future demand for existing products and services. The greater the variability in the future demand estimates the higher the risk of decline in profitability or even loss.

© The Author(s), under exclusive license to Springer Nature Switzerland AG 2022   323
T. S. Cotter, *Engineering Managerial Economic Decision and Risk Analysis*,
Topics in Safety, Risk, Reliability and Quality 39,
https://doi.org/10.1007/978-3-030-87767-5_13

- **Expansion into new products, services, or markets**—high-risk–return category. These capital projects include costs to install a new production line or facility to produce a new product or service or to expand into a geographic area not currently served. These capital budgeting decisions are strategic in that the outcome will change the fundamental operation of the organization. These capital budgeting decisions require incurring costs over long periods for planning, designing, installing, operating, and decommissioning the new operations. As such, these decisions require detailed analyses.
- **Safety and environmental projects**—low-to-medium risk category. Note that this category of capital budgets is not risk–return decisions, because there is no return. As such, these capital budgeting decisions are termed *non-revenue-generating* projects. This category of capital budgeting is necessary to comply with governmental regulations or possibly insurance policy terms. How the capital budgeting decision is made depends on whether it is a low or medium risk decision. The major consideration is minimizing the cost necessary to attain compliance.
- **Other projects**—low-to-medium risk category. Again, these projects fall into the non-revenue-generating category, because they are for supporting infrastructure such as office buildings or a truck fleet to provide on-time delivery of raw materials or finished products.

13.1 Capital Expenditure for Project Proposals

Up until now, we have been given an interest rate for which we selected a project's best mutually exclusive alternative. In doing so, we assumed an *unlimited amount of money* for capital investment. In a free-market economy (purely socialist economies are excluded because capital budgeting is government dictated and may provide no benefit to the general economy), the mechanism of *scarcity of resources* promotes the selection of economically attractive capital projects. The problem of rationing money among competing capital projects is called **capital budgeting**. Conceptually, the capital budgeting process passes through six major steps.

1. Determine the capital cost and source of funding, debt or stock.
2. Estimate expected cash flows from the project including terminating market or salvage values.
3. Estimate the riskiness (structural and random variability) of cash flows and the risk premium necessary to compensate lenders and stockholders.
4. Estimate the IRR of the project given the risk premium as

$$\text{IRR} = i_{\text{project}} = i_{\text{RiskFree}} + i_{\text{RiskPremium}}$$

where inflation is expected to have major impact on project cash flows, the project IRR becomes

$$\text{IRR} = i_{\text{project}} = i_{\text{RiskFree}} + i_{\text{RiskPremium}} + f + (i_{\text{RiskFree}} + i_{\text{RiskPremium}})f$$

5.   Determine the project's rate of return and net present value to obtain an estimate of the project's value to the organization.
6.   Decide on project implementation based on NPV $\geq$ 0 at the MARR or RoR $\geq$ MARR for budgeted funds.

13.2   Rationing Capital by Rate of Return

As noted in the capital budgeting process, one criterion for assessing the value of a capital investment is RoR > MARR for budgeted funds. The budgeting process for rationing capital by the rate of return is:

1.   Compute rate of return (RoR) on each alternative project investment given respective cash flow structures.
2.   Rank the alternatives in order of decreasing RoR.
3.   For a fixed capital budget, select the highest RoR >= MARR investment project alternatives by rank until the allocated capital funds are allocated with the sum of allocated fund less than or equal to the budgeted funds.
4.   Estimate the weighted portfolio RoR = S_i (Project Cost(i)/Sum Allocated Budget) $\times$ ROR(i).
5.   Return unallocated capital budget funds to the organization for distribution among stockholders.

The rate of return on the best-rejected project is called the **cutoff rate of return** or **opportunity cost of capital**.

Example 13.1

An organization is considering the following nine capital investments. The capital budget is $1,250,000. These are medium risk–return projects requiring MARR = 12.0%. Allocate the capital budget using rationing by rate of return.

Budget	$1,400,000			
Project	Cost (1k)	After-tax CF (1k)	Life (years)	Salvage Value
1	$200.00	$47.70	10	$0.00
2	$400.00	$79.70	10	$0.00
3	$100.00	$34.72	4	$0.00
4	$200.00	$40.00	6	$125.00
5	$100.00	$20.00	10	$100.00
6	$300.00	$54.00	10	$150.00
7	$600.00	$189.28	4	$50.00
8	$450.00	$88.88	10	$100.00
9	$100.00	$17.50	10	$50.00
Total	$2,450.00			

Budget	$1,250,000			MARR	12.0%			
Project	Cost (1k)	After-tax CF (1k)	Life (years)	Salvage Value	RoR	Rank	CumBudget
1	$200.00	$47.70	10	$0.00	20.0%	1	$200.00
2	$400.00	$79.70	10	$0.00	15.0%	8	
3	$100.00	$38.20	4	$0.00	19.4%	3	$400.00
4	$200.00	$40.00	6	$125.00	15.8%	6	$1,150.00
5	$100.00	$20.00	10	$100.00	20.0%	2	$300.00
6	$300.00	$54.00	10	$150.00	15.6%	7	Opportunity Cost
7	$600.00	$189.28	4	$50.00	12.5%	9	
8	$450.00	$88.88	10	$100.00	16.1%	5	$950.00
9	$100.00	$19.25	10	$50.00	17.0%	4	$500.00
Total	$2,450.00			Portfolio Weighted RoR	17.4%		

	NPW
	$69.52
	$16.03
	($98.87)
	($19.19)
	$19.99
	($7.33)
Portfolio NPW	($19.86)

Note: In the RoR analysis, the one-life EUAW for a continuing requirement is assumed. The common life is the longest life project, and any project with a shorter life will replace itself until the longest life project.

- Only the first six ranked projects (in rank order 1, 5, 3, 9, 8, and 4) can be funded at a budget of $1,150,000.
- The best-rejected project 6 RoR = 15.6% is the opportunity cost of capital. Weighted portfolio RoR = 17.4%. The portfolio life NPW = (−$19.86) (see example worksheet).

Since the opportunity cost of capital is 15.6% > 12.0% MARR and depending on the weighting of low-to-medium-to-high-risk categories, the selected capital projects will increase the organization's WMCC but, as indicated by NPW analysis, will decrease the organization's net worth. Organizational management has identified high-value capital investments that will increase common stock dividend return.

13.3 Rationing Capital by Net Present Worth

Rationing capital by rate of return addresses stockholders' question of project portfolio rate of return versus required common stock return rate. Rationing capital by present worth addresses management's question of projects portfolio current cash flow value to the organization given limited budgets. However, investment project alternative lives are usually unequal inhibiting comparable estimates of net present worth. To overcome this deficiency, apply the continuing requirement of Chap. 7, and set the common project life = capital budget planning horizon. Estimate the one-life EUAW of each investment project alternative. Since each EUAW is the net of initial cost, uniform annual benefits, and end-of-life salvage value, EUAW incorporates receiving salvage value and repurchasing initial cost of alternative at the end of the specified common project life. Then, set EUAW = A and estimate the present worth of each alternative investment as P = A(P/A, MARR, project life). The budgeting process for rationing capital by the net present worth is:

1. For each project accepted by the rate of return ranking, compute one-life EUAW for each alternative investment project.
2. Set budget life = budgeting planning horizon.
3. Set EUAW = A and estimate P = A(P/A, MARR, budget life) for each alternative.
4. Rank the alternatives in order of decreasing net present worth.
5. For a fixed capital budget, prioritize project value by decreasing ranked order of NPW.
6. Since the weighted portfolio RoR > MARR, stockholder value has been maximized, the NPW ranking indicates the importance of each project to future cash flows.

Example 13.2

For the capital budget of $1,250,000 in Example 13.1, allocate the capital budget by net present worth using the organization's MARR = 12.0%. The capital budget planning horizon is 10 years.

Inputs:					Analysis:				
Budget	$1,250,000		MARR	12.0%	Project life		10 years		
Project	Cost (1k)	After-tax CF (1k)	Life (years)	Salvage Value	EUAW	NPW	Rank	CumBudget	RoR
1	$200.00	$47.70	10	$0.00	$12.30	$69.52	1	$200.00	20.0%
2	$400.00	$79.70	10	$0.00	$8.91	$50.32	2	$600.00	15.0%
3	$100.00	$38.20	4	$0.00	$5.28	$29.81	4	$1,150.00	19.4%
4	$200.00	$40.00	6	$125.00	-$24.05	($135.88)			
5	$100.00	$20.00	10	$100.00	-$3.40	($19.19)			
6	$300.00	$54.00	10	$150.00	-$7.64	($43.18)			
7	$600.00	$189.28	4	$50.00	-$18.72	($105.79)			
8	$450.00	$88.88	10	$100.00	$3.54	$19.99	3	$1,050.00	16.1%
9	$100.00	$19.25	10	$50.00	-$1.30	($7.33)			
Total	$2,450.00				Portfolio NPW	$169.65		Portfolio Wt RoR	16.7%

By rank of decreasing NPW, project importance to future cash flows is 1, 2, 8, and 3. The organization's portfolio NPW = $169.65, increasing organizational worth, and weighted RoR = 16.7% decreasing the return to common stockholders by 0.7%.

Since allocation by NPW is based on estimating P = A(P/A, MARR, project life) where A = EUAW for continuing requirement, the above process for capital budgeting by NPW ranks the set of alternative investment projects by order of current value to the firm. Conversely, ranking by RoR orders the projects within the allocated capital budget fund by future value to stockholders. The selection of the order of ranking by RoR or NPW depends on the organization's strategic priorities of maximizing growth in stock value or maximizing cash flows for operational stability.

13.4 Ranking Project Proposals

Examples 13.1 and 13.2 illustrate the fundamental issue in ordering the selected set of investment project alternatives. Prioritizing the capital budget portfolio by RoR may not maximize future cash flows. Conversely, prioritizing the capital budget portfolio by NPW may not maximize portfolio future value to stockholders. In Example 13.2, the opportunity cost RoR of the best-rejected investment project alternative under the net present worth criterion was 20.0% for project 5, clearly much greater than the 15.6% opportunity cost of project 6 under the RoR criterion in Example 13.1.

Lorie and Savage (1955) note that management must address three questions in rationing capital among competing investment projects.

- Given the organization's cost of capital (MARR), what combination of projects should be selected to maximize discounted cash flows net present worth to the organization. Recall that maximizing IRR may not maximize net present worth of cash flows.
- Given a fixed sum for capital investment, what combination of investment projects should be undertaken? This question has two parts. First, what criteria should management apply to decide on the amount of the capital budget? Second, given the allocated capital budget, what mix of investment projects maximize net present worth under the budget constraint? Too small a capital budget may result in rejecting the net present-worth maximizing investment portfolio for a collection of suboptimal investment projects that fit within the budget constraint. Too large a capital budget may result in accepting projects with low net present worth that downwardly weight the portfolio's net present worth.
- How to select the best combination of investment proposals among the mutually exclusive alternatives? Given the constraint of an optimally sized capital budget, how does management select the combination of projects that maximize portfolio

net present worth? The order by which investment projects are admitted can result in suboptimizing portfolio net present worth.

These problems with portfolio selection are illustrated by comparing the investment alternative projects selected under the net present worth criterion in Example 13.2 to those selected under the rate of return criterion in Example 13.1.

Lori and Savage recommend that the NPW maximizing allocation of capital funds by the net present worth criterion is

$$NPW - p \times PW|initial\ cost| = 0 \qquad (13.1)$$

where $0 < p < 1.0$ is a constant to be determined and NPW is that estimated based on each alternative's useful life. Adjust p until the maximum number of investment alternatives are selected with total capital cost $\leq$ capital budget funds. The final portfolio of investment projects is the penalized maximum NPW of all possible combinations of projects.

Lori and Savage's p-constant approach will select the set of alternative investment projects that maximize the penalized capital budget portfolio NPW. Next, estimating project alternative NPW from the continuing requirement EUAW estimate for a common project life reflects actual continual capital budgeting in application. Capital budgets are set for a life-cycle budgeting period and reviewed continually for changes. All fixed assets in the existing life-cycle budget must have a positive NPW or be capable of being replaced by assets that will have a positive NPW at the current MARR. Selecting alternative projects based on Lori and Savage's p-constant approach assures selection of the set of projects that maximize the penalized NPW over the life-cycle budget. Using the NPW-EUAW common project life estimate.

Example 13.3

For the capital budget of $1,250,000 in Example 13.2, perform sensitivity analysis on the projects' net present worth using the modified Lori and Savage p-constant.

Inputs:							Analysis:	
Budget	$1,250,000		MARR		12.0%			
Project	Cost (1k)	After-tax CF (1k)	Life (years)	Salvage Value	RoR		EUAW	NPW
1	$200.00	$47.70	10	$0.00	20.0%		$12.30	$69.52
2	$400.00	$79.70	10	$0.00	15.0%		$8.91	$50.32
3	$100.00	$38.20	4	$0.00	19.4%		$5.28	$29.81
4	$200.00	$40.00	6	$125.00	15.8%		-$24.05	($135.88)
5	$100.00	$20.00	10	$100.00	20.0%		-$3.40	($19.19)
6	$300.00	$54.00	10	$150.00	15.6%		-$7.64	($43.18)
7	$600.00	$189.28	4	$50.00	12.5%		-$18.72	($105.79)
8	$450.00	$88.88	10	$100.00	16.1%		$3.54	$19.99
9	$100.00	$19.25	10	$50.00	17.0%		-$1.30	($7.33)
Total	$2,450.00						Portfolio NPW	$169.65

Trial p				
0	0.0733	Rank	CumBudget	RoR
$69.52	$84.18	1	$200.00	20.0%
$50.32	$79.64	2	$600.00	15.0%
$29.81	$37.14	4	$1,150.00	19.4%
($135.88)	($121.22)			
($19.19)	($11.86)			
($43.18)	($21.19)			
($105.79)	($61.81)			
$19.99	$52.98	3	$1,050.00	16.1%
($7.33)	($0.00)	5	$1,250.00	17.0%
Wt NPW	$253.94		Portfolio Wt RoR	16.7%

Using Formula (13.1), p-constant $= 0.0733$ admits projects 1, 2, 8, 3, and 9 (by rank order of portfolio life NPW) with a penalized weighted portfolio life NPW $= \$253.94$ and a portfolio weighted RoR $= 16.7\%$ matching the RoR of Example 13.2.

13.5   Summary

$$\text{IRR} = i_{\text{project}} = i_{\text{RiskFree}} + i_{\text{RiskPremium}}$$

$$\text{IRR} = i_{\text{project}} = (i_{\text{RiskFree}} + i_{\text{RiskPremium}}) + f + (i_{\text{RiskFree}} + i_{\text{RiskPremium}})f$$

Rationing Capital by Rate of Return.
Rationing Capital by Net Present Worth.
Lori and Savage's Penalized Net Present Worth.

$$NPW - p \times PW|initial\ cost| = 0$$

13.6   Key Terms

Capital budgeting
Cutoff rate of return
Lori and Savage's p-constant
Opportunity cost of capital
Rationing capital

Problems

1. Nine capital spending proposals have been submitted for management review for the 10-year life-cycle budget. Management has initially budgeted $2,000,000 for the capital budget. Using the rate of return criteria, which projects should be funded, and what is the opportunity cost of capital? What is the portfolio weighted rate of return?

Project	First cost	Ann. benefit	Salvage	Life (yrs)
A	$150,000	$44,300	$0	5
B	$200,000	$61,750	$0	5
C	$300,000	$98,750	$30,000	5
D	$250,000	$62,600	$0	6
E	$400,000	$119,350	$40,000	6
F	$450,000	$115,500	$0	6
G	$350,000	$67,950	$0	10
H	$600,000	$126,900	$60,000	10
I	$750,000	$140,600	$0	10

2. The firm in problem 1 requires MARR = 12.0% on investment alternatives. Allocate the capital budget based on net present worth.
3. Allocate the capital budget in problem 1 using Lorie and Savage's p-criterion.

Chapter 14
Benefit–Cost Ratio Analysis

Abstract The design, implementation, and management of major service systems (agricultural, cybersecurity, energy, health care, information networks, infrastructure, legal, military, public services, safety, etc., as distinguished from small business services) with long lives (10-plus or 20-plus years) and associated intangible benefits and costs are not amenable to singular economic criterion analysis requiring accurate and precise cash flows. Often, the intangible costs and benefits of major systems are not fully known. Major public sector service systems, however, are foundational in that without their existence the private industrial and small business sectors would not be economically viable. Hence, engineering managers need a broader criterion that more completely accounts for the tangible and intangible benefits and costs of major service systems.

14.1 Characteristics of Major Service Systems

The primary purpose of benefit–cost analysis is to determine the true full benefits and costs (tangible and intangible) of major service systems projects. Knowledge gained from benefit–cost analysis can then be utilized to:

- Identify and prioritize short- and long-term benefits.
- Identify and prioritize short- and long-term cost-saving opportunities.
- Identify governmental and public sources of long-term financing needed to implement major service systems.
- Estimate the cash flows necessary to cover the true costs of designing, implementing, and delivering service benefits.
- Price systemic service or product benefits (for paying beneficiaries) at a level that covers the true costs of providing them.

© The Author(s), under exclusive license to Springer Nature Switzerland AG 2022
T. S. Cotter, *Engineering Managerial Economic Decision and Risk Analysis*,
Topics in Safety, Risk, Reliability and Quality 39,
https://doi.org/10.1007/978-3-030-87767-5_14

- Report the true costs of systemic services when claiming government reimbursements.

Knowing the true benefits and costs of a major systemic service program allows determination of the amount of revenue needed from the system to support its long-term existence and ensure that its service is not a net consumer of resources from the organization.

Assessment of a major service system's benefits alignment with organizational mission is the first step toward understanding its true costs and contribution to the organization's mission and overall financial health. This understanding is instrumental to the:

- Effective allocation of financial and human resources.
- Prioritization of core service systems that must be protected especially in economic downturns.
- Identification and elimination of peripheral and financially unhealthy systemic services.
- Design of effective maintenance and long-term growth strategies.
- Continual improvement of the service system's financial health and mission alignment with the organization's mission.

We can understand the characteristics of major service systems by contrasting their service products with those of manufacturing. Compared to manufacturing products:

- Services are intangible in that, in general, they do not have a physical form.
- Services cannot be inventoried. Services exist only at the point and time of demand.
- Service quality requirements are more subjective than those of physical products.
- Services are more subject to and must be customized to the client's preferences at the time of demand. Hence, meeting demanded that service requirements incur greater risk due to on-demand variability.
- Service quality requires collaboration between the server and client to formulate and communicate service requirements at the time of demand.
- Due to variability in service quality requirements, service products cannot be standardized and mass-produced.
- Given the subjectivity and variability of service quality requirements, service processes are labor-intensive as compared to manufactured products.
- Communications and information systems are core to multiplying the human component of service delivery.

14.2 Economic Evaluation of Services

The strategic issues that must be addressed in designing and delivering services are:

- Who is the client and what services are demanded by the client demand?

- What is our service capability and capacity to meet client demand? Capability is the types of service expertise. Service expertise is fulfilled across three dimensions; (1) responsiveness—promptness, helpfulness, and timely delivery of the demanded service; (2) professionalism—knowledge and skills of the service provider; and (3) attentiveness—caring and personalized attention in fulfilling client service demand. Capacity is the number and amount of services that the organization can delivery economically.
- Location of services? Does the client travel to the service location, or does the service travel to the client location?
- Costs associated with providing and improving services?

14.3 Selecting the Appropriate Interest Rate

In the USA, as in many capitalist economies, public sector investment generally involves some concept of promoting the general welfare. Some baseline of public infrastructure, services, and security must be in place to support private sector capitalism. These require public sector investments. Benefit–cost ratio analysis is a decision metric used by public organizations (governments, planning commissions, highway commissions, airport and port authorities, school districts, etc.) to select the best alternatives among major service systems investments. In public sector economic analysis, it is not easy to determine benefits that promote the general welfare or the forecast accuracy of projected benefits. Factors affecting public sector economic analysis include:

- Purpose of the investment—provision of vital goods and services that are either impossible for the private sector to supply efficiently or that require a natural monopoly to achieve economies of scale.
- Viewpoint for the analysis—government analyst, citizens who benefit from an investment, or citizens who are harmed by an investment.
- Financing sources—increased taxes, service fees, usage fees, earnings from the sale of commodities, or public bonds.
- Project duration.
- Major service system life cycle.
- System service capability and capacity versus public demand.
- Economic effects of systemic service location.
- Long-term unintended consequences.

Generally, projected benefits must consider *the interests of all parties involved* and *unintended consequences and costs*. The viewpoint of benefit–cost analysis influences or determines the final recommendation. To avoid suboptimization, the proper approach is *to take a viewpoint at least as broad as those who pay the costs and those who receive the benefits*.

Benefit–cost analysis has three primary objectives: (1) maximize an identified set of benefits for a fixed cost, (2) maximize net benefits with benefits and costs vary, or

(3) minimize the cost to achieve a stated level of benefits. The project interest rate affects the decision outcome for each objective. Several factors influence interest rate selection in the public sector.

- No Time Value of Money—Taxes (the source of public finance) are, to some extent, guaranteed future cash flows and can be spent immediately.
- Cost of Capital—Governments often borrow money or issue bonds to finance projects. Interest on municipal bonds is free from federal taxes. The effective interest rate is (bond rate)/(1—tax rate). General types of municipal bonds are:
  - General obligation bonds—retired through taxes.
  - Revenue bonds—retired through usage fees or other revenues generated by the project.

Theoretically, there are two approaches to determining the appropriate interest rate for public sector investments: (1) internal rate of return and (2) opportunity cost.

14.3.1 Public Sector Internal Rate of Return

In Sect. 6.5, the internal rate of return was specified as the interest rate at which the equivalent benefits are equal to the equivalent costs.

$$PW(\text{Benefits}, i) = PW(\text{Costs}, i)$$
$$EUAB(i) = EUAC(i)$$
$$FW(\text{Benefits}, i) = FW(\text{Costs}, i)$$

For public sector projects, this definition becomes *the interest rate at which the equivalent net user benefits are equal to the life cycle cost of the investment to attain the public asset.*

$$PW(\text{Net User Benefits}, i) = PW(\text{Net Life Cycle Costs}, i) \qquad (14.1)$$

$$EUAB(i) = EUAC(i)$$

Net user benefits are the difference in the benefit and disbenefit cash flows given public asset implementation. *Disbenefits* are *the negative impacts and their resultant effects on some individuals or groups resulting from public sector investments.*

$PW(\text{Net User Benefits})$

$$= \sum_n (\text{Benefit cash flows} - \text{Disbenefit cash flows})_t \, (P/F, i, t) \qquad (14.2)$$

where time $t = -m, -m + 1, ..., 0, 1, 2, ..., n$. Some examples of benefits and disbenefits include:

- Interstate highway construction—Benefits include improved transportation efficiency, reduced fuel costs, increased property values along the new route, and new business opportunities along the new route. Disbenefits include loss of homes and businesses to eminent domain along the new route, loss of business, and lower property values along old highway routes.
- New hospital construction—Benefits include improved health care in the region surrounding the hospital, new opportunities, and revenues for supporting medical practices, and new business opportunities and revenues for supporting information systems and medical supply services. Disbenefits include loss of existing homes and businesses to the hospital site and loss of access to health care in older parts of a city far away from the new hospital.
- Wind electricity generation—Benefits include increased electricity generation efficiency resulting in lower cost per kilowatt hour, creating of new business and employment opportunities, increased tax revenues for local governments, and elimination of the long-term expenses associated with coal fly ash recycling and containment. Disbenefits include high initial cost to install wind turbines, loss of wildlife, decreased local property values resulting from noise and visual pollution, and loss of coal mining jobs.

On the investment sponsor side of Eq. (14.2), net life cycle costs include annual cash flow differences for the future worth of design, development, acquisition, and installation of the public asset, lifetime annual operating and maintenance costs net any revenues realized from the public asset.

$$PW(Asset) = FW(Asset\ costs)p\ (F/P, i, p)$$
$$+ \sum_n PW(O\&M\ costs - Revenues)t\ (P/F, i, t) \qquad (14.3)$$

where time $p = -m, -m + 1, ..., 0$ and time $t = 0, 1, 2, ..., n$ asset lifetime.

14.3.2 Public Sector Opportunity Cost

The internal rate of return approach does not directly account for the opportunity cost to government and taxpayers. In chapter three, opportunity cost was defined as

Opportunity cost—a benefit that is foregone by engaging a resource in a chosen activity instead of engaging that same resource in the foregone activity.

Government opportunity cost arises from having to select among long-term investments in differing public assets given available tax revenues. Investment on one set of public assets often results in long-term delayed investment in other public assets creating long-term disbenefits to public sectors.

Government opportunity cost—Interest rate is set at the financing rate for which funding is not available. Problem: Differing governmental units will have different opportunities and different interest rates which will lead to overall inconsistent funding decisions.

Money for investment in public assets ultimately comes from taxpayers in the form of taxes or fees. This is money that, otherwise, taxpayers could have invested in alternate revenue generating assets.

Taxpayer opportunity cost—The government interest rate paid for public assets should be at least that which taxpayers could have received on the taxes paid to fund the asset.

The United States Office of Management and Budget Circular OMB A94, Section 8 Discount Rate Policy (1) Base Case Analysis specifies:

Constant dollar benefit–cost analyses of proposed investments and regulations should report net present value and other outcomes determined using a real discount rate of 7 percent. This rate approximates the marginal pre-tax rate of return on an average investment in the private sector in recent years.

Given the stipulated 7% real interest rate and 2.5% average inflation rate in the US economy from 1990 to 2018, this yields an approximate market interest rate $i_M = 0.07 + 0.025 + (0.07)(0.025 = 0.097$ or 9.7%. As a rule, *select the larger of the government opportunity cost or taxpayer opportunity cost as the interest rate for funding public investments.* If local taxpayers affected by investment in public sector asset earn less than 9.7%, use the 9.7% interest rate. Conversely, if local taxpayers can earn greater than the 9.7% interest rate, the higher taxpayer interest rate should be used.

14.4   Cost-Effectiveness Analysis

Cost-effectiveness analysis (CEA) is applied when service benefits cannot be monetized. As such, cost-effectiveness analysis is an alternative to benefit–cost analysis. CEA compares the relative costs to the outcome benefits of two or more service designs. CEA measures costs in a common monetary value and the effectiveness of a service design in terms of common outcome benefit. As such, CEA is a comparison tool. CEA will not always yield a unique decision. Rather, CEA provides relative evaluation of service design options for a common service outcome benefit.

CEA distinguishes between service direct and indirect costs. Direct costs result from the application of physical artifacts as part of service delivery (e.g., automobile or truck in transportation, photographic equipment, drugs and medical equipment, tables and chairs in cafeterias, etc.). Indirect costs are costs incurred by the client in seeking and acquiring the service. Intangibles are unintended consequences or adverse effects as a result of the service (time or money lost due to delivery delays, drug or medical side effects, food allergies, or incorrect food delivery).

The cost-effectiveness ratio is simply the sum of all service delivery costs divided by the weighted sum of service benefits.

$$\text{CER} = \frac{\sum \text{direct costs} + \sum \text{indirect costs}}{\sum w_i B_i} \tag{14.4}$$

where w_i = value placed on benefit B_i by the client and $\sum_i w_i = 1.0$. Hence, a service benefit is the inner product of weighted benefit outcomes. As an example, one medical patient may value a reduction in pain over increased mobility where another patient may desire an equivalent reduction to increased mobility.

In service design, CERs can only be ranked relative to each other. CERs must be related in some manner to service budgets to determine the most cost-effective package of service delivery. Since CERs are not commensurable, their direct and indirect costs cannot be added for a selected set of services (i.e., additivity assumes independent services; overlap intersections among services prohibit additivity). Accordingly, development of major system services requires managerial expertise, experience, and judgment within the given service sector.

14.5 Cost-Utility Analysis

Cost-utility analysis is used in health care to compare the cost of alternative interventions versus their respective outcomes. Cost-utility analysis estimates the incremental cost-effectiveness ratio (ICER). ICER is calculated as the difference in the expected cost of two proposed interventions, divided by the difference in the interventions' estimated quality-adjusted life years (QALYs). QALYs measure healthcare intervention outcomes as a combination of the duration and health-related quality of life. ICER estimates are compared with a threshold ICER. Intervention ICER < threshold ICER is funded. ICER comparisons indicate which is the best of identified alternative interventions given the intervention costs. ICER comparisons may not be commensurable across different healthcare areas, because (1) different currencies may be used in measuring intervention outcomes and (2) QALY measures may be sensitive to the quality-of-life dimension. The usefulness of cost-utility analysis is related to QALY measures. The more accurate that a QALY measurement captures the social values of quality of life associated with interventions, the more useful the cost-utility analysis. Given this limitation, cost-utility analysis is an insufficient basis for deciding among alternative intervention projects.

14.6   Benefit–Cost Ratio Analysis

Dividing both sides of Eq. (14.1) by PW(Net Life Cycle Costs, i), the worthiness of a public investment can be expressed as the benefit–cost ratio

$$\frac{B}{C} = \frac{PW(\text{ Net User Benefits, } i)}{PW(\text{ Net Life Cycle Costs, } i)} = \frac{EUAB(i)}{EUAC(i)} \tag{14.5}$$

Equation (14.5) implies that public sector investments with $B/C \geq 1.0$ should be accepted and those with $B/C < 1.0$ should be rejected. Equation (14.5) is termed the **conventional B/C ratio**. The conventional B/C ratio may also be expressed as

$$\frac{B}{C} = \frac{PW(\text{Net User Benefits, } i)}{PW(\text{Initial Costs, } i) + PW(\text{Net O\&M Costs, } i)} \tag{14.6}$$

The conventional B/C ratio compares the benefits to users in the numerator and cost to the public sector agency in the denominator. Economics researchers recognize that taxpayers and users pay the annual net operating and maintenance expenses, rather than the sponsoring public sector agency. They argue that the PW(Net O&M Costs, i) should be subtracted from the B/C numerator and denominator to reflect the life cycle net worth to taxpayers and users in the numerator. This yields the **modified B/C ratio**.

$$\text{Mod}\frac{B}{C} = \frac{PW(\text{Net User Benefits, } i) - PW(\text{Net O\&M Costs, } i)}{PW(\text{Initial Costs, } i)} \tag{14.7}$$

Conventional B/C ratio estimates do not equal the modified B/C ratio estimates, but both estimates yield the same investment decisions.

Example 14.1
City council had requested its city engineering manager to perform benefit–cost analysis for construction of a new interstate interchange and widening of a 2-lane road to 4 lanes with exits for a proposed outlet shopping mall. Federal funding of 50% will be required for the new interstate interchange. Construction costs and employment to the city's citizens and tax benefits to the city are stated below. A conservative interest rate of 10% will be used to be just greater than the OMB A94's 9.7% interest rate accounting for an expected 2.5% inflation rate over the expected 30-year life before the interchange and road will require replacement

Costs.
Land Acquisition: $5,000,000 year -1

Interstate Construction Cost: $3,600,000 year -1

$3,600,000 year 0

Annual Maintenance: $ 156,000/year

Road Expansion

Construction Cost: $2,600,000 year 0

Entrance Traffic Light: $ 250,000 year 0

Annual Maintenance: $ 65,000/year

Benefits Per Year

After-Tax Wages 100 Employees: $3,375,000

Sales Tax Revenues to City: $1,687,500

Future worth of land acquisition and construction costs.
Land acquisition:

$$FW = \$5,000,000\,(F/P, 10\%, 2)$$
$$= \$5,000,000\,(1.210) = \$6,050,000$$

Interstate interchange construction:

$$FW = \$3,600,000\,(F/P, 10\%, 2) + \$3,600,000\,(F/P, 10\%, 1)$$
$$FW = \$3,600,000\,(1.210) + \$3,600,000\,(1.100) = \$8,316,000$$

Road expansion:

$$FW = \$2,600,000\,(F/P, 10\%, 1) = \$2,600,000\,(1.100) = \$2,860,000$$

Traffic light:

$$FW = \$250,000\,(F/P, 10\%, 1) = \$250,000\,(1.100) = \$275,000$$

Present worth estimates.
Wages and tax revenues

$$PW = \$3,375,000\,(P/A, 10\%, 30) + \$1,687,500\,(P/A, 10\%, 30)$$
$$PW = \$3,375,000\,(9.427) + \$1,687,500\,(9.427) = \$47,724,188$$

Annual maintenance costs

$$PW = \$156,000 \, (P/A, 10\%, 30) + \$65,000 \, (P/A, 10\%, 30)$$
$$PW = \$156,000 \, (9.427) + \$65,000 \, (9.427) = \$2,083,367$$

$$\text{Conventional } B/C \text{ ratio } = \$47,724,188 / \left(\$17,501,000 + \$2,083,367\right)$$
$$= 2.437$$

$$\text{Modified } B/C \text{ ratio } = \left(\$47,724,188 - \$2,083,367\right)/\$17,501,000$$
$$= 2.608$$

14.7   Incremental Benefit–Cost Ratio Analysis

Chapter 10 applied incremental analysis to maximize the net present worth of the best alternative selected from multiple mutually exclusive alternative investments. Similarly, when using the B/C criteria to select from among multiple mutually exclusive alternatives, the incremental method should be applied. The process for *incremental B/C* ($\Delta B/\Delta C$) analysis is:

1. Identify all relevant alternatives.
2. Calculate the B/C ratio of each competing alternative in the set. Eliminate individual alternatives with B/C ratio < 1.0.
3. Rank order the projects by increasing denominator cost. The "do-nothing" alternative is always the first on the ordered list.
4. Identify the increment of pairwise projects under consideration.
5. Calculate the $\Delta B/\Delta C$ ratio on the incremental cash flows.
6. Select the next higher cost project if $\Delta B/\Delta C > 1.0$.
7. Iterate steps 4 through 6 until all pairwise comparisons have been considered.
8. Select the highest cost project that has $\Delta B/\Delta C > 1.0$.

   Incremental B/C analysis can be applied with the conventional and modified B/C ratio criteria. The two methods should not be mixed in the same selection problem.

Example 14.2

The City of Coastal California is considering installation of two wind turbine electricity generators to supply its residents electricity needs and minimize the potential for wildfires due to sparks from California Power and Light high-power lines into the city. Four commercial wind turbine manufacturers have submitted the following bids for installation of the wind turbines, power substation, and reconnection of existing underground power lines to the substation. The city engineering has submitted the following cost and benefit estimates to city council for review. The project life is 50 years, and the city uses the interest rate of 10% as an approximation of the OMB A94's 9.7% interest rate. Perform conventional incremental $\Delta B/\Delta C$ analysis.

	Alternatives ($\times$$1000)				
Project costs	Do nothing	A	B	C	D
Installation	$0	(−$25,000)	(−$16,250)	(−$19,000)	(−$27,500)
Annual O&M costs	$0	(−$240)	(−$709)	(−$500)	(−$237)
Project benefits					
Ann utility savings	$0	$1200	$1034	$1444	$2128
Overcapacity rev	$0	$1,400	$813	$304	$659
Annual jobs created	$0	$800	$1,100	$228	$818

			Alternatives			
Project Costs		Do Nothing	A	B	C	D
Turbine/substation installation		$0	$25,000	$16,250	$19,000	$27,500
Annual O & M expenses		$0	$240	$709	$500	$237
	PW Costs	$0	$27,380	$23,280	$23,957	$29,850
Project Benefits						
Annual Savings Utilities Payments		$0	$1,200	$1,034	$1,444	$2,128
Revenue from Overcapacity		$0	$1,400	$813	$304	$659
Annual Effect on Jobs Created		$0	$800	$1,100	$228	$818
	PW Benefits	$0	$33,710	$29,214	$19,592	$35,743
	Project B/C	0.00	1.23	1.25	0.82	1.20

	Incremental Analysis		
	B - DN	A - B	D - A
	$16,250	$8,750	$2,500
	$709	($469)	($3)
	$23,280	$4,100	$2,470
	$1,034	$166	$928
	$813	$588	($741)
	$1,100	($300)	$18
	$29,214	$4,496	$2,033
	1.25	1.10	0.82
Select	B	A	A

Commercial wind turbine manufacturer C is eliminated from consideration since its B/C ratio = 0.82 < 1.0. From incremental B/C ratio analysis, select commercial wind turbine manufacturer A.

Example 14.3

Reconsider the wind turbine electricity generator supplier selection in Example 14.2. This time, use the modified incremental $\Delta B/\Delta C$ analysis.

		Alternatives				
Project Costs		Do Nothing	A	B	C	D
Plant Construction		$0	$25,000	$16,250	$19,000	$27,500
Project Benefits						
Annual Savings Utilities Payments		$0	$1,200	$1,034	$1,444	$2,128
Revenue from Overcapacity		$0	$1,400	$813	$304	$659
Annual Effect on Jobs Created		$0	$800	$1,100	$228	$818
Annual O & M		$0	($240)	($709)	($500)	($237)
	PW Benefits	$0	$31,331	$22,184	$14,634	$33,393
	Project B/C	0.00	1.25	1.37	0.77	1.21

	Incremental Analysis		
	B - DN	A - B	D - A
	$16,250	$8,750	$2,500
	$1,034	$166	$928
	$813	$588	($741)
	$1,100	($300)	$18
	($709)	$469	$3
	$36,244	$9,146	$2,003
	2.23	1.05	0.80
Select	B	A	A

Commercial wind turbine manufacturer C is eliminated from consideration since its modified *B/C* ratio $= 0.77 < 1.0$. The same pairwise comparison result is: B—do nothing, A–B, and D–A. From incremental $\Delta B/\Delta C$ ratio analysis, select commercial wind turbine manufacturer A.

14.8 Factors Affecting Benefit–Cost Ratio Analysis Decisions

Four other factors affect public sector investment *B/C* ratio estimates and alternative selection: investment financing, investment life cycle, quantifying benefits and disbenefits, and public sector politics and policies.

14.8.1 Public Sector Investment Financing

The primary sources of public sector financing are taxes or municipal bonds. Governments (federal, state, and local) tend to spend tax revenues on immediate service needs. As an example, in its 2019 fiscal year the US federal government took in approximately $3.5 trillion in taxes from corporations and individuals.

Individual income taxes	$1.700 trillion
Payroll taxes	$1.200
Corporate income taxes	$0.230
Excise taxes	$0.099
Other sources	$0.271
Total revenue	$3.500 trillion

For the same fiscal year, the US federal government incurred about $4.4 trillion in expenses.

Social security	$1.000 trillion
National defense	$0.676
Non-defense	$0.661
Medicare	$0.644
Medicaid	$0.409
Interest on debt	$0.375
Other expenses	$0.642
Total expenses	$4.407 trillion

Since 49 of the 50 states have balanced budget provisions, states tend to spend their tax revenues on immediate service needs. Although state budgets vary, the average state expenses are as follows:

K-12 education	26%
Medicaid & children's health	17%
Higher education	15%
Transportation	6%
Corrections	5%
Public assistance	1%
All other	30%

Further, to prevent excessive borrowing and assure repayment, the US government restricts local governmental debt financing.

- Local governments' borrowing is restricted to a specified percentage of assessed property value in the taxation district.
- Borrowed funds through the sale of bonds for new construction require local voter approval (sometimes by 2/3 majority).
- Public debt repayment must be made in a preset period in accordance with a specific plan.

Hence, governments must borrow money for large-scale public investments, which means that most public infrastructure and service projects are 100% financed with bonds.

14.8.2 Public Sector Investment Life Cycle

Public sector investment life cycles are longer than those in the private sector. Private sector investments are tied to the finite planning horizon life cycle budget of three to ten or twelve years and long-term business-cycle budget of two or three life cycle budgets. Public sector investments tend to have 20- to 50-year life cycles and corresponding budget horizon. The investments, likewise, have longer lives with some that could be considered as permanent investment of infinite lives.

Unlike private sector investments (Fig. 1.5), public sector investments require substantial funding in the needs assessment, conceptual design, and detailed design phases of the product life cycle. Hence, it is advantageous to advocates of public sector projects to spread initial costs of investment into the latter part of the conceptual design phase or into the detailed design phase. Doing so reduces $PW_0 = FW$(initial costs, i) in the denominator of Eq. (14.5) increasing the resultant B/C ratio. Likewise, if advocates can secure interest rates less than the OMB A94 7.0% real interest rate in the early life cycle phases, $PW_0 = FW$(initial costs, i) in the denominator of Eq. (14.5) will be reduced, increasing the resultant B/C ratio.

14.8.3 Quantifying Benefits and Disbenefits

Public sector benefits and disbenefits tend to be difficult to quantify, and resultant estimates will tend to exhibit large variances. As examples, in a flood control project, what is the value of a lost human life? In a K-12 school investment, what is the differential value to students in investing in a regular classroom or a computerized, media-enabled, advanced classroom? In investing in a new airport addition or runway, how accurately and precisely can benefit cash flows be estimated 50 years into the future versus the accuracy and precision of current disbenefit cash flows? Hence, it is advantageous to advocates of public sector investments to inflate early benefit cash flows and deflate later benefit cash flows to offset current disbenefit cash flows.

14.8.4 Public Sector Politics and Policies

Politics affect all public sector investments, because these investments are large scale affecting a large proportion of citizens and have the potential to affect supporting private sector industries. Politics put further pressure on advocates to modify the timing and amount (discussed above) of benefit and cost cash flows and using lower interest rates in the early life cycle phases to yield a higher B/C ratio favoring the investment. Further, the lengthy and complex processes required for many public sector investments are vulnerable to political bias to favor public sector investments at each budgeting step. As an example, the federal budgeting process for public sector investments consists of the following major steps.

- Managers develop requirements.

  - "Funded" requirements—within existing budgets.
  - "Unfunded" requirements—more funds required.

- Managers submit requirements to program managers.
- Programs prioritize requirements by B/C.
- Subprograms submit to higher programs—often many steps here.
- At each step, requirements are prioritized, and program budgets are adjusted.
- The president receives all budgets and applies a final prioritization.
- Federal government sets tax rates and funding rates.
- Spent by the executive branch.
- Set by federal law:

  - The president submits a budget proposal.
  - Congress passes budget/appropriations bills.
  - President signs or vetoes.
  - Congress can override a veto.
  - Highly political process.

- Federal Reserve Bank sets interest rates.
- Specific funds appropriated for specific purposes.
- Each funding type can only be used in certain ways.
- Appropriated funds.

  - Research, development, testing, and evaluation.
  - Procurement.
  - Operations and maintenance.
  - Construction.
  - Military salaries/compensation.

- Non-appropriated funds.

  - Agencies charge fees for services—largest: US Postal Service.
  - Working capital managed and reported.

Allocated federal funds are tightly tied to the federal fiscal year.

- October 1 to September 30.
- Funds must be at least obligated (on contract) before the end of the fiscal year.
- Expired funds are "lost" to the program and go back to the general fund. Hence, there is pressure to spend allocated funds.

Some categories of allocated funds can be extended into later years.

- Operations and maintenance—1 year only.
- RDT&E—2 years.
- Procurement—3 years.

- Ship construction—5 years.
- Building construction—5 years.

At some point in the budgeting cycle, there is pressure to spend allocated funds or lose them.

14.9 Summary

Optimal budgeting in the public sector requires public employees *to take a viewpoint at least as broad as those who pay the costs and those who receive the benefits.*

$$PW(\text{Asset}) = FW(\text{Asset costs})p\,(F/P, i, p)$$
$$+ \sum_n PW(\text{O\&M costs} - \text{Revenues})t\,(P/F, i, t)$$

Conventional *B/C* ratio:

$$\frac{B}{C} = \frac{PW(\text{Net User Benefits, } i)}{PW(\text{Initial Costs, } i) + PW(\text{Net O\&M Costs, } i)}$$

Modified *B/C* ratio:

$$\text{Mod}\frac{B}{C} = \frac{PW(\text{Net User Benefits, } i) - PW(\text{Net O\&M Costs, } i)}{PW(\text{Initial Costs, } i)}$$

Incremental B/C($\Delta B/\Delta C$) ratio.

14.10 Key Terms

Benefits
Benefit–cost ratio analysis
Conventional *B/C* ratio
Cost-effectiveness analysis
Cost-utility analysis
Disbenefits
Government opportunity cost
Incremental *B/C* ratio analysis
Modified *B/C* ratio analysis
Net benefits to users
Public sector investment life cycle
Public sector investment financing
Politics

Revenue bonds
Taxpayer opportunity cost
User fees.

Problems

1. Calculate the conventional and modified benefit–cost ratios for the following project.

Category	Cash flow
Initial cost	$1,875,000
Annual benefits to users	$625,000
Annual disbenefits to users	$43,750
Annual government maintenance cost	$225,000
Project life	50
Government interest rate	5.0%

2. A city's water treatment department has a budget of $1,080,000 for projects to improve its treatment and distribution operations. The following improvement projects have been proposed.

Project	Initial cost	Annual benefit	Salvage value	Life
A	$180,000	$63,000	$10,000	5
B	$360,000	$91,450	$60,000	10
C	$450,000	$168,650	$2,000	5
D	$360,000	$91,250	$12,000	8
E	$180,000	$37,600	$0	12

(a) What is the city's opportunity cost of capital? Based on the opportunity cost of capital, which projects should be recommended?

(b) If the city uses market rate of 9.7%, which projects are recommended using the conventional B/C ratio?

Using the $\Delta B/\Delta C$ criteria, which alternative is preferred?

3. The regional electric cooperative needs additional electricity capacity. REC is committed to changing to clean energy sources to reduce its long-term costs and liability for coal fly ash storage and containment. REC's engineering manager has selected two plans for general management consideration. REC's interest rate is 5.0%, and this project has a 40-year life.

	Solar cell	Wind turbine
Initial cost	$5,320,000	$2,660,000
10th year cost	$0	$380,000
O&M, replace/year	$95,000	$47,500
Avg annual pwr cost		
First 10 years	$0	$47,500
Next 30 years	$0	$190,000

4. Evaluate the three mutually exclusive alternatives in the following table with a 25-year project life and MARR = 12.0%. Use (a) conventional B/C ratio, (b) modified B/C ratio, and (c) net present worth.

Cash flow	A	B	C
Initial cost	$15,000	$22,200	$26,400
Annual savings	$3840	$6000	$11,750
Annual costs	$1200	$3300	$7680
Salvage value	$7200	$5040	$16,800

Chapter 15
Introduction to Management Economic Decision Theory and Risk Analysis

Abstract Except for the introductory discussion of risk–return fundamentals in Sect. 12.1, discounted cash flow analyses have assumed that estimates of future cash flows are known with certainty. This is never the case in applied engineering economic analyses. Macroeconomic, business economic, and organizational internal variables induce structural and random variation in future cash flow outcomes. Hence, it is essential that the engineering manager accounts for the riskiness of structural and random variation in future cash flow estimates and report probabilities of not achieving required returns on organizational investments.

15.1 Elements of Engineering Managerial Economic Decisions

Selecting the best from among competing investment alternatives is hard due to variability in future cash flows resulting from variances in macroeconomic, business economic, and organizational internal variables. Variances in macroeconomic variables drive short- and long-term risk of loss. International macroeconomic variables include political, social, and economic instability, trade imbalances, trade wars, currency instability, government regulations, and pandemics. Business economic risk variables include changes in consumer preferences or demand, competitive per-unit costs and price, interest rates, technology, and availability of raw materials, components, and suppliers. Organizational internal variables include management capability and tolerance for risk, engineering and technological capability, organizational capability and knowledge, cash flows from existing products and services, and time horizon for required returns on investment. The joint effects of variances in macroeconomic, business economic, and organizational internal variables can be summarized into four basic sources of decision-making difficulty.

© The Author(s), under exclusive license to Springer Nature Switzerland AG 2022
T. S. Cotter, *Engineering Managerial Economic Decision and Risk Analysis*,
Topics in Safety, Risk, Reliability and Quality 39,
https://doi.org/10.1007/978-3-030-87767-5_15

- Complexity due to interacting variances in driving variables. Given just the macroeconomic, business economic, and organizational internal variables listed above, there are 33,554,432 combinations of statistically significant main effect variances plus 5.62949E+14 possible two-way interactions. Essentially, there is a countably infinite number of variable interactions that can (and do) affect future cash flows.
- Inherent risk in predictor variable variances. The magnitude and timing of predictor variable future variances can never be known with certainty.
- Whether an organization is working toward a single objective or toward multiple objectives. As previously noted in Chap. 13 on capital budgeting, an organization can seek to maximize the present worth of future cash flows, to maximize the rate of return to investors, or to maximize some weighted combination of the two in mixed investment portfolios.
- Competing management perspectives of investment objectives may lead to different future cash flow estimates and different investment decisions. Even if a single management perspective dominates, small changes in the variances of certain predictor variables may lead to changes in the relative ordering of investment criterion or weighted criteria and different investment decisions.

Figure 15.1 sets forth a general engineering managerial economic investment decision analysis process.

The most important and difficult step in the economic investment decision analysis process is the problem statement. To the degree that the problem statement is

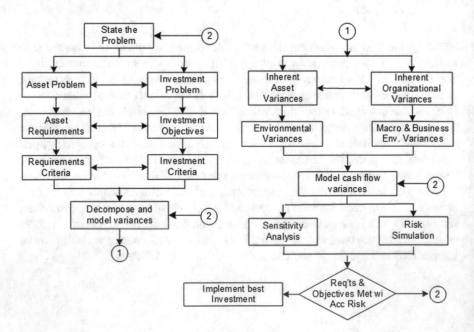

Fig. 15.1 General economic investment decision analysis process

misaligned with the asset requirements and investment objectives, the asset performance and investment outcomes will be misaligned with desired results. Stated conversely, a decision outcome's quality is limited to the degree that the investment problem statement yields a model that approximates required asset physical and economic performance. Why a model? Resultant detailed asset design is only a model (mathematical and graphical representation). Discounted cash flow equations are likewise only mathematical models. It is only when a decision-maker selects an investment alternative, physically implements it, and realizes actual life cycle cash flows that the decision-maker can assess model quality in terms of variances from designed physical and economic performance.

In the problem statement step, the asset problem must be decomposed and assessed separately from the investment problem, but their interactive effects on each other must be incorporated into the requirements, objectives, and criteria. Asset and investment decomposition allow the engineering manager to identify optimal asset performance design and optimal discounted cash flows. Assessment of the interactive effects requires the engineering manager to recognize cash flow constraints imposed on asset performance variances from optimum and asset performance constraints imposed on cash flows from optimum. These constraint differences become the inputs to model variances in the second phase of the economic decision analysis process.

In phase two, the engineering manager must consider additional constraints imposed on asset performance and economics by organizational constraints (capacities, capabilities, technologies, and social dynamics), production environment constraints, end-use environmental constraints, and projected macroeconomic and business economic constraints over the asset's life cycle. These impose additional variances from optimum performance and cash flow outcomes.

In phase three, the engineering manager uses the knowledge gained to develop mathematical and graphical models of asset performance and cash flow. The models must reflect realistic asset performance and cash flows within identified constraints. The models may be assessed through sensitivity analyses to determine the ranges over which the best asset performance and cash flows hold, or constraint variances assessed through simulations to develop performance risk profiles and economic risk profiles. The asset design or alternative that in some sense optimizes performance or economic criteria relative to risk is selected as the best investment.

15.2 Probability and Statistics Concepts for Management Economic Decisions

To assess economic risk (we will forgo performance risk assessment because that is a development and design engineering task beyond the scope of this discussion), we need some fundamental probability and statistics concepts. Parametric statistical models partition structure from error. Structure is measured by some metric of central

location, and error is measured by the variance of outcomes around the central location. Some additional statistical metrics measure the shape and sharpness of the peak
of the distribution around the central location.

Fundamental Probability Theory

Estimates of probabilities are the foundation of statistical analyses. All probabilities
must exhibit three properties.

> **Definition**: Probability is a set function $P(x \in A)$ or $f(x \in A)$ that assigns to each
> event $x \in A$ in the sample space S a number $P(x \in A) = a$ or $f(x \in A) = a$, called
> the probability of event A, such that the following properties are satisfied:

Discrete distributions

Property 1: $P(x = A) \geq 0$.
Property 2: $P(S = A + \overline{A}) = 1.0$
Property 3: If $A_1, A_2, \ldots, A_n$ are events $A_i \cap A_j = \emptyset$, then

$$P(A_1 \cup A_2 \cup \cdots \cup A_n) = P(A_1) + P(A_2) + \cdots + P(A_n)$$

Continuous distributions

Property 1: $f(X = A) > 0, A \in \Re$.
Property 2: $\int_{\Re} f(X) \mathrm{d}x = 1.0$
Property 3: The probability of event $X \in A$ is $P(X \in A) = \int_A f(x)\mathrm{d}x$

Probability Theorems

$$P(A) = 1 - P(\overline{A})$$
$$P(\emptyset) = 0$$

For each event A, $P(A) \leq 1.0$
If A and B are any two events, $P(A \cup B) = P(A) + P(B) - P(A \cap B)$.
Note: If $A \cap B = \emptyset$, $P(A \cup B) = P(A) + P(B)$
If any two events A and B are independent, $P(A \cap B) = P(A) \times P(B)$.

Conditional Probabilities

Suppose you are interested in the probability of event B given that event A also
occurs. Then,

$$P(B|A) = \frac{P(B \cap A)}{P(A)}$$

Independence—Events B and A are independent if and only if $P(B|A) = P(B)$.
Conditional independence—Events B and A are conditionally independent if and
only if

$$P(B|A, C) = P(A|B)P(B)$$

Because of the symmetry of the definition of conditional probability,

$$P(B|A)P(A) = P(A|B)P(B)$$

from which **Bayes theorem** can be derived.

$$P(B|A) = \frac{P(A|B)P(B)}{P(A)}.$$

Expanding $P(A)$ with the formula for total probabilities,

$$P(B|A) = \frac{P(A|B)P(B)}{P(A|B)P(B) + P\left(A|\overline{B}\right)P\left(\overline{B}\right)}$$

Discrete Probability Distributions

When estimating proportions of individuals falling into one of two categories (i.e., pass–fail, male–female, citizen–non-citizen, etc.) or estimating counts per unit (customers per capita, defects per unit, species per square kilometer, etc.), we apply probability theory to discrete units. Since X is a countable event, it is termed a **random variable**.

Definition: A *random variable* X is a numerical function specified over some sample space S that can take on only a countable number of values in S.

The probabilities of $P(X = x)$ are termed a probability mass function (pmf) because the probabilities are associated with point values.

The mean and variance are required to specify the location and spread of any probability distribution.

Definition: Let X be a discrete random variable with probability distribution function $P(X = x)$. The *mean* or *expected* value of X is

$$\mu = E[X] = \sum_N x_i P(x_i) \tag{15.1}$$

where N = number of individuals in the population. In cases where all individuals of a population cannot be enumerated, the population mean is estimated from a random sample by the sample mean.

$$\overline{x} = \frac{\sum_n x_i}{n} \tag{15.2}$$

Definition: Let X be a discrete random variable with probability distribution function $P(X = x)$. The *variance* of X is

$$\sigma^2 = V[X] = \sum_N (x_i - \mu)^2 P(x_i) \qquad (15.3)$$

where N = number of individuals in the population. In cases where all individuals of a population cannot be enumerated, the population variance is estimated from a random sample by the sample variance

$$s^2 = \frac{\sum_n (x_i - \overline{x})^2}{n - 1} \qquad (15.4)$$

where $n - 1$ adjusts for the loss of 1 degree of freedom using x-bar to estimate the population mean μ, because $E[x\text{-bar}] = \mu$. The standard deviation of a random variable is defined as

$$s = \sqrt{s^2} \qquad (15.5)$$

The **Bernoulli** distribution is the most basic of the discrete distributions. A Bernoulli random variable is defined as the dichotomous of a trial where $X = 1$ if a success occurs and $X = 0$ if a failure occurs. The Bernoulli pmf distribution is

$$p(x = 0, 1) = p^x (1 - p)^x \qquad (15.6)$$

Characteristics of a Bernoulli random variable are:

- The trial results in one of two mutually exclusive outcomes.
- The outcomes are exhaustive.
- The probabilities of the outcomes are assigned $P(X = 1) = p$ and $P(X = 0) = 1 - p, 0 < p \le 1$.

The mean and variance of a Bernoulli random variable are

$$E[X] = p$$

$$\sigma^2 = p(1 - p)$$

The sum of a sequence of Bernoulli trials results in the random variable being binomially distributed. The **binomial** pmf distribution is

$$P(X = x) = \binom{n}{x} p^x (1 - p)^{n-x} \qquad (15.7)$$

where $\binom{n}{x}$ = combination of x out of n. Characteristics of a binomial random variable are:

- Each trial consists of identical Bernoulli random variables.
- $P(X = 1) = p$ remains constant from trial to trial.
- Each trail is independent.
- The binominal random variable $X = 1$ is the number of successes in n trials.

The mean and variance of a binomial random variable are

$$E[X] = np$$

$$\sigma^2 = np(1 - p)$$

Example 15.1

A quality engineer at Vapor Phase Products, Inc., is investigating impurities in the last two 100-batch runs of 5-L reactants. In the first 100-batch run, 5 batches were found contaminated. In the last two 100-batch runs, 2 batches and 5 batches, respectively, were found contaminated. Estimate the probability distribution of batch contamination.

Each batch may be classified as not contaminated, 0, or contaminated, 1. Such 0–1 classification arises from a Bernoulli process, where $0 < p < 1$ is the expected proportion of units contaminated and n is the number of units in the sample. For this case, the best estimate of the proportion of units contaminated per 100-batch run is $p = (5 + 2 + 5)/(100 + 100 + 100) = 0.04$.

Since this is continuous process producing a sequence of 100-batch runs, the distribution of contaminated batches X per 100-batch run is the sum of a Bernoulli process resulting in a binomial pmf. Letting $n = 100$, the resultant distribution-contaminated batches per 100-batch run are

x	P(x)	CumP(x)
0	0.01687	0.01687
1	0.07029	0.08716
2	0.14498	0.023214
3	0.19773	0.042948
4	0.19939	0.62886

(continued)

(continued)

x	P(x)	CumP(x)
5	0.15951	0.78837
6	0.10523	0.89361
7	0.05888	0.65249
8	0.02852	0.98101
9	0.01215	0.99316
10	0.00161	0.99776
11	0.00157	0.99933
12	0.00049	0.99982

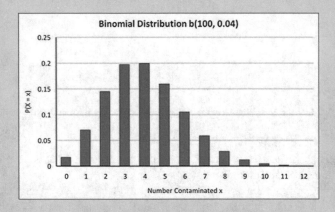

The mean and variance of the contamination rate are

$$E[X] = 4.0$$

$$V[X] = 3.84$$

A contamination rate of just 0.04, or 4.0%, represents a threat to the reactant product's profit margin, because only 0.01687, or approximately one 100-batch run out of 100 100-batch runs, will yield zero lost batches due to contamination.

If we set $\lambda = E[X] = np$, hold λ constant as $n \to \infty$ as $p \to 0$, and perform some algebraic manipulations, we can derive the **Poisson** distribution. The Poisson pmf distribution is

$$P(X = x) = \frac{\lambda^x e^{\lambda-}}{x!}. \qquad (15.8)$$

Characteristics of Poisson random variable are:

- In each trial, count the number of times a particular event X occurs over a $\sigma^2 = E[X] =$ constant unit area, volume, weight, or any unit of measure.
- The probability that a number of events X in any unit area remain constant across all unit areas sampled.
- The number of events X that occur in one unit area is independent of the number that occurs in all other unit areas sampled.

The mean and variance of a Poisson random variable are

$$E[X] = \lambda$$

$$\sigma^2 = \lambda$$

Example 15.2

The Environmental Protection Agency (EPA) limits ethylene oxide emissions to 10 ppm after sterilization. Suppose that the safety engineer sampling for the last fiscal month of operation indicated an average of 4 ppm. If the ppm random variable follows a Poisson distribution, (a) what is the standard deviation in ppm and (b) what is the probability that an air sample will exceed 10 ppm?

Since the standard deviation is the square root of the variance and for this Poisson random variable $E[X] = \sigma^2 = \lambda = 4$, the estimated standard deviation is 2.

The Poisson pmf for $\lambda = 4$ is

x	$P(x)$	CumP(x)
0	0.01832	0.01832
1	0.07326	0.09158
2	0.14653	0.23810
3	0.19537	0.43347
4	0.19537	0.62884
5	0.15629	0.78513
6	0.10420	0.88933
7	0.05954	0.94887
8	0.02977	0.97864

(continued)

(continued)

x	$P(x)$	CumP(x)
9	0.01323	0.99187
10	0.00529	0.99716
11	0.00192	0.99908
12	0.00064	0.99973
13	0.00020	0.99992
14	0.00006	0.99998
15	0.00002	1.00000

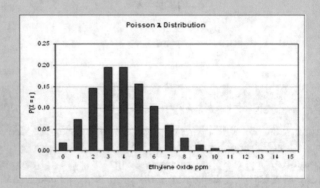

With $\lambda = 4$, the probability that an air sample will exceed 10 ppm is $P(X > 10) = 1 - P(X \leq 10) = 1 - 0.99716 = 0.00284$ or approximately 0.3%.

Continuous Probability Distributions

When a random variable can be measured on the real line, it is termed a **continuous random variable**.

Definition: A *continuous random variable* X has the following properties.

- X takes on an uncountable infinite number of values in the $\Re$ interval $(-\infty, \infty)$.
- The cumulative distribution function $F(X)$ is continuous.
- The $P(X = x) = 0$.

Definition: The *cumulative distribution function* $F(x)$ for a random variable X is equal to the probability

$$F(x) = P(X \leq x), -\infty < x < \infty \tag{15.9}$$

Definition: If $F(x)$ is the cumulative distribution function for a continuous random variable X, then the density function $f(x)$ for X is

$$f(x) = \frac{dF(x)}{dx} \tag{15.10}$$

The mean and variance of a continuous random variable are defined as follows.

Definition: Let X be a continuous random variable with density function $f(x)$. Then, the *expected value* or *mean* of X is

$$\mu = E[X] = \int_{-\infty}^{\infty} xf(x)dx \tag{15.11}$$

Definition: Let X be a continuous random variable with $E[X] = \mu$. Then, the *variance* of X is

$$\sigma^2 = \int_{-\infty}^{\infty} (x - \mu)^2 f(x)dx \tag{15.12}$$

The sample mean and variance of a continuous random variable are estimated the same as in Eqs. (15.2) and (15.4).

Suppose a number X is randomly selected from a population such that X is a point in the interval $a < X \le b$. The density function of X is the **uniform distribution** shown in Fig. 15.2.

The uniform distribution probability density function (pdf) is

$$f(x) = \frac{1}{b-a}, \quad a \le x \le b$$
$$f(x) = 0 \qquad \text{elsewhere} \tag{15.13}$$

Fig. 15.2 Uniform density function

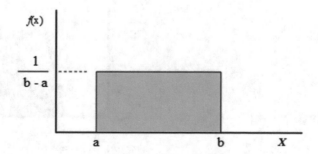

Each $a \leq X = x \leq b$ has an equal probability of occurrence for the continuous uniform distribution. Accordingly, the continuous uniform distribution is an important reference distribution. It is most widely used in the generation of random numbers in the uniform interval (0, 1). The mean and variance of the continuous uniform distribution are

$$E[X] = \frac{a+b}{2}$$

$$\sigma^2 = \frac{(b-a)^2}{12}$$

The **triangular distribution** is a second important reference distribution in statistical modeling. Forms of the triangular density function are illustrated in Fig. 15.3.

The triangular distribution probability density function (pdf) is

$$f(x) = \frac{(x-a)^2}{(b-a)(c-a)}, a \leq x \leq c \qquad (15.14)$$

$$f(x) = 1 - \frac{(b-x)^2}{(b-a)(b-c)}, c \leq x \leq b$$

The triangular distribution is used to model population distributions for which there is only limited sample data or where the distributional form of a variable is known but data is insufficient to fit a model. The standard triangular distribution corresponds to $a = 0$, $b = 1$ with median SQRT($c/2$) for $c \leq 1/2$ and $1 - $ SQRT(($1 - c)/2$) for $c \geq 1/2$. The mean and variance of the triangular distribution are

$$E[X] = \frac{a+b+c}{3}$$

Fig. 15.3 Triangular density function

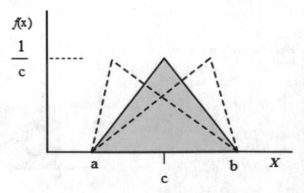

$$\sigma^2 = \frac{a^2 + b^2 + c^2 - ab - ac - bc}{18}$$

Perhaps, the normal distribution is the most widely applied of all continuous distributions. The **normal** (Gaussian) **density function** describes the relative frequency distribution of errors arising from additive processes. The normal probability density function is

$$f(x) = \frac{1}{\sigma\sqrt{2\pi}} e^{-\frac{1}{2}\left(\frac{x-\mu}{\sigma}\right)^2} \tag{15.15}$$

The term $Z = (x - \mu)/\sigma$ in the exponent transforms any normally distributed variate into the **standard normal distribution**. Visually, the standard normal probability density function is a symmetric, unimodal, bell-shaped curve illustrated in Fig. 15.4 with associated $\pm 1\sigma$, $\pm 2\sigma$, and $\pm 3\sigma$ probabilities. Since the $\pm 3\sigma$ interval covers 99.73% of the error probabilities, it is often referred to as the **natural tolerance limits** of a process.

The mean and variance of a normal random variable $N(\mu, \sigma^2)$ are

$$E[X] = \mu$$

$$V[X] = \sigma^2$$

Hence, a normal variate is completely described by its first two moments. The normal distribution has many useful properties.

- The **standard normal variate** $N(0,1)$ and the normal variate $N(\mu, \sigma^2)$ are related by

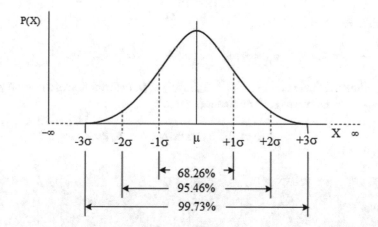

Fig. 15.4 Standard normal probability density function

$$N(0, 1) = [N(\mu, \sigma) - \mu]/\sigma$$

- **Central Limit Theorem**: If $x_1, x_2, \ldots, x_n$ are identically, independently distributed (IID) random variables from any unimodal distribution with mean μ and variance σ^2, and if $y = x_1 + x_2 + \ldots + x_n$, then the distribution of y

$$\frac{y - \sum_n \mu_i}{\sqrt{\sum_n \sigma_i^2}}$$

approaches $N(0, 1)$ as $n \to \infty$. Since y is the summation of x_i IID variables, the variance of the mean is

$$V[\mu] = V\left[\frac{y}{n}\right] = \frac{n\sigma^2}{n^2} = \frac{\sigma^2}{n}$$

Taking the square root yields the **standard error of the mean** SEM $= \sigma/\sqrt{n}$. Hence, the mean of a sample of x_n $\sigma/\sqrt{n}$ from any IID random variable is distributed $N(\mu, \sigma/\sqrt{n})$ as the $n \to \infty$. In application, sample sizes in the range of 12–15 are sufficient to yield means that are $N(\mu, \sigma/\sqrt{n})$ distributed.

- **Linear combinations**: If $x_{i,j}$, where $i = 1$ to a units and $j = 1$ to jth variable, are $N(\mu, \sigma^2)$ distributed, then the linear combination of

$$y = a_1 x_1 + a_2 x_2 + \cdots + a_j x_j$$

is normally distributed with mean

$$\mu_y = a_1 \mu_1 + a_2 \mu_2 + \cdots + a_j \mu_j$$

and variance

$$\sigma_y^2 = a_1^2 \sigma_1^2 + a_2^2 \sigma_2^2 + \cdots + a_j^2 \sigma_j^2$$

- Extending the linear combination rule, any fixed transformation of a normal variate is also a normal variate. For constants b and c,

$$b + c(N : \mu, \sigma) \sim N : b + \mu, c\sigma$$

Example 15.3

A firm is considering investing in an automated weighing system to weigh metal powders for an alloy molding process. The proposed automated weighing system has an initial cost of $N(\$32,000, \$500^2)$ and an estimated market value of $N(\$3000, \$1000^2)$ at the end of the contract life of 5 years. The initial cost variance is due to uncertainties about the costs of the system's COTS components. The market value uncertainty is due to uncertainty about future inflation rates. Industrial engineering has estimated a uniform annual savings of $N(\$8000, \$300^2)$ per year. For the risk level of this investment, the firm requires MARR $= 8.0\%$. What is the probability that the investment will have a negative NPW (i.e., $P(\text{NPW} \leq \$0)$)?

Using the linear combinations of normally distributed variates, the expected mean NPW is

$$\begin{aligned} E[\text{NPW}] = &\left(-\$32,000\right) + 1 \times \$8000\,(P/F, 8\%, 1) \\ &+ 1 \times \$8000\,(P/F, 8\%, 2) + 1 \times \$8000\,(P/F, 8\%, 3) \\ &+ 1 \times \$8000\,(P/F, 8\%, 4) + 1 \times \$8000\,(P/F, 8\%, 5) \\ &+ 1 \times \$3000\,(P/F, 8\%, 5) \end{aligned}$$

$$\begin{aligned} E[\text{NPW}] = &\left(-\$32,000\right) + 1 \times \$8000\,(0.9259) \\ &+ 1 \times \$8000\,(0.8573) + 1 \times \$8000\,(0.7938) \\ &+ 1 \times \$8000\,(0.7350) + 1 \times \$8000\,(0.6806) \\ &+ 1 \times \$3000\,(0.6806) \end{aligned}$$

$$E[\text{NPW}] \quad \$1983$$

The variance estimate is

$$\begin{aligned} V[\text{NPW}] = &500^2 + 1^2 \times \$300^2\,(P/F, 8\%, 1) + 1^2 \times \$300^2\,(P/F, 8\%, 2) \\ &+ 1^2 \times \$300^2\,(P/F, 8\%, 3) + 1^2 \times \$300^2\,(P/F, 8\%, 4) \\ a \quad &+ 1^2 \times \$300^2\,(P/F, 8\%, 5) + 1^2 \times \$1000^2\,(P/F, 8\%, 5) \end{aligned}$$

$$\begin{aligned} V[\text{NPW}] = &500^2 + 1^2 \times \$300^2\,(0.9259) + 1^2 \times \$300^2\,(0.8573) \\ &+ 1^2 \times \$300^2\,(0.7938) + 1^2 \times \$300^2\,(0.7350) \\ &+ 1^2 \times \$300^2\,(0.6806) + 1^2 \times \$1000^2\,(0.6806) \end{aligned}$$

$$V[\text{NPW}] = \$1,289,934$$

The standard deviation is

$$\sigma = \sqrt{\$1,289,934} = \$1,135.75.$$

Transforming to a standard normal variable,

$$Z = \frac{\$0 - \$1,983}{\$1,135.75} = -1.74589$$

For $Z = -1.74589$, the cumulative probability of $P(\text{NPW} \leq \$0) = 0.040374$ or about 4.04%.

Many processes exhibit exponential growth or decay. An exponential process occurs when the rate of change with respect to a variable is proportional to the quantity of the variable itself. Examples of exponential processes include radioactivity, chain reactions, cooling, charging and discharging, spread of information, propagation of defects through a medium or system, population growth in biomechanical processes, interest compounding, and departure from or arrival at a systemic equilibrium state. If the variable of proportionality is random, then the process exhibits an *exponential probability distribution function*.

$$f(x) = \lambda e^{-\lambda x} \text{ or } f(x) = \lambda e^{\lambda x} \tag{15.16}$$

Exponential decay occurs when λ is negative, and exponential growth occurs when λ is positive. The mean and variance of the exponential distribution are

$$\mu = \frac{1}{\lambda}$$

$$\sigma^2 = \frac{1}{\lambda^2}$$

In reliability engineering, the exponential distribution is used to model the random time to failure of a component or system. In this use, λ is termed the **failure rate**, which yields $1/\lambda = $ **mean time to failure**. As an example, suppose that a component has a failure rate of 0.001/hour of continuous operation. That is, $\lambda = 0.0001$ and the mean time to failure is $1/\lambda = 1/0.0001 = 1000$ h. What is the probability that any randomly selected component will not fail before 2000 h of operation?

$$P(x \geq 2000) = \int_{2000}^{\infty} \lambda e^{-\lambda x} dx = e^{-2} = 0.1353$$

There is an important relationship between the Poisson and exponential distributions. If the Poisson probability density function is restated to model the number of occurrences of an event in the interval $(0, t]$, then

$$P(X = t) = \frac{(\lambda t)^x e^{-\lambda t}}{x!}$$

For $x = 0$, there are no occurrences in $(0, t]$, and $P(X = 0) = P(X \geq t) = e^{-\lambda t}$. Since

$$F(t) = P(X \leq t) = 1 - e^{-\lambda}$$

and applying $f(x) = dF(t)/dx$, we derive

$$f(x) = \lambda e^{-\lambda x}$$

as the interval between successive occurrences. Therefore, if the number of occurrences of an event is Poisson distributed with parameter λ, then the distribution of the intervals between each successive occurrence is exponentially distributed with parameter λ.

Many processes are multiplicative (with no time lags or feedback loops). Examples include (1) crushing boulders into sequentially smaller rocks, pebbles, and power; (2) sequentially blending and purifying chemicals and pharmaceutical powders; (3) sequentially separating and refining gases; and (4) software coding processes. In multiplicative processes, product characteristics grow or decline exponentially $y = e^{\pm \lambda x}$, where $x = $ the sequential step. A special case occurs when the x is a normally distributed random variable (i.e., the processing step intervals are normally distributed rather than constant). In this case, the distribution of x is termed a **lognormal distribution** with probability density function (Fig. 15.5).

$$f(x) = \frac{1}{x \sigma_N \sqrt{2\pi}} e^{-\frac{1}{2}\left(\frac{\ln(x) - \mu_N}{\sigma_N}\right)^2} \tag{15.17}$$

The mean and variance of the lognormal distribution are

$$\mu = m e^{\sigma_N^2 / 2}$$

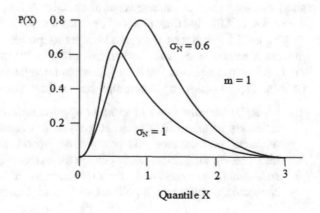

Fig. 15.5 Standard lognormal probability density function

$$V[X] = m^2 e^{\sigma_N^2}\left(e^{\sigma_N^2} - 1\right)$$

The parameter m = median. Equation (15.17) implies that the $\ln(x)$ of a lognormally distributed variable is normally distributed. Thus, if data x is lognormally distributed, $\ln(x)$ will be normally distributed, and the standard normal variate Z can be used to estimate probabilities of occurrence.

Example 15.4

In cybersecurity, the "intrusion kill chain" is a model of the sequence of counter events that must occur to effectively detect and deny, disrupt, or degrade and contain a cyberintrusion. The intrusion kill chain for a particular SQL injection attack is known to be lognormally distributed with $\mu_N = 3.2484$ and $\sigma_N = 0.10837$. A recent SQL injection kill chain required 30 h to disrupt and contain the intrusion. What was the intrusion kill chain's probability of containment (i.e., $1 - P(X \leq x)$)?

$$P(\text{containment}) = 1 - \Phi\left(\frac{\ln(30) - 3.2484}{0.10837}\right) = 1 - \Phi(1.41)$$

$$= 1 - 0.9207 = 0.0793$$

For this particular SQL intrusion, the kill chain was 7.93% effective.

15.3   Structuring Decision Problems

The engineering management investment decision is the selection of the portfolios of short- and long-term productive assets that maximizes net present worth cash flows at the current MARR and maximizes rate of return when WMCC > MARR (Fig. 15.6).

Chapter 13 introduced the capital budgeting problem. Capital budgeting is the process of selecting the asset investment portfolio that maximizes the rate of return on asset investments over the life cycle budget and business-cycle budget planning periods. Capital budgeting is accomplished in six major steps:

1. *Identify Potential Asset Investment Opportunities*. Asset investment opportunities are identified from market analysis of demand trends, gaps in existing products' performance, and new product opportunities. Competitive analysis identifies existing and emerging competitors, their strategies, and their strengths and weaknesses relative to the organizations. Identify macro-environmental economic, technological, political, and social trends that may modify variable

Fig. 15.6 Engineering management investment decision

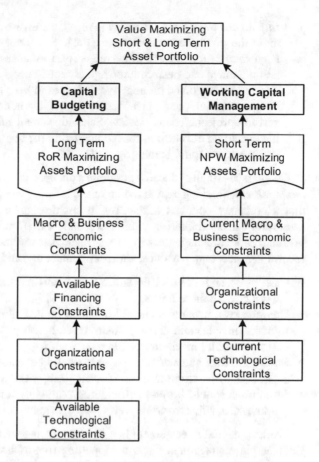

relationships in the market analysis. Likewise, identify organizational socio-technical capacities, capabilities, and potential for transformation to support successful implementation of identified opportunities. Select asset investment opportunities with the highest probability of increasing return on investment.

2. ***Develop the Capital Budget Investment Portfolio***. Categorize fixed asset investment opportunities according to their purpose: market expansion, new product, replacement, upgrade, regulatory requirement, or social welfare. Within each category, rank opportunities by estimated rate of return or net present worth.

3. ***Decision Making***. Select those opportunities that maximize rate of return, net present worth, or are a weighted mix to achieve a target rate of return and cash flow.

4. ***Prepare the Capital Budget and Appropriations***. Classify fixed asset outlays into economic risk value categories. Higher-value outlays are approved by upper management for long-term investment. Medium- and lower-value outlays are approved by middle management and usually covered by blanket appropriations for rapid implementation.

5. ***Implement the Capital Budget***. Project teams are formed and authorized to make the authorized outlays to acquire and implement the fixed assets.
6. ***Performance Review***. Establish project performance reviews to track implementation outlays and performance against actual costs and performance. Corrective actions are taken on negative variances to keep project implementation on budget, schedule, and performance targets. Upon completion of each project, perform project closeout review. Schedule annual audits to compare fixed asset life cycle rate of return to that forecasted in the opportunity identification step. Begin a new capital budgeting cycle.

Chapters 6–10 introduced economic concepts necessary for ***working capital management***. Working capital management is a *business strategy* designed to ensure that a company operates efficiently by monitoring and maintaining its current assets and current liabilities within acceptable current ratio and quick ratio ranges. Working capital management involves balancing movements among cash, accounts receivables, accounts payables, short-term financing, and inventories.

- Cash levels should be sufficient to cover ordinary and small, unexpected needs but not beyond these levels.
- Credit sales (accounts receivable) should be just sufficient to balance the need to maintain sales and healthy business relationships while limiting exposure to customers with low creditworthiness.
- Short-term debt and accounts payable should be just sufficient to allow liquidity for ordinary operations and unexpected needs without increasing financial risk.
- Inventories should be just sufficient to maintain current production levels and product sales while avoiding losses due to accumulation and obsolescence.

Working capital management is operations management's primary responsibility. Engineering management plays a supporting role in assisting operations management in maintaining working capital. For engineering management, working capital management is the process of managing current fixed assets over the zero-budget period and life-cycle budget period to maximize the net present worth of cash flows.

Influence Diagrams

An influence diagram is a graphical representation of the key elements of a decision problem: decisions, alternatives, constraints, risk, criteria, scope, outcome, and stakeholders.

Definition: A *decision* is the selection of a sequence of actions designed to resolve a problem or attain an objective.

Definition: *Alternatives* are two or more available actions.

Definition: *Constraints*—(1) Limitations or restrictions on selecting an alternative or on a selected alternative's range of outcome values. (2) Decision constraints are a set of rules for acceptable values of decision variables.

Definition: *Risk* is observed in those situations in which the potential outcomes can be described by well-known probability distributions.

Definition: *Criteria* are acceptable values of decision variables.

Definition: An *outcome* of a decision is the realized value of the decision criterion or criteria vector.

Definition: *Stakeholders* are all persons impacted by or with interest in the outcome of a decision.

Influence diagrams are constructed from the outcome backward through the variables, constraints, and decisions necessary to achieve the outcome.

Step 1.   Define the problem: Capital budget management—maximize the weighted rate of return > MARR. Working capital management—maximize the net present worth of cash flows from current fixed assets for the required MARR.

Step 2.   Identify the outcome and the random independent variables that directly determine its value.

>   2(a)   Identify each stakeholder impacted by or with interest in the outcome.
>
>   2(b)   Identify the constraints on the values of each independent variable.

Step 3.   Identify the decision node that determines each random independent variable and its probability of occurrence.

Step 4.   Continue decomposition for each random variable and decision node until no variable or decision remains.

Influence diagram symbols are illustrated in Fig. 15.7.

The elementary risky decision is maximizing the net present worth of working capital cash flows for a required MARR is investment in a single asset. For this decision, there is one decision to make; one uncertain event, the variation is actual cash flows versus predicted cash flows; and one stakeholder, management. The influence diagram for this decision is illustrated in Fig. 15.8.

The decision node is to invest or not invest given the net present worth of future annual revenue and expense cash flows and net cash receipt from market value or

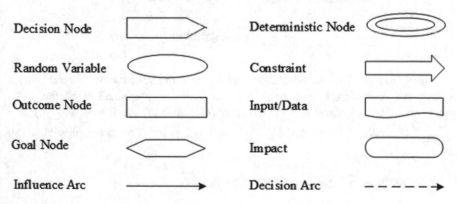

Fig. 15.7 Influence diagram symbols

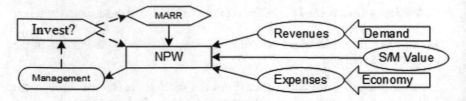

Fig. 15.8 Single asset investment, working capital influence diagram

salvage value upon disposal of the asset at the end of the project life. Operating annual revenue cash flows are constrained by demand for the asset's product or service, which, in turn, is constrained by customer preferences and demand-side business and macroeconomic factors. Hence, annual revenue cash flows can be modeled as a regression equation.

$$\text{Revenue}(t) = \beta_{R(0,t)} + \beta_{C,t}\text{CustPref}(t) + \sum_{j} \beta_{R(j)}\text{BusFactor}(j, t)$$

$$+ \sum_{k} \beta_{R(k)}\text{MacFactor}(k, t) + \varepsilon_{R(j,k,t)}$$

This regression representation incorporates only a weighted sum of customer preferences CustPref(t), business economic factors main effects BusFactor(t), and macroeconomic factors main effects MacFactor(t). Possible two-factor interactions are omitted. The annual revenue error term is a linear combination of the customer preference, business economic factors, and macroeconomic factors' variances $\varepsilon_{R(j,k,t)} = \sum_{j,k,t} \left((\text{Actual}_R(j, k, t) - E_R[X_{j,k,t}])^2 + \sigma^2_{R(j,k,t)} \right)$. Correspondingly, the annual operating expense cash flows are constrained by supply-side business and macroeconomic factors and can be represented as a regression equation.

$$\text{Expense}(t) = \beta_{E(0,t)} + \sum_{j} \beta_{E(j)}\text{BusFactor}(j, t)$$

$$+ \sum_{k} \beta_{E(k)}\text{MacFactor}(k, t) + \varepsilon_{E(j,k,t)}$$

Again, only business economic factors and macroeconomic factors' main effects are considered. The error term is a linear combination of the business economic factors and macroeconomic factors' variances $\varepsilon_{E(j,k,t)} = \sum_{j,k,t} \left((\text{Actual}_R(j, k, t) - E_E[X_{j,k,t}])^2 + \sigma^2_{E(j,k,t)} \right)$. Investment net present worth is

$$\text{NPW} = \sum_{n} \text{Revenue}(t)(P/F, \text{MARR}, t)$$

$$+ \sum_n \text{Expenses}(t)(P/F, \text{MARR}, t) + \varepsilon_{R(j,k,n)} + \varepsilon_{E(j,k,n)}$$

where Revenue(t) and Expenses(t) are the realized revenues and expenses, and the sum of the variances $\varepsilon_{R(j,k,t)}$ and $\varepsilon_{E(j,k,t)}$ results in the linear combination of additive variances.

15.4   Sensitivity Analyses

Section 10.1 introduced sensitivity analysis in the context of incremental analysis and applied sensitivity analysis to determine the ranges over which alternatives cash flows maximized NPW or EUAW. Given an influence diagram structured representation of a decision problem, sensitivity analysis allows an engineering manager to determine the range over which an optimal solution holds for variation in some influential predictor variable.

Definition: *Sensitivity analysis* is a determination of the amount of variation ($\pm$) in an objective function predictor variable or constraint necessary to change the decision to select a particular alternative. The point at which the decision changes from one alternative to another is the *point of indifference*.

Figure 15.8 illustrates input variables annual revenues and expenses and the market and salvage value of a zero-budget period or life cycle budget period investment in which the objective is to maximize cash flow (NPW). Input variables are those over which management can exercise some level of control. The figure also illustrates constraints. Constraints are conditions or limitations on input variables that affect the latter's average or variance. Management can exercise limited or no control over constraints and must respond to their effects on input variable and the resultant objective function (NPW in Fig. 15.8). Sensitivity analysis informs management of the points of indifference at which the decision changes and management response is required.

Assuming random errors are identically and independently normally distributed, $N(E[X], \sigma^2)$, the expected mean NPW of an investment alternative is

$$E[\text{NPW}] = \sum_n E[\text{Revenue}(t)](P/F, \text{MARR}, t)$$

$$+ \sum_n E\big[\text{Expenses}(t)\big](P/F, \text{MARR}, t)$$

Hence, the expected NPW of an investment alternative is just its accepted predicted value. The variance of an investment alternative's NPW is obtained by applying the variance operator resulting in

$$V[\text{NPW}] = \sum_n \left(\text{Actual}_R(j, k, t) - E_R\left(X_{j,k,t}\right)\right)^2$$
$$+ \sum_n \left(\text{Actual}_E(j, k, t) - E_E\left(X_{j,k,t}\right)\right)^2.$$
$$+ \sigma^2_{R(j,k,t)} + \sigma^2_{E(j,k,t)}$$

Sensitivity analysis informs engineering management about variances from expected NPW due to bias variances $(\text{Actual}_R(j,k,t) - E_R[X_{j,k,t}])^2$ and $(\text{Actual}_E(j,k,t) - E_E[X_{j,k,t}])^2$.

Example 15.5

Following the six-step graphical incremental rate of return sensitivity analysis in Sect. 10.1, consider variances in competing investment alternatives. An engineering manager is considering two alternatives for improving throughput on a production line. Both alternatives have useful lives of 12 years to end of the expected product life.

Alternative 1: A partial reconfiguration will cost $7,200,000 and require annual maintenance expenses of $120,000. During construction in year one, disruption to production is expected to cost $720,000. The reconfiguration is predicted to save $1,480,000 annually.

Alternative 2: A complete reconfiguration will cost $9,600,000. The completely reconfigured line will require $60,000 annual maintenance expenses. This construction is completely disruptive shutting down the production line at a cost of $1,730,000 the first year. Once complete, the full reconfiguration is predicted to save $1,960,000 annually.

Although construction and disruption costs are certain due to their short-term prediction, annual saving for each alternative could vary ±5% and annual maintenance expenses could vary ±10%. The MARR for this investment is 12.0%. Estimate the expected NPW for each alternative, and perform sensitivity analyses for the expected variances in annual savings and maintenance expenses.

Expected NPW:
Alternative 1:

$$E[\text{NPW}] = (-\$7,200,000) + (-\$720,000)(P/F, 12\%, 1)$$
$$+ (\$1,480,000 - \$120,000)(P/F, 12\%, 12)$$

$$E[\text{NPW}] = (-\$7,200,000) + (-\$720,000)(0.8929)$$
$$+ (\$1,480,000 - \$120,000)(6.194)$$

$$E[\text{NPW}] = (-\$7,200,000) + (-\$642,888) + \$8,423,840$$

$$E[NPW] = \$580,952$$

Alternative 2:

$$E[NPW] = (-\$9,600,000) + (-\$1,730,000)(P/F, 12\%, 1)$$
$$+ (\$1,960,000 - \$60,000)(P/F, 12\%, 12)$$

$$E[NPW] = (-\$9,600,000) + (-\$1,730,000)(0.8929)$$
$$+ (\$1,960,000 - \$60,000)(6.194)$$

$$E[NPW] = (-\$9,600,000) + (-\$1,544,717) + \$11,768,600$$

$$E[NPW] = \$623,888 \rightarrow \text{Prefer Alternative 2.}$$

Since Alternative 2 is preferred, consider the case where Alternative 2's annual savings decreases 5% to $1,862,000 but Alternative 1's remains at $1,480,000.
Alternative 2:

$$E[NPW] = (-\$9,600,000) + (-\$1,730,000)(P/F, 12\%, 1)$$
$$+ (\$1,862,000 - \$60,000)(P/F, 12\%, 12)$$

$$E[NPW] = (-\$9,600,000) + (-\$1,730,000)(0.8929)$$
$$+ (\$1,862,000 - \$60,000)(6.194)$$

$$E[NPW] = (-\$9,600,000) + (-\$1,544,717) + \$11,161,588$$

$$E[NPW] = \$16,871 \quad \text{Now prefer Alternative 1 with NPW} = \$580,952$$

Now, assume Alternative 2's savings remains at $1,960,000 but its annual maintenance cost increases 10% to (−$66,000).
Alternative 2:

$$E[NPW] = (-\$9,600,000) + (-\$1,730,000)(P/F, 12\%, 1)$$
$$+ (\$1,960,000 - \$60,000)(P/F, 12\%, 12)$$

$$E[NPW] = (-\$9,600,000) + (-\$1,730,000)(0.8929)$$

$$+ (\$1,960,000 - \$66,000)(6.194)$$

$$E[NPW] = (-\$9,600,000) + (-\$1,544,717) + \$11,731,436$$

$$E[NPW] = \$586,719 \quad \text{Prefer Alternative 2.}$$

Now, we know that change in either alternative's annual savings is the variance driver in the decision point of indifference. Evaluating different scenarios in % change annual savings and maintenance expenses in Excel® yields the following approximate points of indifference and decisions.

Expected NPW	$581,492	$624,668				
Alternative 1			**Alterative 2**			
Ann Savings	Ann M Expense	NPW	Ann Savings	Ann M Expense	NPW	Decision
0.0%	0.0%	$581,492	0.0%	0.0%	$624,668	2
0.00%	-5.81%	$624,679	0.0%	0.0%	$624,668	1
0.50%	0.35%	$624,729	0.0%	0.0%	$624,668	1
1.00%	6.52%	$624,704	0.0%	0.0%	$624,668	1
1.50%	12.69%	$624,679	0.0%	0.0%	$624,668	1
0.00%	0.00%	$581,492	0.0%	11.61%	$581,518	2
0.00%	0.00%	$581,492	-0.5%	-4.72%	$581,506	2
0.00%	0.00%	$581,492	-1.00%	-21.05%	$581,493	2
0.00%	0.00%	$581,492	-1.50%	-37.39%	$581,518	2

If NPW is replaced with RoR, Fig. 15.7 represents the generic elements of a capital budgeting decision in which the objective is to maximize rate of return over the business-cycle budget period. Sensitivity analysis can be performed for the internal rate of return > MARR at NPW = EUAW = \$0 on capital budgeting projects. Investment rate of return is

$$IRR(NPW = \$0) = \sum_n Revenue(t)(P/F, i, t)$$

$$+ \sum_n Expenses(t)(P/F, i, t) + \varepsilon_{R(j,k,n)} + \varepsilon_{E(j,k,n)}$$

Again, assuming random errors are identically and independently normally distributed, $N(E[X], \sigma^2)$, the expected $IRR[NPW = \$0]$ of an investment alternative is

$$IRR[NPW = \$0] = \sum_n E[Revenue(t)](P/F, i, t)$$

$$+ \sum_n E[\text{Expenses}(t)](P/F, i, t)$$

with variance

$$V[\text{IRR}] = \sum_n \left(\text{Actual}_R(j, k, t) - E_R(X_{j,k,t})\right)^2$$

$$+ \sum_n \left(\text{Actual}_E(j, k, t) - E_E(X_{j,k,t})\right)^2$$

$$+ \sigma^2_{R(j,k,t)} + \sigma^2_{E(j,k,t)}$$

Example 15.6 illustrates rate of return sensitivity analysis for the simple case of a portfolio of two capital projects.

Example 15.6

For the capital budget of $1,250,000 in Example 13.1, conduct sensitivity analysis to determine the increase in rejected projects 2, 6, and 7 after-tax benefit cash flows necessary for alternatives to achieve the 15.8% rate of return of project 4 (last accepted project) and be admitted as acceptable alternatives. Given that projects 2, 6, and 7 become acceptable at a 15.8% rate of return, how would the $1,250,000 be re-allocated to maximize value to the organization?

Determine (P/A, 15.8%, 4), (P/A, 15.8%, 6), (P/A, 15.8%, 10), (P/F, 15.8%, 4), (P/F, 15.8%, 6), and (P/F, 15.8%, 10) by linear interpolation.

$$(P/A, 15.8\%, 4) = (P/A, 15\%, 4) + [(P/A, 18\%, 4) - (P/A, 15\%, 4)]$$
$$[(15.8\% - 15\%)/(18\% - 15\%)]$$

$$(P/A, 15.8\%, 4) = 2.855 + [2.690 - 2.855][0.26666] = 2.811$$

$$(P/A, 15.8\%, 6) = (P/A, 15\%, 6) + [(P/A, 18\%, 6)$$
$$- (P/A, 15\%, 6)][0.26666]$$
$$(P/A, 15.8\%, 6) = 3.784 + [3.498 - 3.784][0.26666] = 3.708$$

$$(P/A, 15.8\%, 10) = (P/A, 15\%, 10) + [(P/A, 18\%, 10)$$
$$- (P/A, 15\%, 10)][0.26666]$$

$$(P/A, 15.8\%, 6) = 5.019 + [4.494 - 5.019][0.26666] = 4.879$$

$$(P/F, 15.8\%, 4) = (P/F, 15\%, 4) + [(P/F, 18\%, 4)$$
$$- (P/F, 15\%, 4)][0.26666]$$
$$(P/F, 15.8\%, 4) = 0.5718 + [0.5158 - 0.5718]$$
$$[0.26666] = 0.5569$$

$$(P/F, 15.8\%, 6) = (P/F, 15\%, 6) + [(P/F, 18\%, 6)$$
$$-(P/F, 15\%, 6)][0.26666]$$
$$(P/F, 15.8\%, 6) = 0.4323 + [0.3704 - 0.4323]$$
$$[0.26666] = 0.4158$$

$$(P/F, 15.8\%, 10) = (P/F, 15\%, 10) + [(P/F, 18\%, 10)$$
$$- (P/F, 15\%, 10)][0.26666]$$
$$(P/F, 15.8\%, 10) = 0.2472 + [0.1911 - 0.2472]$$
$$[0.26666] = 0.2322$$

Manual solutions—Set projects 2, 6, and 7 cash flow equations equal to that of project 6 and at 15.6% and solve for the unknown after-tax benefit cash flow. For project 4, 15.8% = IRR; therefore, NPW(Project 4) = $0.
Project 2 = Project 4

$$(-\$400) + \text{ATCF}(P/A, 15.8\%, 10) = (-\$200) + \$40(P/A, 15.8\%, 6)$$
$$+ \$125(P/F, 15.8\%, 6)$$

$$(-\$400) + \text{ATCF}(4.879) = (-\$200) + \$40(3.708)$$
$$+ \$125(0.4158)(-\$400) + \text{ATCF}(4.879) = \$0$$

ATCF = $400/4.879 = $81.98 increase from $79.70 or delta = $2.190 K
New capital budget with project 2 replacing project 4: $1150 − $200 + $400 = $1350.
Over the set budget = $1250. Not feasible.
Project 6 = Project 4

$$(-\$300) + \text{ATCF}(P/A, 15.8\%, 10) + \$150(P/F, 15.8\%, 10) = \$0$$
$$(-\$300) + \text{ATCF}(4.879) + \$150(0.2322) = \$0$$

ATCF = ($300 − $34.83)/4.879 = $54.349 increase from $54.00 or delta = $0.349 K.

New capital budget with project 6 replacing project 4: $1150 − $200 + $300 = $1250.

On the set budget = $1250. Feasible.

$$NPW = (-\$300) + \$54.35\,(P/A, 12.0\%, 10)$$
$$+ \$150\,(P/F, 12.0\%, 10)$$
$$NPW = (-\$300) + \$54.35\,(5.650) + \$150\,(0.3220)$$
$$= \$55.38 \text{ New Portfolio}$$
$$NPW = \$267.78 - \$27.79 + \$55.38$$
$$= \$295.37;\ \text{net increase} + \$27.59\,K$$

Project 7 = Project 4

$$(-\$600) + ATCF\,(P/A, 15.8\%, 4) + \$50\,(P/F, 15.8\%, 4) = \$0$$
$$(-\$600) + ATCF\,(2.811) + \$50\,(0.5569) = \$0$$

ATCF = ($600 − $27.845)/2.811 = $203.54 increase from $189.28 or delta = $14.260 K

New capital budget with project 7 replacing project 4. $1150 − $200 + $600 = $1550.

Over the set budget = $1250. Not feasible.

15.5 Decision Analysis

Influence diagrams are excellent for representing the key elements of a decision; however, influence diagrams do not represent the risk structure of a decision. Decision trees are used to represent the probabilistic risk structure of decisions.

Definition: A *decision tree* is a static graphical representation of the probabilistic outcomes of a decision.

A decision tree represents the sequential relationships of decision nodes, random variables, and outcomes that make up a risky decision. A decision tree describes the problem by starting at the decision that must be made, the root node, and adding chance and decision nodes in the proper logic sequence. Both root and leaf nodes contain questions or criteria to be answered. Branches are arrows connecting nodes, showing the flow from question to answer. Each node typically has two or more nodes extending from it. Best practices for creating a decision tree are:

- **Start the decision tree**—draw the root decision node near the left edge of the drawing area. Write the question regarding the investment to be made and the economic criteria (net present worth, equivalent annual worth, or rate of return) to be used to make the decision.
- **State the goal**—above the root decision node, add a goal node with a statement of the investment objective in terms of the economic criteria (i.e., NPW or EUAW > 0 at MARR = 12%, RoR ≥ 14.5% opportunity cost of capital, etc.). Link the goal node to the root decision node with an arc (solid line).
- **Add branches**—moving to the right in time order from each node, draw an arc for each possible alternative path reflecting the decision logic of the investment's possible future cash flows. Each branch should reflect the decision sequence of chance nodes, deterministic nodes, intermediate decision and goal nodes, and outcomes that reflect the future cash flow outcomes of the investment.
- **Mutual exclusivity and exhaustivity**—all decision and chance nodes with their associated probabilities must be mutually exclusive and exhaustive. Each decision node must have the number of arcs emanating from it equal to the number of possible decision routes from the node. Each chance node must have the number of arcs emanating from it equal to the number of possible outcome nodes. Each chance node arc must state the probability of its respective outcome, and the sum of the probabilities across all chance node arcs must equal 1.0.

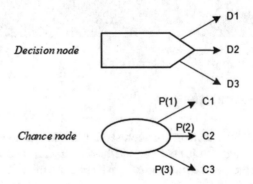

- **Terminate in outcome nodes**—decision logic branch must terminate in an outcome node.

- **Add a timeline**—each chance node, deterministic node, intermediate decision and goal node, and outcome node must be associated with its time interval of occurrence from 0 to n periods of the investment or project life.

A rudimentary decision tree example is illustrated in Fig. 15.8.

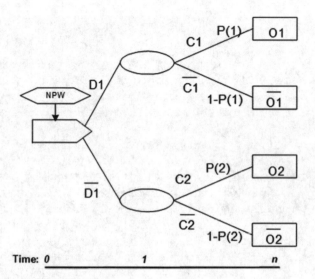

Time: 0   1   n

To solve a decision tree, work backward from the outcome nodes at time n to the decision root node.

- Add predicted cash flows at each intermediate and time n outcome node. For a NPW criteria goal, restate each predicted future cash flow to its present worth at the MARR. For an EUAW or a rate of return criteria goal, use the predicted future cash flows.
- Add cash flow requirements to each intermediate decision node.
- Add mutually exclusive and exhaustive $P(\text{outcome}(i))$ to each branch from a chance node. $\text{Sum}(P(\text{outcome}(i)) = 1.0$.
- Starting at each time n outcome node, work backward through the decision logic structure estimating either expected present worth cash flows at each time $t = 1$, 2, ..., n for a NPW criteria goal or expected monetary cash flows for an EUAW or a rate of return criteria goal. For each chance node, the expected cash flow is estimated as

$$E[\text{CF}] = \sum_n P(\text{state}_i) \times \text{CF}_i$$

- Select the decision network path that maximizes the expected NPW of future present worth cash flows at the root decision node. For and EUAW or a rate of return goal criteria, select the decision network path that maximizes the expected monetary value at the root decision node, and then estimate the life EUAW or internal rate of return for the cash flows along that decision network path.

Example 15.7

Management of a building supply manufacturer is deciding on whether to open a new warehouse outlet for local building contractors. There are two options for the new warehouse. A warehouse building can be erected for a cost $8,750,000, and new warehousing fixtures and equipment and office fixtures and equipment can be purchased for $3,500,000. Option two is to purchase an existing empty warehouse for $9,450,000 with the building market value $7,000,000 and equipment $2,450.000. Due to logistics of the existing warehouse location, expected Real$ revenues will be 90% of those set forth in the following table. Either way, the building, fixtures, and equipment will be depreciated under MACRS beginning the first day of the next fiscal year. Forecasted annual cash flow net revenues (revenues–operating expenses–maintenance expenses) are set forth in the following table along with their probabilities of occurrence. If demand meets or exceeds expectations, the outlet will remain in service indefinitely. If demand falls below expectations, outlet operations will be terminated at the end of eight years. All chase flows are in Real$, and the market MARR for this project is 12.0%. The organization operates in a state with 6.0% corporate income tax rate. Based on expected value, which alternative should be pursued?

Demand	Predicted annual cash flow revenue		
	Low	Expected	High
P(demand)	0.3	0.6	0.1
Revenue	$3,050,000	$3,660,000	$5,850,000

For each warehouse alternative, estimate:

- Building depreciation—year 1, initial cost $\times$ 0.02461; year 2, initial cost $\times$ 0.02564.
- Equipment and fixture depreciation—initial cost $\times$ GDS 7-year property percentages.

- Combined income tax rate $= 0.06 + 0.21(1 - 0.06) = 0.2574$ or 25.74%.
- Taxable income $=$ CF Revenue $-$ Building Depreciation $-$ Equipment Depreciation
- Taxes paid $=$ Taxable income $\times 0.2574$
- After-tax Cash Flow $=$ CF Revenue $-$ Taxes paid
- NPW $=$ (Building cost $+$ Equipment cost) $+$ Sum(PV(After-tax Cash Flows, 12.0%, n).

The eight-year NPW estimates are set forth from the associated Excel® worksheet.

New Warehouse - Low Demand

Year	Building Cost	Equipment	CF Revenue	Build Depr	Eq Depr	Taxable Income	Income Tax Paid	After Tax Cash Flow	NPW
0	($8,750,000)	($3,500,000)						($12,250,000)	($100,376)
1			$3,050,000	($215,338)	($500,000)	$2,334,663	($600,942)	$2,449,058	
2			$3,050,000	($224,350)	($857,143)	$1,968,507	($506,694)	$2,543,306	
3			$3,050,000	($224,350)	($612,245)	$2,213,405	($569,730)	$2,480,270	
4			$3,050,000	($224,350)	($437,318)	$2,388,332	($614,757)	$2,435,243	
5			$3,050,000	($224,350)	($312,370)	$2,513,280	($646,918)	$2,403,082	
6			$3,050,000	($224,350)	($312,370)	$2,513,280	($646,918)	$2,403,082	
7			$3,050,000	($224,350)	($312,370)	$2,513,280	($646,918)	$2,403,082	
8			$3,050,000	($224,350)	($156,185)	$2,669,465	($687,120)	$2,362,880	

New Warehouse - Expected Demand

CF Revenue	Taxable Income	Income Tax Paid	After Tax Cash Flow	NPW
			($12,250,000)	$2,149,895
$3,660,000	$2,944,663	($757,956)	$2,902,044	
$3,660,000	$2,578,507	($663,708)	$2,996,292	
$3,660,000	$2,823,405	($726,744)	$2,933,256	
$3,660,000	$2,998,332	($771,771)	$2,888,229	
$3,660,000	$3,123,280	($803,932)	$2,856,068	
$3,660,000	$3,123,280	($803,932)	$2,856,068	
$3,660,000	$3,123,280	($803,932)	$2,856,068	
$3,660,000	$3,279,465	($844,134)	$2,815,866	

New Warehouse - High Demand

CF Revenue	Taxable Income	Income Tax Paid	After Tax Cash Flow	NPW
			($12,250,000)	$10,228,738
$5,850,000	$5,134,663	($1,321,662)	$4,528,338	
$5,850,000	$4,768,507	($1,227,414)	$4,622,586	
$5,850,000	$5,013,405	($1,290,450)	$4,559,550	
$5,850,000	$5,188,332	($1,335,477)	$4,514,523	
$5,850,000	$5,313,280	($1,367,638)	$4,482,362	
$5,850,000	$5,313,280	($1,367,638)	$4,482,362	
$5,850,000	$5,313,280	($1,367,638)	$4,482,362	
$5,850,000	$5,469,465	($1,407,840)	$4,442,160	

Existing Warehouse - Low Demand

Year	Building Cost	Equipment	CF Revenue	Build Depr	Eq Depr	Taxable Income	Income Tax Paid	After Tax Cash Flow	NPW
0	($7,000,000)	($2,450,000)						($9,450,000)	$1,333,488
1			$2,745,000	($172,270)	($350,000)	$2,222,730	($572,131)	$2,172,869	
2			$2,745,000	($179,480)	($600,000)	$1,965,520	($505,925)	$2,239,075	
3			$2,745,000	($179,480)	($428,571)	$2,136,949	($550,051)	$2,194,949	
4			$2,745,000	($179,480)	($306,122)	$2,259,398	($581,569)	$2,163,431	
5			$2,745,000	($179,480)	($218,659)	$2,346,861	($604,082)	$2,140,918	
6			$2,745,000	($179,480)	($218,659)	$2,346,861	($604,082)	$2,140,918	
7			$2,745,000	($179,480)	($218,659)	$2,346,861	($604,082)	$2,140,918	
8			$2,745,000	($179,480)	($109,329)	$2,456,191	($632,223)	$2,112,777	

Existing Warehouse - Expected Demand

CF Revenue	Taxable Income	Income Tax Paid	After Tax Cash Flow	NPW
			($9,450,000)	$3,358,732
$3,294,000	$2,771,730	($713,443)	$2,580,557	
$3,294,000	$2,514,520	($647,237)	$2,646,763	
$3,294,000	$2,685,949	($691,363)	$2,602,637	
$3,294,000	$2,808,398	($722,882)	$2,571,118	
$3,294,000	$2,895,861	($745,395)	$2,548,605	
$3,294,000	$2,895,861	($745,395)	$2,548,605	
$3,294,000	$2,895,861	($745,395)	$2,548,605	
$3,294,000	$3,005,191	($773,536)	$2,520,464	

Existing Warehouse - High Demand				
	Taxable	Income	After Tax	
CF Revenue	Income	Tax Paid	Cash Flow	NPW
			($12,250,000)	$7,829,690
$5,265,000	$4,742,730	($1,220,779)	$4,044,221	
$5,265,000	$4,485,520	($1,154,573)	$4,110,427	
$5,265,000	$4,656,949	($1,198,699)	$4,066,301	
$5,265,000	$4,779,398	($1,230,217)	$4,034,783	
$5,265,000	$4,866,861	($1,252,730)	$4,012,270	
$5,265,000	$4,866,861	($1,252,730)	$4,012,270	
$5,265,000	$4,866,861	($1,252,730)	$4,012,270	
$5,265,000	$4,976,191	($1,280,871)	$3,984,129	

Decision Tree:

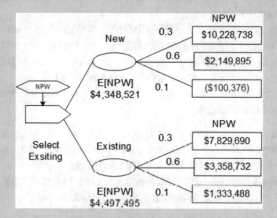

$$E[\text{NPW New}] = 0.3(\$10,228,738) + 0.6(\$2,149,895)$$
$$+ 0.1(-\$100,376) = \$4,348,521$$

$$E[\text{NPW Existing}] = 0.3(\$7,829,690) + 0.6(\$3,358,732)$$
$$+ 0.1(\$1,333,488) = \$4,497,495$$

Select the existing warehouse with expected NPW = $4,497,495.

15.6 Decision Tree Risk Simulation

In application, many decision tree variables are subject to random variation. The combinatorics of modeling even simple decision trees with random variables, as in the prior example, would create a bush of decision paths. The goal of any decision problem is to estimate expected outcomes (NPW, EUAW, IRR, etc.). Risk analysis simulations allow the estimation of expected outcomes within a range of prediction error.

A decision tree risk simulation is a mathematical model (a system of formulas) that specifies how the structural and random components of an economic decision change over time. Decision tree risk simulation models are used to investigate a wide variety of "what if" questions about risky decisions.

- The long-run expected average values of variables and metrics.
- The prediction error of the expected average values.
- The points at which a decision changes from one alternative to another are due to random variation (we examined structural points of indifference in Sect. 10.0 graphical sensitivity analysis).
- The breakeven points where NPW or EUAW ≤ 0 given a stated MARR, or the points where IRR $\leq$ MARR or stockholders' required return on equity.
- How changes in input factor or variable values affect the expected values and prediction error of output metrics (NPW, EUAW, or RoI).
- Which input factors or variables dominate (are probabilistically important components of) the decision.

To decide which of the alternatives is preferred, the engineering analyst assigns random probability estimates to chance node probabilities and random cash flow estimates (NPW, EUAW, RoI, etc.) to outcome nodes representing their structural and random values based on the best available information. In Excel, uniform distribution probabilities can be generated using the following functions.

=RAND() returns uniformly distributed random numbers between 0 and 1.

=RANDBETWEEN(bottom, top) returns a random integer number between the bottom integer and the top integer numbers.

Simulation of chance node probabilities requires observing their rank order. As an example, if a chance node has three outcomes $P(\text{low}) = 0.2$, $P(\text{expected}) = 0.5$, and $P(\text{high}) = 0.3$, then the rank order of Rank(low) < Rank(high) < Rank(expected) must not be violated in random number generation. To maintain rank order,

1. Set $T() = P() \times 100$.
2. Beginning with lowest Rank(1), assign $V(1) = \text{RANDBETWEEN}(1, \text{ROUND}((T(2) + T(1))/2,0))$.
3. For each intermediate Rank(2, ..., Max − 1), assign $V(2+) = \text{RANDBETWEEN}(V(1+) + 1, \text{ROUND}(((T(2+) + T(3 +))/2 - \text{SUM}(V1+)),0))$.
4. Repeat step 3 until all $V(2, ..., \text{Max} - 1)$ have been assigned. Assign $V(\text{Max}) = 1 - \text{Sum}(V1(), V2(), ..., V(\text{Max} - 1))$.
5. Assign random chance probabilities $\text{Pc}(1) = V(1)/100$, $\text{Pc}(2) = V(2)/100$, ..., $P(\text{Max}) = V(\text{Max})/100$.

To generate cash flow, NPW, EUAW, or RoI random values, the specific population probability density function form must be known or estimated from sample data. The general form of the distribution may be approximated by plotting a histogram. If the data is reasonably symmetric, the normal, lognormal, or triangular distribution may provide a good distributional fit. If the data is positively skewed, the lognormal or triangular distribution may provide a good approximation of the fit. If the data is positively skewed with the mode at the minimum value, the exponential distribution may provide the best fit. If the data is negatively skewed, the triangular distribution may provide a good approximation of the fit. One of the parametric distribution fitting methods (probability plotting with the Anderson–Darling or Kolmogorov–Smirnov test, method of moments, maximum likelihood, etc.) will be required to determine the parameters of the probability mass function (covered in statistics textbooks). This discussion will assume that the probability density function is known or has been estimated from sample data.

In Microsoft® Excel, random values may be generated for a normal distribution using

=NORM.INV(probability, mean, standard deviation)

with RAND() used for the probability argument. Similarly, random values may be generated for a lognormal distribution by using

=LOGNORM.INV(probability, mean, standard deviation)

where RAND() used for the probability argument, mean = AVERAGE(LN(data)), standard deviation = STDEV(LN(data)).

Microsoft® Excel® does not have a comparable inverse function for the triangular distribution. Random values may be generated for a triangular distribution using the following observations about the triangular distribution. The area of the triangular distribution may be found as the sum of the areas of two right-angle distributions.

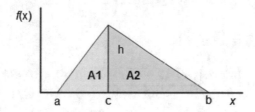

From basic geometry for any triangle, the area $= 1/2 \times$ base $\times$ height,

$$A1 = 1/2 \times (c-a) \times h \text{ and } A2 = 1/2 \times (b-c) \times h$$

Now, the total area $= A1 + A2$ or

$$A = 1/2 \times (c-a) \times h + 1/2 \times (b-c) \times h$$
$$= 1/2((c-a) + (b-c))h$$

Setting the $A = 1$, we get $h = 2/(b - a)$. Evaluating the A1 yellow triangle,

$$P(a < x \le c) = 1/2 \times (c-a) \times h = 1/2 \times (c-a)$$
$$\times 2/(b-a) = (c-a)/(b-a)$$

Evaluating the A2 blue triangle,

$$P(c < x \le b) = 1/2 \times (b-c) \times h = 1/2 \times (b-c)$$
$$4 \times 2/(b-a) = (b-c)/(b-a)$$

From the above figure of the triangular distribution and utilizing conformity of the triangles, for $a \le x \le c$,

$$P(X \le x) = \left(\frac{x-a}{c-a}\right)^2 \left(\frac{c-a}{b-a}\right)$$

Taking the derivative with respect to x,

$$f(x) = \frac{x-a}{c-a}\left(\frac{2}{b-a}\right)$$

From the complement rule for the case $c \le x \le b$

$$P(X \le x) = 1 - \left(\frac{b-x}{b-c}\right)^2 \left(\frac{b-c}{b-a}\right)$$

Taking the derivative with respect to x,

$$f(x) = \frac{b-x}{b-c}\left(\frac{2}{b-a}\right)$$

The inverse triangle cdf of X is

$$F^{-1}(x) = \begin{cases} a + \sqrt{U(b-a)(c-a)} & \text{for} 0 < U < F(c) \\ b - \sqrt{(1-U)(b-a)(b-c)} & \text{for} F(c) \le U < 1 \end{cases}$$

where U is the standard $U(0, 1)$ distribution. The inverse triangle cdf of X can be incorporated into the cells of an Excel® spreadsheet as follows.

Parameter	Value	Assumed cell
Lower limit	a	A1
Most likely	c	A2
Upper limit	B	A3
$U(0, 1)$	U	A4

(continued)

(continued)

Parameter	Value	Assumed cell
$X = x$	x	See formula below
$= \text{ROUND(IF}(A4 \le (A2 - A1)/(A3 - A1), A1 + \text{SQRT}(A4 * (A3 - A1) * (A2 - A1)), A3 - \text{SQRT}(1 - A4) * (A3 - A1) * (A3 - A2))),0)$		

Similarly, Microsoft® Excel® does not have a comparable inverse function for the exponential distribution. Random values may be generated for an exponential distribution using the following observations about the exponential distribution.

Exponential probability density function:

$$f(x) = \lambda e^{-\lambda}$$

Exponential cumulative density function:

$$F(x) = 1 - e^{-\lambda x}$$

The inverse exponential cdf of X is

$$F^{-1}(x) = -\lambda Ln(1 - x)$$

where $0 < x \le 1$. The inverse exponential cdf of X can be incorporated into the cells of an Excel® spreadsheet as follows.

Parameter	Value	Assumed cell
λ	Average (data)	A1
U	$U(0, 1)$	
$X = x$	x	See formula below
$= -A1 * \text{LN(1-RAND())}$		

Decision tree simulations are iterative in that the initial decision tree values are randomly altered, outcome economic criteria (NPW, EUAW, or RoI) are recorded, the altered decision tree becomes the decision outcomes under study, and the cycle repeats until a sufficient number of random criteria values are recorded to permit sensitivity analysis. Figure 15.9 presents a schematic of an economic decision tree simulation process flow.

1. Identify the Economic Investment Problem—This was the central discussion of the first eleven chapters of this text: new product, new process, upgraded product or process, new facility or facility expansion, alternative investment comparisons (incremental analysis), replacement analysis, etc.

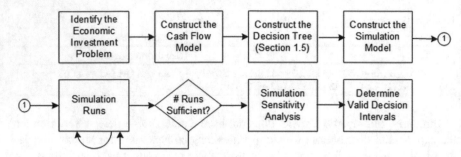

Fig. 15.9 Decision tree simulation process flow

2. Construct the Cash Flow Model—A mathematical model that estimates the current value of an investment based on its expected future cash flows.
3. Construct the Decision Tree—The static graphical representation of the probabilistic outcomes of an investment decision.
4. Construct the Simulation Model—Translation of the graphical decision tree structure into a discounted cash flow model that can be executed in the selected simulation software application.
5. Simulation Runs—Determine the number of runs necessary to achieve a required statistical prediction interval (subject of statistical or simulation textbooks) and conduct those runs. After each run, record the values of economic criteria (NPW, EUAW, or RoI).
6. Simulation Sensitivity Analysis—Given the observed random variation in the economic criteria, determine the statistical prediction interval and points of indifferences in interval unions where decision changes occur.

Example 15.8

Conduct a simulation sensitivity analysis of the warehouse investment decision in Example 15.8 using the data in the following table. Revenues are normally distributed with E[Revenue] means and Stdev[Rev] standard deviations in the following table. Conduct 100 simulation runs for each of the following scenarios: (a) P(Demand) random variation only with E[Revenue] held constant. (b) Revenue random variation using $N(E$[Revenue], Stdev[Rev]) population values with P(Demand) held constant.

	Predicted annual cash flow revenue		
Demand	Low	Expected	High
P(Demand)	0.3	0.6	0.1
E[Revenue]	$3,050,000	$3,660,000	$5,850,000
Stdev[Rev]	$122,000	$190,000	$558,000

P(Demand) simulation

Add a demand simulation matrix of cells to the worksheet, and encode the rank order simulation functions into cells (relative to the worksheet example below).

	fx	=J$16/SUM($H$16:$J$16)						
C	D	E	F	G	H	I	J
	Predicted Annual Cash Flow Revenue				Demand Simulation		
Demand	Low	Expected	High		Low	Expected	High
P(Demand)	0.3	0.6	0.1	T	30	60	10
E[Revenue]	$3,050,000	$3,660,000	$5,850,000	V	25	55	20
Stdev[Rev]	$122,000	$190,000	$558,000	P	0.25	0.55	0.2

Cell H15: $= \$D\$15 * 100$ Cell I15: $= \$E\$15 * 100$ Cell J15: $= \$F\$15 * 100$
In rank order:
Cell J16: $= \text{RANDBETWEEN}(1,\text{ROUND}((\$H\$15 + \$J\$15)/2,0))$
Cell H16: $= \text{RANDBETWEEN}(\$J\$16 + 1,\text{ROUND}(((\$I\$15 + \$H\$15)/2-\$J\$16),0))$
Cell I16: $= 100\text{-SUM}(\$H\$16,\$J\$16)$
Cell H17: $= \text{H}\$16/\text{SUM}(\$H\$16:\$J\$16)$
Cell I17: $= \text{I}\$16/\text{SUM}(\$H\$16:\$J\$16)$
Cell J17: $= \text{J}\$16/\text{SUM}(\$H\$16:\$J\$16)$
In the decision tree, reference the P(Demand) probability cells to the random demand cells.

	Predicted Annual Cash Flow Revenue				Demand Simulation		
Demand	Low	Expected	High		Low	Expected	High
P(Demand)	0.3	0.6	0.1	T	30	60	10
E[Revenue]	$3,050,000	$3,660,000	$5,850,000	V	14	75	11
Stdev[Rev]	$122,000	$190,000	$558,000	P	0.14	0.75	0.11
Decision Tree							
	Warehouse						
Decision	E[NPW]	P(Demand)	Outcome(CF) NPW				
	New	0.11 High	$10,228,738				
	$2,723,530	0.75 Expected	$2,149,895				
		0.14 Low	($100,376)				
Existing Warehouse							
	Existing	0.11 High	$7,829,690				
	$3,567,003	0.75 Expected	$3,358,732				
		0.14 Low	$1,333,488				

Run simulation samples (100 for this example) of the variable demand by clicking on any cell with a formula, clicking in the formula bar, repeatedly clicking on the √ button to generate a new random sample.

EXPON.DIST ▾ : × ✓ *fx* =$X$23

▲	A	B	C	D	E	F	G
45							
46			Decision Tree				
47				Warehouse			
48			Decision	E[NPW]	P(Demand)	Outcome(CF)	NPW
49							
50				New	0.12	High	=$X$23
51				$2,759,313	0.72	Expected	$2,149,895
52					0.16	Low	($100,376)
53			Existing Warehouse				
54				Existing	0.12	High	$7,829,690
55				$3,571,208	0.72	Expected	$3,358,732
56					0.16	Low	$1,333,488

Copy and paste each new simulated E[NPW] New and Existing to columns in the spreadsheet.

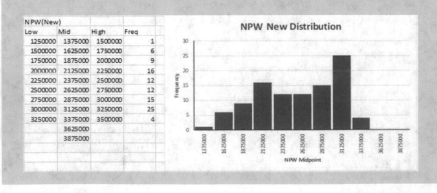

Plot histogram distributions of E[NPW|New], E[NPW|Existing], and E[NPW|Existing]—E[NPW|New], the last to test for negative differences indicating a change in decision from the existing warehouse to the new warehouse.

NPW(New)						
Low	Mid	High	Freq			
1250000	1375000	1500000	1			
1500000	1625000	1750000	6			
1750000	1875000	2000000	9			
2000000	2125000	2250000	16			
2250000	2375000	2500000	12			
2500000	2625000	2750000	12			
2750000	2875000	3000000	15			
3000000	3125000	3250000	25			
3250000	3375000	3500000	4			
	3625000					
	3875000					

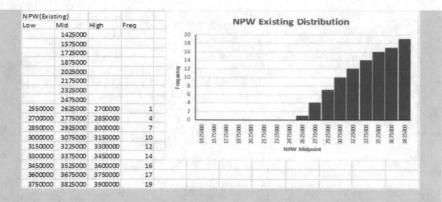

N PW(Existing)			
Low	Mid	High	Freq
	1425000		
	1575000		
	1725000		
	1875000		
	2025000		
	2175000		
	2325000		
	2475000		
2550000	2625000	2700000	1
2700000	2775000	2850000	4
2850000	2925000	3000000	7
3000000	3075000	3150000	10
3150000	3225000	3300000	12
3300000	3375000	3450000	14
3450000	3525000	3600000	16
3600000	3675000	3750000	17
3750000	3825000	3900000	19

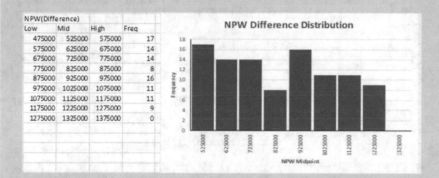

N PW(Difference)			
Low	Mid	High	Freq
475000	525000	575000	17
575000	625000	675000	14
675000	725000	775000	14
775000	825000	875000	8
875000	925000	975000	16
975000	1025000	1075000	11
1075000	1125000	1175000	11
1175000	1225000	1275000	9
1275000	1325000	1375000	0

Since the NPW difference distribution is always > \$0, E[NPW|Existing] > E[NPW|New] everywhere, and no random variation in the ranked proportion demand will change the decision from purchasing the existing warehouse to building the new warehouse.

Revenue simulation

Since revenue cash flows are determined by the mutually exclusive demand outcomes, conduct simulation runs for demand outcome (low, expected, and high). For low demand, set the cash flow revenue to

New Warehouse: = NORM.INV(RAND(),\$D\$16,\$D\$17)
Existing Warehouse: = NORM.INV(RAND(),\$D\$16*\$F\$7,\$D\$17).

where \$F\$7 = 90%

| F24 | | ▼ | : | × | ✓ | fx | =NORM.INV(RAND(),$D$16,$D$17) |

◢ A	B	C	D	E	F	G
12						
13			Predicted Annual Cash Flow Revenue			
14		Demand	Low	Expected	High	
15		P(Demand)	0.3	0.6	0.1	
16		E[Revenue]	$3,050,000	$3,660,000	$5,850,000	
17		Stdev[Rev]	$122,000	$190,000	$558,000	
18						
19	Analysis:	Eight Year NPW				
20		New Warehouse - Low Demand				
21						
22		Year	Building Cost	Equipment	CF Revenue	Build [
23		0	($8,750,000)	($3,500,000)		
24		1			$3,081,938	($21
25		2			$3,081,370	($22
26		3			$3,044,194	($22
32						
33		Existing Warehouse - Low Demand				
34						
35		Year	Building Cost	Equipment	CF Revenue	Build D
36		0	($7,000,000)	($2,450,000)		
37		1			$2,509,365	($172
38		2			$2,745,000	($179
39		3			$2,745,000	($179

Run simulation samples (100 for this example) of the variable revenue for demand = low by clicking on any cell with a formula, clicking in the formula bar, repeatedly clicking on the ✓ button to generate a new random sample.

Copy and paste each new simulated E[NPW] New and Existing to columns in the spreadsheet.

Repeat the simulation runs for new and existing warehouse–expected demand and new and existing warehouse–high demand.

Plot histogram distributions of

- NPW(New|Low) versus NPW(Exist|Low)
- NPW(New|Expected) versus NPW(Exist|Expected).

to assess for distribution overlaps indicating ranges of indifference (change in decision) and possibly dominance.

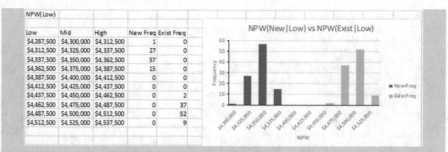

NPW(Low)

Low	Mid	High	New Freq	Exist Freq
$4,287,500	$4,300,000	$4,312,500	1	0
$4,312,500	$4,325,000	$4,337,500	27	0
$4,337,500	$4,350,000	$4,362,500	57	0
$4,362,500	$4,375,000	$4,387,500	15	0
$4,387,500	$4,400,000	$4,412,500	0	0
$4,412,500	$4,425,000	$4,437,500	0	0
$4,437,500	$4,450,000	$4,462,500	0	2
$4,462,500	$4,475,000	$4,487,500	0	37
$4,487,500	$4,500,000	$4,512,500	0	52
$4,512,500	$4,525,000	$4,537,500	0	9

For low demand with variable revenues, NPW(Existing|Low) dominates NPW(New|Low), and the decision will be to always invest in the existing warehouse.

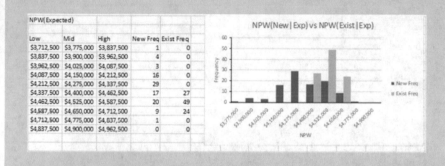

NPW(Expected)

Low	Mid	High	New Freq	Exist Freq
$3,712,500	$3,775,000	$3,837,500	1	0
$3,837,500	$3,900,000	$3,962,500	4	0
$3,962,500	$4,025,000	$4,087,500	3	0
$4,087,500	$4,150,000	$4,212,500	16	0
$4,212,500	$4,275,000	$4,337,500	29	0
$4,337,500	$4,400,000	$4,462,500	17	27
$4,462,500	$4,525,000	$4,587,500	20	49
$4,587,500	$4,650,000	$4,712,500	9	24
$4,712,500	$4,775,000	$4,837,500	1	0
$4,837,500	$4,900,000	$4,962,500	0	0

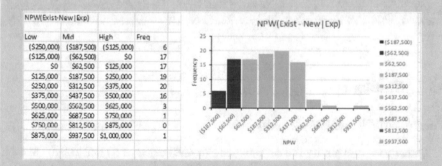

NPW(Exist-New|Exp)

Low	Mid	High	Freq
($250,000)	($187,500)	($125,000)	6
($125,000)	($62,500)	$0	17
$0	$62,500	$125,000	17
$125,000	$187,500	$250,000	19
$250,000	$312,500	$375,000	20
$375,000	$437,500	$500,000	16
$500,000	$562,500	$625,000	3
$625,000	$687,500	$750,000	1
$750,000	$812,500	$875,000	0
$875,000	$937,500	$1,000,000	1

For expected demand variable revenues less than about $4,756,000, invest in the new warehouse at $P(\text{New}|\$4,377,500 < \text{Exp}[\text{Rev}] < \$4,756,000) \sim 0.23$. Invest in the existing warehouse at $P(\text{Existing}|\$3,55,300 < E[\text{Rev}]) \sim 0.77$.

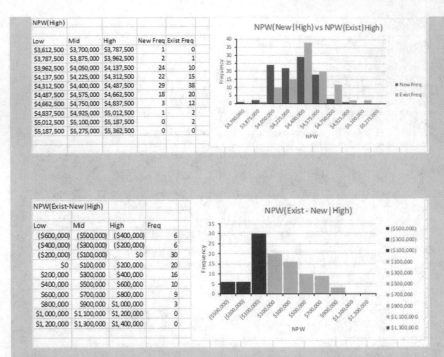

For high demand variable revenues less than about \$4,928,100, invest in the new warehouse at $P(\text{New}|\$3,612,500 < E[\text{Revenue}] < \$4,928,100) \sim 0.42$. Invest in the existing warehouse at $P(\text{Existing}|\$3,933,200 > E[\text{Revenue}]) \sim 0.58$.

15.7 Summary

Fundamental Probability Theory

Definition: Probability is a set function $P(x \in A)$ or $f(x \in A)$ that assigns to each event $x \in A$ in the sample space S a number $P(x \in A) = a$ or $f(x \in A) = a$, called the probability of event A, such that the following properties are satisfied:

Discrete distributions

Property 1: $P(x = A) \geq 0$
Property 2: $P(S = A + \overline{A}) = 1.0$
Property 3: If $A_1, A_2, \ldots, A_n$ are events $A_i \cap A_j = \emptyset$, then

$$P(A_1 \cup A_2 \cup \cdots \cup A_n) = P(A_1) + P(A_2) + \cdots + P(A_n)$$

Continuous distributions

Property 1: $f(X = A) > 0, A \in \mathfrak{R}$.
Property 2: $\int_{\mathfrak{R}} f(X)\, dx = 1.0$
Property 3: The probability of event $X \in A$ is $P(X \in A) = \int_A f(x)\, dx$.

Probability Theorems

$$P(A) = 1 - P(\bar{A})$$
$$P(\emptyset) = 0.$$

For each event A, $P(A) \le 1.0$
If A and B are any two events, $P(A \cup B) = P(A) + P(B) - P(A \cap B)$.
Note: If $A \cap B = \emptyset$, $P(A \cup B) = P(A) + P(B)$.
If any two events A and B are independent, $P(A \cap B) = P(A) \times P(B)$.

Discrete Probability Distributions

Definition: A ***random variable*** X is a numerical function specified over some sample space S that can take on only a countable number of values in S.
Definition: The ***probability distribution*** for a discrete random variable X is a table, graph, or formula that gives the probability $P(X = x)$ of each possible value of $X = x$.
Definition: Let X be a discrete random variable with probability distribution function $P(X = x)$. The ***mean*** or ***expected*** value of X is

$$\mu = E[X] = \sum_N x_i P(x_i)$$

where N = number of individuals in the population. In cases where all individuals of a population cannot be enumerated, the population mean is estimated from a random sample by the sample mean.

$$\bar{x} = \frac{\sum_n x_i}{n}$$

Definition: Let X be a discrete random variable with probability distribution function $P(X = x)$. The ***variance*** of X is

$$\sigma^2 = V[X] = \sum_N (x_i - \mu)^2 P(x_i)$$

where N = number of individuals in the population. In cases where all individuals of a population cannot be enumerated, the population variance is estimated from a random sample by the sample variance

$$s^2 = \frac{\sum_n (x_i - \overline{x})^2}{n - 1}$$

where $n - 1$ adjusts for the loss of 1 degree of freedom using x-bar to estimate the population mean μ, because $E[x\text{-bar}] = \mu$. The standard deviation of a random variable is defined as

$$s = \sqrt{s^2}$$

Bernoulli Distribution

$$p(x = 0, 1) = P^x (1 - p)^x$$

$$E[X] = p$$

$$\sigma^2 = p(1 - p)$$

Binomial Distribution

$$P(X = x) = \binom{n}{x} p^x (1 - p)^{n-x}$$

$$E[X] = np$$

$$\sigma^2 = np(1 - p)$$

Poisson Distribution

$$P(X = x) = \frac{\lambda^x e^{-\lambda}}{x!}$$

$$E[X] = \lambda$$

$$\sigma^2 = \lambda$$

Continuous Probability Distributions

Definition: A *continuous random variable* X has the following properties.

- The cumulative distribution function F
  X takes on an uncountable infinite number of values in the $\Re$ interval $(-\infty, \infty)$.
- (X) is continuous.
- The $P(X = x) = 0$.

Definition: The *cumulative distribution function* $F(x)$ for a random variable X is equal to the probability

$$F(x) = P(X \le x), -\infty < x < \infty$$

Definition: If $F(x)$ is the cumulative distribution function for a continuous random variable X, then the density function $f(x)$ for X is

$$f(x) = \frac{dF(x)}{dx}$$

Definition: Let X be a continuous random variable with density function $f(x)$. Then, the *expected value* or *mean* of X is

$$\mu = E[X] = \int_{-\infty}^{\infty} x f(x) dx$$

Definition: Let X be a continuous random variable with $E[X] = \mu$, Then, the *variance* of X is

$$\sigma^2 = \int_{-\infty}^{\infty} (x - \mu)^2 f(x) dx$$

Uniform Distribution

$$f(x) = \frac{1}{b - a}, a \le x \le b$$

$$f(x) = 0 \text{ elsewhere}$$

$$E[X] = \frac{a+b}{2}$$

$$\sigma^2 = \frac{(b-a)^2}{12}$$

Triangular Distribution

$$f(x) = \frac{(x-a)^2}{(b-a)(c-a)}, a \leq x \leq c$$

$$f(x) = 1 - \frac{(b-x)^2}{(b-a)(b-c)}, c \leq x \leq b$$

$$E[X] = \frac{a+b+c}{3}$$

$$\sigma^2 = \frac{a^2 + b^2 + c^2 - ab - ac - bc}{18}$$

Normal Distribution

$$f(x) = \frac{1}{\sigma\sqrt{2\pi}}e^{-\frac{1}{2}\left(\frac{x-\mu}{\sigma}\right)^2}$$

$$E[X] = \mu$$

$$V[X] = \sigma^2$$

The **standard normal variate** $N(0, 1)$ and the normal variate $N(\mu, \sigma^2)$ are related by

$$N(0, 1) = [N(\mu, \sigma) - \mu]/\sigma$$

Central Limit Theorem: If $x_1, x_2, \ldots, x_n$ are identically, independently distributed (IID) random variables from any unimodal distribution with mean μ and variance σ^2, and if $y = x_1 + x_2 + \cdots + x_n$, then the distribution of y

$$\frac{y - \sum_n \mu_i}{\sqrt{\sum_n \sigma_i^2}}$$

approaches $N(0, 1)$ as $n \to \infty$. Since y is the summation of x_i IID variables, the variance of the mean is

$$V[\mu] = V\left[\frac{y}{n}\right] = \frac{n\sigma^2}{n^2} = \frac{\sigma^2}{n}$$

Taking the square root yields the **standard error of the mean** SEM $= \sigma/\sqrt{n}$.

Linear Combinations: If $x_{i,j}$, where $i = 1$ to a units and $j = 1$ to jth variable, are $N(\mu, \sigma^2)$ distributed, then the linear combination of

$$y = a_1 x_1 + a_2 x_2 + \cdots + a_j x_j$$

is normally distributed with mean

$$\mu_y = a_1 \mu_1 + a_2 \mu_2 + \cdots + a_j \mu_j$$

and variance

$$\sigma_y^2 = a_1^2 \sigma_1^2 + a_2^2 \sigma_2^2 + \cdots + a_j^2 \sigma_j^2$$

Exponential Distribution

$$f(x) = \lambda e^{-\lambda x} \text{ or } f(x) = \lambda e^{\lambda x}$$

$$\mu - \frac{1}{\lambda}$$

$$\sigma^2 - \frac{1}{\lambda^2}$$

Lognormal Distribution

$$f(x) = \frac{1}{x \sigma_N \sqrt{2\pi}} e^{-\frac{1}{2}\left(\frac{\ln(x) - \mu_N}{\sigma_N}\right)^2}$$

$$\mu = m e^{\sigma_N^2/2}$$

$$V[X] = m^2 e^{\sigma_N^2}\left(e^{\sigma_N^2} - 1\right)$$

Influence Diagrams

Definition: A *decision* is the selection of a sequence of actions designed to resolve a problem or attain an objective.

Definition: *Alternatives* are two or more available actions.

Definition: *Constraints*—(1) Limitations or restrictions on selecting an alternative or on a selected alternative's range of outcome values. (2) Decision constraints are a set of rules for acceptable values of decision variables.

Definition: *Risk* is observed in those situations in which the potential outcomes can be described by well-known probability distributions.

Definition: *Criteria* are acceptable values of decision variables.

Definition: An *outcome* of a decision is the realized value of the decision criterion or criteria vector.

Definition: *Stakeholders* are all persons impacted by or with interest in the outcome of a decision.

Definition: *Sensitivity analysis* is a determination of the amount of variation ($\pm$) in an objective function predictor variable or constraint necessary to change the decision to select a particular alternative. The point at which the decision changes from one alternative to another is the *point of indifference*.

Decision Analysis

Definition: A *decision tree* is a static graphical representation of the probabilistic outcomes of a decision.

Decision Tree Risk Simulation

Used to investigate a wide variety of "what if" questions about risky decisions.

- The long-run expected average values of variables and metrics.
- The prediction error of the expected average values.
- The points at which a decision changes from one alternative to another are due to random variation (we examined structural points of indifference in Sect. 10.0 graphical sensitivity analysis).
- The breakeven points where NPW or EUAW ≤ 0 given a stated MARR, or the points where IRR $\leq$ MARR or stockholders' required return on equity.
- How changes in input factor or variable values affect the expected values and prediction error of output metrics (NPW, EUAW, or RoI).
- Which input factors or variables dominate (are probabilistically important components of) the decision.

Process

1. Identify the Economic Investment Problem—This was the central discussion of the first eleven chapters of this text: new product, new process, upgraded product or process, new facility or facility expansion, alternative investment comparisons (incremental analysis), replacement analysis, etc.
2. Construct the Cash Flow Model—A mathematical model that estimates the current value of an investment based on its expected future cash flows.
3. Construct the Decision Tree—The static graphical representation of the probabilistic outcomes of an investment decision.

4. Construct the Simulation Model—Translation of the graphical decision tree structure into a discounted cash flow model that can be executed in the selected simulation software application.
5. Simulation Runs—Determine the number of runs necessary to achieve a required statistical prediction interval (subject of statistical or simulation textbooks) and conduct those runs. After each run, record the values of economic criteria (NPW, EUAW, or RoI).
6. Simulation Sensitivity Analysis—Given the observed random variation in the economic criteria, determine the statistical prediction interval and points of indifferences in interval unions where decision changes occur.

15.8 Key Terms

Asset criteria
Asset problem
Bernoulli distribution
Binomial distribution
Decision tree
Economic risk
Exponential distribution
Influence diagram
Investment objective
Lognormal distribution
Mean
Normal distribution
Probability
Probability distribution
Probability mass function
Poisson distribution
Population
Random variation
Requirements criteria
Risk simulation
Sensitivity analysis
Simulation
Standard deviation
Statistics
Structural variation
Triangular distribution
Uniform distribution
Variance

Problems

1. Heat recirculatory exchangers are being installed in a factory heating system to increase heat exchange efficiency in the facility. The exchangers' initial cost is $96,000 with an expected life of 12 years. Since the facility is in a Midwest state with variable year-to-year winter temperatures, the predicted annual savings in natural gas is triangularly distributed with P(low savings|mild = $12,000) = 0.10, P(moderate savings|normal = $16,000) = 0.6, and P(high savings|severe = $18,000) = 0.3.

    (a)   Estimate the expected (mean) savings.
    (b)   Estimate the variance and standard deviation of savings.
    (c)   Estimate the expected internal rate of return.
    (d)   If MARR = 9.0% for this type of investment, should the recirculatory exchangers be installed?

2. For the heat recirculatory exchangers' installation in problem 1, assume that the distribution of annual savings can be estimated by a triangular distribution with low(a) = $12,000, mode(c) = $16,000, and high (b) = $18,000.

    (a)   Estimate the expected (mean) savings.
    (b)   Estimate the variance and standard deviation of savings.
    (c)   Estimate the expected internal rate of return.
    (d)   If MARR = 9.0% for this type of investment, should the recirculatory exchangers be installed?

3. A new development project has a life of 7 years and no salvage value. The organization uses MARR = 18% for risky new development projects. The project has independent uncertain initial cost and annual net revenue as shown in the following table.

P (Init cost)	Initial cost	P (Ann Rev)	Annual Rev
0.2	$450,000	0.1	$176,000
0.7	$600,000	0.5	$212,000
0.3	$900,000	0.4	$240,000

    (a)   Since each initial cost and annual net revenue are independent, estimate the joint probability distribution P(Cost ∩ Revenue) = P(Cost) × P(Revenue).
    (b)   Estimate the expected value, variance, and standard deviation of the NPW of the joint probability distribution.
    (c)   Defining pessimistic as initial cost = $900,000 and annual revenue = $176,000, most likely as initial cost = $600,000 and annual revenue = $212,000, and most likely as optimistic as initial cost = $450,000 and annual revenue= $240,000, estimate the pessimistic, most likely, and optimistic NPW.

4. Construct a decision tree of the new development project in problem 3 and find the expected NPW.

5. Using the decision tree for problem 4, by what percentage can the joint annual revenues decrease before the decision changes to not invest? (b) By what percentage can the joint initial costs increase (go more negative) before the decision changes to not invest?

6. Assume that initial costs are known with a high degree of certainty. Given the expected means and historical standard deviations of normally distributed annual revenues, conduct a 100-trial simulation of the project (NPW) using the decision tree from problem 4. What percentage of the outcomes will yield the decision to not invest?

P (Init cost)	Initial cost	P (Ann Rev)	E[Ann Rev]	Std deviation
0.2	$450,000	0.1	$176,000	$19,950
0.7	$600,000	0.5	$212,000	$24,000
0.3	$900,000	0.4	$240,000	$27,200

Appendix A
Use of the Microsoft® Excel® Compound Interest Calculator Spreadsheet

Engineering economics textbooks typically provide an appendix with multiple pages of compound interest tables indexed by interest percentage. Before the advent of personal computers and spreadsheet programs such as Excel®, this was the only effective means of avoiding manual computation of the correct compound interest formula each time it was used in a cash flow equation. Spreadsheet programs supply built-in discrete compounding functions and provide the ability to code formulas specifying geometric gradient and continuous compounding functions. In addition to setting forth multiple pages of discrete payment compound amount factors for functional notation (i.e., (P/F, i, n), (A/P, i, n), etc.) in Appendix B, this work provides a one-spreadsheet Excel® workbook as a supplement.

The spreadsheet requires only two inputs:

- Interest rate i. This interest rate is used to update the spreadsheet factors for both discrete and continuous compounding.

	A	B	C	D	E	F	G	H	I
1				*Engineering Managerial Economics*					
2				*Compound Interest Table Calculator*					
3									
4		**Discrete Compounding**							
5		Interest Rate	i	7.00%					
6									
7		Single Payment		Uniform Payment Series				Arithmetic Gradient	
8	n	F/P	P/F	A/F	A/P	F/A	P/A	A/G	P/G
9	1	1.07000	0.93458	1.00000	1.07000	1.00000	0.93458	0.00000	0.00000
10	2	1.14490	0.87344	0.48309	0.55309	2.07000	1.80802	0.48309	0.87344
11	3	1.22504	0.81630	0.31105	0.38105	3.21490	2.62432	0.95493	2.50603
12	4	1.31080	0.76290	0.22523	0.29523	4.43994	3.38721	1.41554	4.79472
13	5	1.40255	0.71299	0.17389	0.24389	5.75074	4.10020	1.86495	7.64666

© The Editor(s) (if applicable) and The Author(s), under exclusive license to Springer Nature Switzerland AG 2022
T. S. Cotter, *Engineering Managerial Economic Decision and Risk Analysis*, Topics in Safety, Risk, Reliability and Quality 39, https://doi.org/10.1007/978-3-030-87767-5

M	N	O	P	Q	R
Continuous Compounding					
Single Payment		Uniform Payment Series			
F/P	P/F	A/F	A/P	F/A	P/A
1.072508	0.932394	1.00000	1.07251	1.00000	0.93239
1.150274	0.869358	0.48251	0.55502	2.07251	1.80175
1.233678	0.810584	0.31029	0.38280	3.22278	2.61234
1.32313	0.755784	0.22439	0.29690	4.45646	3.36812
1.419068	0.704688	0.17302	0.24553	5.77959	4.07281

- Geometric growth rate g. The combination of I and g is required to update the geometric gradient (P/A, i, g, n) and (F/A, i, g, n) compound amount factors.

J	K
g	2.00%
	** i >= g
Geometric Gradient	
P/A,i,g	F/A,I,g
0.93458	1.00000
1.82549	2.09000
2.67476	3.27670
3.48435	4.56728
4.25611	5.96942

Appendix B
Discrete Payment Compound Interest Factors

Interest Rate: 0.25%

n	Single payment		Uniform payment series			
	F/P	P/F	A/F	A/P	F/A	P/A
1	1.00250	0.99751	1.00000	1.00250	1.00000	0.99751
2	1.00501	0.99502	0.49938	0.50188	2.00250	1.99252
3	1.00752	0.99254	0.33250	0.33500	3.00751	2.98506
4	1.01004	0.99006	0.24906	0.25156	4.01503	3.97512
5	1.01256	0.98759	0.19900	0.20150	5.02506	4.96272
6	1.01509	0.98513	0.16563	0.16813	6.03763	5.94785
7	1.01763	0.98267	0.14179	0.14429	7.05272	6.93052
8	1.02018	0.98022	0.12391	0.12641	8.07035	7.91074
9	1.02273	0.97778	0.11000	0.11250	9.09053	8.88852
10	1.02528	0.97534	0.09888	0.10138	10.11325	9.86386
11	1.02785	0.97291	0.08978	0.09228	11.13854	10.83677
12	1.03042	0.97048	0.08219	0.08469	12.16638	11.80725
13	1.03299	0.96806	0.07578	0.07828	13.19680	12.77532
14	1.03557	0.96565	0.07028	0.07278	14.22979	13.74096
15	1.03816	0.96324	0.06551	0.06801	15.26537	14.70420
16	1.04076	0.96084	0.06134	0.06384	16.30353	15.66504
17	1.04336	0.95844	0.05766	0.06016	17.34429	16.62348
18	1.04597	0.95605	0.05438	0.05688	18.38765	17.57953
19	1.04858	0.95367	0.05146	0.05396	19.43362	18.53320
20	1.05121	0.95129	0.04882	0.05132	20.48220	19.48449

(continued)

© The Editor(s) (if applicable) and The Author(s), under exclusive license
to Springer Nature Switzerland AG 2022
T. S. Cotter, *Engineering Managerial Economic Decision and Risk Analysis*,
Topics in Safety, Risk, Reliability and Quality 39,
https://doi.org/10.1007/978-3-030-87767-5

(continued)

n	Single payment		Uniform payment series			
	F/P	P/F	A/F	A/P	F/A	P/A
21	1.05383	0.94892	0.04644	0.04894	21.53341	20.43340
22	1.05647	0.94655	0.04427	0.04677	22.58724	21.37995
23	1.05911	0.94419	0.04229	0.04479	23.64371	22.32414
24	1.06176	0.94184	0.04048	0.04298	24.70282	23.26598
25	1.06441	0.93949	0.03881	0.04131	25.76457	24.20547
26	1.06707	0.93714	0.03727	0.03977	26.82899	25.14261
27	1.06974	0.93481	0.03585	0.03835	27.89606	26.07742
28	1.07241	0.93248	0.03452	0.03702	28.96580	27.00989
29	1.07510	0.93015	0.03329	0.03579	30.03821	27.94004
30	1.07778	0.92783	0.03214	0.03464	31.11331	28.86787
35	1.09132	0.91632	0.02738	0.02988	36.52924	33.47243
40	1.10503	0.90495	0.02380	0.02630	42.01320	38.01986
45	1.11892	0.89372	0.02102	0.02352	47.56606	42.51088
50	1.13297	0.88263	0.01880	0.02130	53.18868	46.94617
60	1.16162	0.86087	0.01547	0.01797	64.64671	55.65236
70	1.19099	0.83964	0.01309	0.01559	76.39444	64.14385
80	1.22110	0.81894	0.01131	0.01381	88.43918	72.42595
90	1.25197	0.79874	0.00992	0.01242	100.78845	80.50382
100	1.28362	0.77904	0.00881	0.01131	113.44996	88.38248

Interest rate: 0.25%

n	Arithmetic grad	
	A/G	P/G
1	0.00000	0.00000
2	0.49938	0.99502
3	0.99834	2.98009
4	1.49688	5.95028
5	1.99501	9.90065
6	2.49272	14.82630
7	2.99001	20.72235
8	3.48689	27.58391
9	3.98335	35.40614
10	4.47940	44.18420
11	4.97503	53.91328
12	5.47025	64.58858

(continued)

(continued)

n	Arithmetic grad	
	A/G	P/G
13	5.96504	76.20532
14	6.45943	88.75874
15	6.95339	102.24409
16	7.44694	116.65666
17	7.94008	131.99172
18	8.43279	148.24459
19	8.92510	165.41059
20	9.41698	183.48508
21	9.90845	202.46341
22	10.39951	222.34096
23	10.89014	243.11313
24	11.38036	264.77534
25	11.87017	287.32301
26	12.35956	310.75160
27	12.84853	335.05657
28	13.33709	360.23340
29	13.82523	386.27760
30	14.31296	413.18468
35	16.74535	560.50760
40	19.16735	728.73988
45	21.57895	917.34000
50	23.98016	1,125.77667
60	28.75142	1,600.08454
70	33.48117	2,147.61109
80	38.16942	2,764.45681
90	42.81623	3,446.86997
100	47.42163	4,191.24173

Interest rate: 0.50%

n	Single payment		Uniform payment series			
	F/P	P/F	A/F	A/P	F/A	P/A
1	1.00500	0.99502	1.00000	1.00500	1.00000	0.99502
2	1.01003	0.99007	0.49875	0.50375	2.00500	1.98510
3	1.01508	0.98515	0.33167	0.33667	3.01502	2.97025
4	1.02015	0.98025	0.24813	0.25313	4.03010	3.95050

(continued)

(continued)

n	Single payment		Uniform payment series			
	F/P	P/F	A/F	A/P	F/A	P/A
5	1.02525	0.97537	0.19801	0.20301	5.05025	4.92587
6	1.03038	0.97052	0.16460	0.16960	6.07550	5.89638
7	1.03553	0.96569	0.14073	0.14573	7.10588	6.86207
8	1.04071	0.96089	0.12283	0.12783	8.14141	7.82296
9	1.04591	0.95610	0.10891	0.11391	9.18212	8.77906
10	1.05114	0.95135	0.09777	0.10277	10.22803	9.73041
11	1.05640	0.94661	0.08866	0.09366	11.27917	10.67703
12	1.06168	0.94191	0.08107	0.08607	12.33556	11.61893
13	1.06699	0.93722	0.07464	0.07964	13.39724	12.55615
14	1.07232	0.93256	0.06914	0.07414	14.46423	13.48871
15	1.07768	0.92792	0.06436	0.06936	15.53655	14.41662
16	1.08307	0.92330	0.06019	0.06519	16.61423	15.33993
17	1.08849	0.91871	0.05651	0.06151	17.69730	16.25863
18	1.09393	0.91414	0.05323	0.05823	18.78579	17.17277
19	1.09940	0.90959	0.05030	0.05530	19.87972	18.08236
20	1.10490	0.90506	0.04767	0.05267	20.97912	18.98742
21	1.11042	0.90056	0.04528	0.05028	22.08401	19.88798
22	1.11597	0.89608	0.04311	0.04811	23.19443	20.78406
23	1.12155	0.89162	0.04113	0.04613	24.31040	21.67568
24	1.12716	0.88719	0.03932	0.04432	25.43196	22.56287
25	1.13280	0.88277	0.03765	0.04265	26.55912	23.44564
26	1.13846	0.87838	0.03611	0.04111	27.69191	24.32402
27	1.14415	0.87401	0.03469	0.03969	28.83037	25.19803
28	1.14987	0.86966	0.03336	0.03836	29.97452	26.06769
29	1.15562	0.86533	0.03213	0.03713	31.12439	26.93302
30	1.16140	0.86103	0.03098	0.03598	32.28002	27.79405
35	1.19073	0.83982	0.02622	0.03122	38.14538	32.03537
40	1.22079	0.81914	0.02265	0.02765	44.15885	36.17223
45	1.25162	0.79896	0.01987	0.02487	50.32416	40.20720
50	1.28323	0.77929	0.01765	0.02265	56.64516	44.14279
60	1.34885	0.74137	0.01433	0.01933	69.77003	51.72556
70	1.41783	0.70530	0.01197	0.01697	83.56611	58.93942
80	1.49034	0.67099	0.01020	0.01520	98.06771	65.80231
90	1.56655	0.63834	0.00883	0.01383	113.31094	72.33130
100	1.64667	0.60729	0.00773	0.01273	129.33370	78.54264

Interest rate: 0.50%

n	Arithmetic grad	
	A/G	P/G
1	0.00000	0.00000
2	0.49875	0.99007
3	0.99667	2.96037
4	1.49377	5.90111
5	1.99003	9.80260
6	2.48545	14.65519
7	2.98005	20.44933
8	3.47382	27.17552
9	3.96675	34.82436
10	4.45885	43.38649
11	4.95013	52.85264
12	5.44057	63.21360
13	5.93018	74.46023
14	6.41896	86.58346
15	6.90691	99.57430
16	7.39403	113.42380
17	7.88031	128.12311
18	8.36577	143.66343
19	8.85040	160.03602
20	9.33419	177.23221
21	9.81716	195.24341
22	10.29929	214.06109
23	10.78060	233.67676
24	11.26107	254.08203
25	11.74072	275.26856
26	12.21953	297.22805
27	12.69751	319.95231
28	13.17467	343.43317
29	13.65099	367.66255
30	14.12649	392.63241
35	16.49153	528.31226
40	18.83585	681.33469
45	21.15947	850.76312
50	23.46242	1,035.69659
60	28.00638	1,448.64580
70	32.46796	1,913.64274

(continued)

(continued)

| n | Arithmetic grad | |
	A/G	P/G
80	36.84742	2,424.64551
90	41.14508	2,976.07688
100	45.36126	3,562.79343

Interest rate: 0.75%

| n | Single payment | | Uniform payment series | | | |
	F/P	P/F	A/F	A/P	F/A	P/A
1	1.00750	0.99256	1.00000	1.00750	1.00000	0.99256
2	1.01506	0.98517	0.49813	0.50563	2.00750	1.97772
3	1.02267	0.97783	0.33085	0.33835	3.02256	2.95556
4	1.03034	0.97055	0.24721	0.25471	4.04523	3.92611
5	1.03807	0.96333	0.19702	0.20452	5.07556	4.88944
6	1.04585	0.95616	0.16357	0.17107	6.11363	5.84560
7	1.05370	0.94904	0.13967	0.14717	7.15948	6.79464
8	1.06160	0.94198	0.12176	0.12926	8.21318	7.73661
9	1.06956	0.93496	0.10782	0.11532	9.27478	8.67158
10	1.07758	0.92800	0.09667	0.10417	10.34434	9.59958
11	1.08566	0.92109	0.08755	0.09505	11.42192	10.52067
12	1.09381	0.91424	0.07995	0.08745	12.50759	11.43491
13	1.10201	0.90743	0.07352	0.08102	13.60139	12.34235
14	1.11028	0.90068	0.06801	0.07551	14.70340	13.24302
15	1.11860	0.89397	0.06324	0.07074	15.81368	14.13699
16	1.12699	0.88732	0.05906	0.06656	16.93228	15.02431
17	1.13544	0.88071	0.05537	0.06287	18.05927	15.90502
18	1.14396	0.87416	0.05210	0.05960	19.19472	16.77918
19	1.15254	0.86765	0.04917	0.05667	20.33868	17.64683
20	1.16118	0.86119	0.04653	0.05403	21.49122	18.50802
21	1.16989	0.85478	0.04415	0.05165	22.65240	19.36280
22	1.17867	0.84842	0.04198	0.04948	23.82230	20.21121
23	1.18751	0.84210	0.04000	0.04750	25.00096	21.05331
24	1.19641	0.83583	0.03818	0.04568	26.18847	21.88915
25	1.20539	0.82961	0.03652	0.04402	27.38488	22.71876
26	1.21443	0.82343	0.03498	0.04248	28.59027	23.54219
27	1.22354	0.81730	0.03355	0.04105	29.80470	24.35949
28	1.23271	0.81122	0.03223	0.03973	31.02823	25.17071

(continued)

(continued)

n	Single payment		Uniform payment series			
	F/P	P/F	A/F	A/P	F/A	P/A
29	1.24196	0.80518	0.03100	0.03850	32.26094	25.97589
30	1.25127	0.79919	0.02985	0.03735	33.50290	26.77508
35	1.29890	0.76988	0.02509	0.03259	39.85381	30.68266
40	1.34835	0.74165	0.02153	0.02903	46.44648	34.44694
45	1.39968	0.71445	0.01877	0.02627	53.29011	38.07318
50	1.45296	0.68825	0.01656	0.02406	60.39426	41.56645
60	1.56568	0.63870	0.01326	0.02076	75.42414	48.17337
70	1.68715	0.59272	0.01091	0.01841	91.62007	54.30462
80	1.81804	0.55004	0.00917	0.01667	109.07253	59.99444
90	1.95909	0.51044	0.00782	0.01532	127.87899	65.27461
100	2.11108	0.47369	0.00675	0.01425	148.14451	70.17462

Interest rate: 0.75%

n	Arithmetic grad	
	A/G	P/G
1	0.00000	0.00000
2	0.49813	0.98517
3	0.99502	2.94083
4	1.49066	5.85250
5	1.98506	9.70581
6	2.47821	14.48660
7	2.97011	20.18084
8	3.46077	26.77467
9	3.95019	34.25438
10	4.43836	42.60641
11	4.92529	51.81736
12	5.41097	61.87398
13	5.89541	72.76316
14	6.37860	84.47197
15	6.86055	96.98758
16	7.34126	110.29735
17	7.82072	124.38875
18	8.29894	139.24940
19	8.77592	154.86708
20	9.25165	171.22969

(continued)

(continued)

	Arithmetic grad	
n	A/G	P/G
21	9.72614	188.32527
22	10.19939	206.14200
23	10.67139	224.66820
24	11.14216	243.89233
25	11.61168	263.80295
26	12.07996	284.38879
27	12.54701	305.63869
28	13.01281	327.54162
29	13.47737	350.08668
30	13.94069	373.26310
35	16.23872	498.24714
40	18.50583	637.46933
45	20.74209	789.71734
50	22.94756	953.84863
60	27.26649	1,313.51888
70	31.46337	1,708.60649
80	35.53908	2,132.14723
90	39.49462	2,577.99605
100	43.33112	3,040.74530

Interest rate: 1.00%

	Single payment		Uniform payment series			
n	F/P	P/F	A/F	A/P	F/A	P/A
1	1.01000	0.99010	1.00000	1.01000	1.00000	0.99010
2	1.02010	0.98030	0.49751	0.50751	2.01000	1.97040
3	1.03030	0.97059	0.33002	0.34002	3.03010	2.94099
4	1.04060	0.96098	0.24628	0.25628	4.06040	3.90197
5	1.05101	0.95147	0.19604	0.20604	5.10101	4.85343
6	1.06152	0.94205	0.16255	0.17255	6.15202	5.79548
7	1.07214	0.93272	0.13863	0.14863	7.21354	6.72819
8	1.08286	0.92348	0.12069	0.13069	8.28567	7.65168
9	1.09369	0.91434	0.10674	0.11674	9.36853	8.56602
10	1.10462	0.90529	0.09558	0.10558	10.46221	9.47130
11	1.11567	0.89632	0.08645	0.09645	11.56683	10.36763
12	1.12683	0.88745	0.07885	0.08885	12.68250	11.25508

(continued)

(continued)

n	Single payment		Uniform payment series			
	F/P	P/F	A/F	A/P	F/A	P/A
13	1.13809	0.87866	0.07241	0.08241	13.80933	12.13374
14	1.14947	0.86996	0.06690	0.07690	14.94742	13.00370
15	1.16097	0.86135	0.06212	0.07212	16.09690	13.86505
16	1.17258	0.85282	0.05794	0.06794	17.25786	14.71787
17	1.18430	0.84438	0.05426	0.06426	18.43044	15.56225
18	1.19615	0.83602	0.05098	0.06098	19.61475	16.39827
19	1.20811	0.82774	0.04805	0.05805	20.81090	17.22601
20	1.22019	0.81954	0.04542	0.05542	22.01900	18.04555
21	1.23239	0.81143	0.04303	0.05303	23.23919	18.85698
22	1.24472	0.80340	0.04086	0.05086	24.47159	19.66038
23	1.25716	0.79544	0.03889	0.04889	25.71630	20.45582
24	1.26973	0.78757	0.03707	0.04707	26.97346	21.24339
25	1.28243	0.77977	0.03541	0.04541	28.24320	22.02316
26	1.29526	0.77205	0.03387	0.04387	29.52563	22.79520
27	1.30821	0.76440	0.03245	0.04245	30.82089	23.55961
28	1.32129	0.75684	0.03112	0.04112	32.12910	24.31644
29	1.33450	0.74934	0.02990	0.03990	33.45039	25.06579
30	1.34785	0.74192	0.02875	0.03875	34.78489	25.80771
35	1.41660	0.70591	0.02400	0.03400	41.66028	29.40858
40	1.48886	0.67165	0.02046	0.03046	48.88637	32.83469
45	1.56481	0.63905	0.01771	0.02771	56.48107	36.09451
50	1.64463	0.60804	0.01551	0.02551	64.46318	39.19612
60	1.81670	0.55045	0.01224	0.02224	81.66967	44.95504
70	2.00676	0.49831	0.00993	0.01993	100.67634	50.16851
80	2.21672	0.45112	0.00822	0.01822	121.67152	54.88821
90	2.44863	0.40839	0.00690	0.01690	144.86327	59.16088
100	2.70481	0.36971	0.00587	0.01587	170.48138	63.02888

Interest rate: 1.00%

n	Arithmetic grad	
	A/G	P/G
1	0.00000	0.00000
2	0.49751	0.98030
3	0.99337	2.92148
4	1.48756	5.80442

(continued)

(continued)

n	Arithmetic grad	
	A/G	P/G
5	1.98010	9.61028
6	2.47098	14.32051
7	2.96020	19.91681
8	3.44777	26.38120
9	3.93367	33.69592
10	4.41792	41.84350
11	4.90052	50.80674
12	5.38145	60.56868
13	5.86073	71.11263
14	6.33836	82.42215
15	6.81433	94.48104
16	7.28865	107.27336
17	7.76131	120.78340
18	8.23231	134.99569
19	8.70167	149.89501
20	9.16937	165.46636
21	9.63542	181.69496
22	10.09982	198.56628
23	10.56257	216.06600
24	11.02367	234.18002
25	11.48312	252.89446
26	11.94092	272.19566
27	12.39707	292.07016
28	12.85158	312.50472
29	13.30444	333.48630
30	13.75566	355.00207
35	15.98711	470.15831
40	18.17761	596.85606
45	20.32730	733.70372
50	22.43635	879.41763
60	26.53331	1,192.80614
70	30.47026	1,528.64744
80	34.24920	1,879.87710
90	37.87245	2,240.56748
100	41.34257	2,605.77575

Interest rate: 1.25%

n	Single payment		Uniform payment series			
	F/P	P/F	A/F	A/P	F/A	P/A
1	1.01250	0.98765	1.00000	1.01250	1.00000	0.98765
2	1.02516	0.97546	0.49689	0.50939	2.01250	1.96312
3	1.03797	0.96342	0.32920	0.34170	3.03766	2.92653
4	1.05095	0.95152	0.24536	0.25786	4.07563	3.87806
5	1.06408	0.93978	0.19506	0.20756	5.12657	4.81784
6	1.07738	0.92817	0.16153	0.17403	6.19065	5.74601
7	1.09085	0.91672	0.13759	0.15009	7.26804	6.66273
8	1.10449	0.90540	0.11963	0.13213	8.35889	7.56812
9	1.11829	0.89422	0.10567	0.11817	9.46337	8.46234
10	1.13227	0.88318	0.09450	0.10700	10.58167	9.34553
11	1.14642	0.87228	0.08537	0.09787	11.71394	10.21780
12	1.16075	0.86151	0.07776	0.09026	12.86036	11.07931
13	1.17526	0.85087	0.07132	0.08382	14.02112	11.93018
14	1.18995	0.84037	0.06581	0.07831	15.19638	12.77055
15	1.20483	0.82999	0.06103	0.07353	16.38633	13.60055
16	1.21989	0.81975	0.05685	0.06935	17.59116	14.42029
17	1.23514	0.80963	0.05316	0.06566	18.81105	15.22992
18	1.25058	0.79963	0.04988	0.06238	20.04619	16.02955
19	1.26621	0.78976	0.04696	0.05946	21.29677	16.81931
20	1.28204	0.78001	0.04432	0.05682	22.56298	17.59932
21	1.29806	0.77038	0.04194	0.05444	23.84502	18.36969
22	1.31429	0.76087	0.03977	0.05227	25.14308	19.13056
23	1.33072	0.75147	0.03780	0.05030	26.45737	19.88204
24	1.34735	0.74220	0.03599	0.04849	27.78808	20.62423
25	1.36419	0.73303	0.03432	0.04682	29.13544	21.35727
26	1.38125	0.72398	0.03279	0.04529	30.49963	22.08125
27	1.39851	0.71505	0.03137	0.04387	31.88087	22.79630
28	1.41599	0.70622	0.03005	0.04255	33.27938	23.50252
29	1.43369	0.69750	0.02882	0.04132	34.69538	24.20002
30	1.45161	0.68889	0.02768	0.04018	36.12907	24.88891
35	1.54464	0.64740	0.02295	0.03545	43.57087	28.20786
40	1.64362	0.60841	0.01942	0.03192	51.48956	31.32693
45	1.74895	0.57177	0.01669	0.02919	59.91569	34.25817
50	1.86102	0.53734	0.01452	0.02702	68.88179	37.01288
60	2.10718	0.47457	0.01129	0.02379	88.57451	42.03459
70	2.38590	0.41913	0.00902	0.02152	110.87200	46.46968

(continued)

(continued)

n	Single payment		Uniform payment series			
	F/P	P/F	A/F	A/P	F/A	P/A
80	2.70148	0.37017	0.00735	0.01985	136.11880	50.38666
90	3.05881	0.32692	0.00607	0.01857	164.70501	53.84606
100	3.46340	0.28873	0.00507	0.01757	197.07234	56.90134

Interest rate: 1.25%

	Arithmetic grad	
n	A/G	P/G
1	0.00000	0.00000
2	0.49689	0.97546
3	0.99172	2.90230
4	1.48447	5.75687
5	1.97516	9.51598
6	2.46377	14.15685
7	2.95032	19.65715
8	3.43479	25.99494
9	3.91720	33.14870
10	4.39754	41.09733
11	4.87581	49.82011
12	5.35202	59.29670
13	5.82616	69.50717
14	6.29824	80.43196
15	6.76825	92.05186
16	7.23620	104.34806
17	7.70208	117.30207
18	8.16591	130.89580
19	8.62767	145.11145
20	9.08738	159.93161
21	9.54502	175.33919
22	10.00062	191.31742
23	10.45415	207.84986
24	10.90564	224.92039
25	11.35507	242.51321
26	11.80245	260.61282
27	12.24778	279.20402
28	12.69106	298.27192

(continued)

(continued)

	Arithmetic grad	
n	A/G	P/G
29	13.13230	317.80191
30	13.57150	337.77969
35	15.73688	443.90369
40	17.85148	559.23198
45	19.91557	682.27103
50	21.92950	811.67385
60	25.80834	1,084.84285
70	29.49130	1,370.45134
80	32.98225	1,661.86513
90	36.28548	1,953.83026
100	39.40577	2,242.24109

Interest rate: 1.50%

	Single payment		Uniform payment series			
n	F/P	P/F	A/F	A/P	F/A	P/A
1	1.01500	0.98522	1.00000	1.01500	1.00000	0.98522
2	1.03023	0.97066	0.49628	0.51128	2.01500	1.95588
3	1.04568	0.95632	0.32838	0.34338	3.04522	2.91220
4	1.06136	0.94218	0.24444	0.25944	4.09090	3.85438
5	1.07728	0.92826	0.19409	0.20909	5.15227	4.78264
6	1.09344	0.91454	0.16053	0.17553	6.22955	5.69719
7	1.10984	0.90103	0.13656	0.15156	7.32299	6.59821
8	1.12649	0.88771	0.11858	0.13358	8.43284	7.48593
9	1.14339	0.87459	0.10461	0.11961	9.55933	8.36052
10	1.16054	0.86167	0.09343	0.10843	10.70272	9.22218
11	1.17795	0.84893	0.08429	0.09929	11.86326	10.07112
12	1.19562	0.83639	0.07668	0.09168	13.04121	10.90751
13	1.21355	0.82403	0.07024	0.08524	14.23683	11.73153
14	1.23176	0.81185	0.06472	0.07972	15.45038	12.54338
15	1.25023	0.79985	0.05994	0.07494	16.68214	13.34323
16	1.26899	0.78803	0.05577	0.07077	17.93237	14.13126
17	1.28802	0.77639	0.05208	0.06708	19.20136	14.90765
18	1.30734	0.76491	0.04881	0.06381	20.48938	15.67256
19	1.32695	0.75361	0.04588	0.06088	21.79672	16.42617
20	1.34686	0.74247	0.04325	0.05825	23.12367	17.16864

(continued)

(continued)

n	Single payment		Uniform payment series			
	F/P	P/F	A/F	A/P	F/A	P/A
21	1.36706	0.73150	0.04087	0.05587	24.47052	17.90014
22	1.38756	0.72069	0.03870	0.05370	25.83758	18.62082
23	1.40838	0.71004	0.03673	0.05173	27.22514	19.33086
24	1.42950	0.69954	0.03492	0.04992	28.63352	20.03041
25	1.45095	0.68921	0.03326	0.04826	30.06302	20.71961
26	1.47271	0.67902	0.03173	0.04673	31.51397	21.39863
27	1.49480	0.66899	0.03032	0.04532	32.98668	22.06762
28	1.51722	0.65910	0.02900	0.04400	34.48148	22.72672
29	1.53998	0.64936	0.02778	0.04278	35.99870	23.37608
30	1.56308	0.63976	0.02664	0.04164	37.53868	24.01584
35	1.68388	0.59387	0.02193	0.03693	45.59209	27.07559
40	1.81402	0.55126	0.01843	0.03343	54.26789	29.91585
45	1.95421	0.51171	0.01572	0.03072	63.61420	32.55234
50	2.10524	0.47500	0.01357	0.02857	73.68283	34.99969
60	2.44322	0.40930	0.01039	0.02539	96.21465	39.38027
70	2.83546	0.35268	0.00817	0.02317	122.36375	43.15487
80	3.29066	0.30389	0.00655	0.02155	152.71085	46.40732
90	3.81895	0.26185	0.00532	0.02032	187.92990	49.20985
100	4.43205	0.22563	0.00437	0.01937	228.80304	51.62470

Interest rate: 1.50%

n	Arithmetic grad	
	A/G	P/G
1	0.00000	0.00000
2	0.49628	0.97066
3	0.99007	2.88330
4	1.48139	5.70985
5	1.97023	9.42289
6	2.45658	13.99560
7	2.94046	19.40176
8	3.42185	25.61574
9	3.90077	32.61248
10	4.37721	40.36748
11	4.85118	48.85681
12	5.32267	58.05708

(continued)

(continued)

	Arithmetic grad	
n	A/G	P/G
13	5.79169	67.94540
14	6.25824	78.49944
15	6.72231	89.69736
16	7.18392	101.51783
17	7.64306	113.93999
18	8.09973	126.94349
19	8.55394	140.50842
20	9.00569	154.61536
21	9.45497	169.24532
22	9.90180	184.37976
23	10.34618	200.00058
24	10.78810	216.09009
25	11.22758	232.63103
26	11.66460	249.60654
27	12.09918	267.00017
28	12.53132	284.79585
29	12.96102	302.97790
30	13.38829	321.53101
35	15.48820	419.35212
40	17.52773	524.35682
45	19.50739	635.01098
50	21.42772	749.96361
60	25.09296	988.16739
70	28.52901	1,231.16582
80	31.74228	1,473.07411
90	34.73987	1,709.54387
100	37.52953	1,937.45061

Interest rate: 1.75%

	Single payment		Uniform payment series			
n	F/P	P/F	A/F	A/P	F/A	P/A
1	1.01750	0.98280	1.00000	1.01750	1.00000	0.98280
2	1.03531	0.96590	0.49566	0.51316	2.01750	1.94870
3	1.05342	0.94929	0.32757	0.34507	3.05281	2.89798
4	1.07186	0.93296	0.24353	0.26103	4.10623	3.83094

(continued)

(continued)

n	Single payment		Uniform payment series			
	F/P	P/F	A/F	A/P	F/A	P/A
5	1.09062	0.91691	0.19312	0.21062	5.17809	4.74786
6	1.10970	0.90114	0.15952	0.17702	6.26871	5.64900
7	1.12912	0.88564	0.13553	0.15303	7.37841	6.53464
8	1.14888	0.87041	0.11754	0.13504	8.50753	7.40505
9	1.16899	0.85544	0.10356	0.12106	9.65641	8.26049
10	1.18944	0.84073	0.09238	0.10988	10.82540	9.10122
11	1.21026	0.82627	0.08323	0.10073	12.01484	9.92749
12	1.23144	0.81206	0.07561	0.09311	13.22510	10.73955
13	1.25299	0.79809	0.06917	0.08667	14.45654	11.53764
14	1.27492	0.78436	0.06366	0.08116	15.70953	12.32201
15	1.29723	0.77087	0.05888	0.07638	16.98445	13.09288
16	1.31993	0.75762	0.05470	0.07220	18.28168	13.85050
17	1.34303	0.74459	0.05102	0.06852	19.60161	14.59508
18	1.36653	0.73178	0.04774	0.06524	20.94463	15.32686
19	1.39045	0.71919	0.04482	0.06232	22.31117	16.04606
20	1.41478	0.70682	0.04219	0.05969	23.70161	16.75288
21	1.43954	0.69467	0.03981	0.05731	25.11639	17.44755
22	1.46473	0.68272	0.03766	0.05516	26.55593	18.13027
23	1.49036	0.67098	0.03569	0.05319	28.02065	18.80125
24	1.51644	0.65944	0.03389	0.05139	29.51102	19.46069
25	1.54298	0.64810	0.03223	0.04973	31.02746	20.10878
26	1.56998	0.63695	0.03070	0.04820	32.57044	20.74573
27	1.59746	0.62599	0.02929	0.04679	34.14042	21.37173
28	1.62541	0.61523	0.02798	0.04548	35.73788	21.98695
29	1.65386	0.60465	0.02676	0.04426	37.36329	22.59160
30	1.68280	0.59425	0.02563	0.04313	39.01715	23.18585
35	1.83529	0.54487	0.02095	0.03845	47.73084	26.00725
40	2.00160	0.49960	0.01747	0.03497	57.23413	28.59423
45	2.18298	0.45809	0.01479	0.03229	67.59858	30.96626
50	2.38079	0.42003	0.01267	0.03017	78.90222	33.14121
60	2.83182	0.35313	0.00955	0.02705	104.67522	36.96399
70	3.36829	0.29689	0.00739	0.02489	135.33076	40.17790
80	4.00639	0.24960	0.00582	0.02332	171.79382	42.87993
90	4.76538	0.20985	0.00465	0.02215	215.16462	45.15161
100	5.66816	0.17642	0.00375	0.02125	266.75177	47.06147

Interest rate: 1.75%

n	Arithmetic grad	
	A/G	P/G
1	0.00000	0.00000
2	0.49566	0.96590
3	0.98843	2.86447
4	1.47832	5.66334
5	1.96531	9.33099
6	2.44941	13.83671
7	2.93062	19.15057
8	3.40895	25.24345
9	3.88439	32.08698
10	4.35695	39.65354
11	4.82662	47.91623
12	5.29341	56.84886
13	5.75733	66.42596
14	6.21836	76.62270
15	6.67653	87.41495
16	7.13182	98.77919
17	7.58424	110.69257
18	8.03379	123.13283
19	8.48048	136.07832
20	8.92431	149.50799
21	9.36529	163.40134
22	9.80341	177.73847
23	10.23868	192.49999
24	10.67111	207.66706
25	11.10069	223.22138
26	11.52744	239.14512
27	11.95135	255.42098
28	12.37243	272.03215
29	12.79069	288.96226
30	13.20613	306.19544
35	15.24123	396.38241
40	17.20665	492.01087
45	19.10318	591.55397
50	20.93167	693.70101
60	24.38848	901.49545
70	27.58564	1,108.33334

(continued)

(continued)

n	Arithmetic grad	
	A/G	P/G
80	30.53289	1,309.24819
90	33.24089	1,500.87981
100	35.72112	1,681.08862 lePara>

Interest rate: 2.00%

n	Single payment		Uniform payment series			
	F/P	P/F	A/F	A/P	F/A	P/A
1	1.02000	0.98039	1.00000	1.02000	1.00000	0.98039
2	1.04040	0.96117	0.49505	0.51505	2.02000	1.94156
3	1.06121	0.94232	0.32675	0.34675	3.06040	2.88388
4	1.08243	0.92385	0.24262	0.26262	4.12161	3.80773
5	1.10408	0.90573	0.19216	0.21216	5.20404	4.71346
6	1.12616	0.88797	0.15853	0.17853	6.30812	5.60143
7	1.14869	0.87056	0.13451	0.15451	7.43428	6.47199
8	1.17166	0.85349	0.11651	0.13651	8.58297	7.32548
9	1.19509	0.83676	0.10252	0.12252	9.75463	8.16224
10	1.21899	0.82035	0.09133	0.11133	10.94972	8.98259
11	1.24337	0.80426	0.08218	0.10218	12.16872	9.78685
12	1.26824	0.78849	0.07456	0.09456	13.41209	10.57534
13	1.29361	0.77303	0.06812	0.08812	14.68033	11.34837
14	1.31948	0.75788	0.06260	0.08260	15.97394	12.10625
15	1.34587	0.74301	0.05783	0.07783	17.29342	12.84926
16	1.37279	0.72845	0.05365	0.07365	18.63929	13.57771
17	1.40024	0.71416	0.04997	0.06997	20.01207	14.29187
18	1.42825	0.70016	0.04670	0.06670	21.41231	14.99203
19	1.45681	0.68643	0.04378	0.06378	22.84056	15.67846
20	1.48595	0.67297	0.04116	0.06116	24.29737	16.35143
21	1.51567	0.65978	0.03878	0.05878	25.78332	17.01121
22	1.54598	0.64684	0.03663	0.05663	27.29898	17.65805
23	1.57690	0.63416	0.03467	0.05467	28.84496	18.29220
24	1.60844	0.62172	0.03287	0.05287	30.42186	18.91393
25	1.64061	0.60953	0.03122	0.05122	32.03030	19.52346
26	1.67342	0.59758	0.02970	0.04970	33.67091	20.12104
27	1.70689	0.58586	0.02829	0.04829	35.34432	20.70690
28	1.74102	0.57437	0.02699	0.04699	37.05121	21.28127

(continued)

(continued)

n	Single payment		Uniform payment series			
	F/P	P/F	A/F	A/P	F/A	P/A
29	1.77584	0.56311	0.02578	0.04578	38.79223	21.84438
30	1.81136	0.55207	0.02465	0.04465	40.56808	22.39646
35	1.99989	0.50003	0.02000	0.04000	49.99448	24.99862
40	2.20804	0.45289	0.01656	0.03656	60.40198	27.35548
45	2.43785	0.41020	0.01391	0.03391	71.89271	29.49016
50	2.69159	0.37153	0.01182	0.03182	84.57940	31.42361
60	3.28103	0.30478	0.00877	0.02877	114.05154	34.76089
70	3.99956	0.25003	0.00667	0.02667	149.97791	37.49862
80	4.87544	0.20511	0.00516	0.02516	193.77196	39.74451
90	5.94313	0.16826	0.00405	0.02405	247.15666	41.58693
100	7.24465	0.13803	0.00320	0.02320	312.23231	43.09835

Interest rate: 2.00%

n	Arithmetic grad	
	A/G	P/G
1	0.00000	0.00000
2	0.49505	0.96117
3	0.98680	2.84581
4	1.47525	5.61735
5	1.96040	9.24027
6	2.44226	13.68013
7	2.92082	18.90349
8	3.39608	24.87792
9	3.86805	31.57197
10	4.33674	38.95510
11	4.80213	46.99773
12	5.26424	55.67116
13	5.72307	64.94755
14	6.17862	74.79992
15	6.63090	85.20213
16	7.07990	96.12881
17	7.52564	107.55542
18	7.96811	119.45813
19	8.40732	131.81388
20	8.84328	144.60033

(continued)

(continued)

| n | Arithmetic grad | |
	A/G	P/G
21	9.27599	157.79585
22	9.70546	171.37947
23	10.13169	185.33090
24	10.55468	199.63049
25	10.97445	214.25924
26	11.39100	229.19872
27	11.80433	244.43113
28	12.21446	259.93924
29	12.62138	275.70639
30	13.02512	291.71644
35	14.99613	374.88264
40	16.88850	461.99313
45	18.70336	551.56519
50	20.44198	642.36059
60	23.69610	823.69753
70	26.66323	999.83432
80	29.35718	1,166.78677
90	31.79292	1,322.17008
100	33.98628	1,464.75275

Interest rate: 2.50%

| n | Single payment | | Uniform payment series | | | |
	F/P	P/F	A/F	A/P	F/A	P/A
1	1.02500	0.97561	1.00000	1.02500	1.00000	0.97561
2	1.05063	0.95181	0.49383	0.51883	2.02500	1.92742
3	1.07689	0.92860	0.32514	0.35014	3.07563	2.85602
4	1.10381	0.90595	0.24082	0.26582	4.15252	3.76197
5	1.13141	0.88385	0.19025	0.21525	5.25633	4.64583
6	1.15969	0.86230	0.15655	0.18155	6.38774	5.50813
7	1.18869	0.84127	0.13250	0.15750	7.54743	6.34939
8	1.21840	0.82075	0.11447	0.13947	8.73612	7.17014
9	1.24886	0.80073	0.10046	0.12546	9.95452	7.97087
10	1.28008	0.78120	0.08926	0.11426	11.20338	8.75206
11	1.31209	0.76214	0.08011	0.10511	12.48347	9.51421
12	1.34489	0.74356	0.07249	0.09749	13.79555	10.25776

(continued)

(continued)

n	Single payment		Uniform payment series			
	F/P	P/F	A/F	A/P	F/A	P/A
13	1.37851	0.72542	0.06605	0.09105	15.14044	10.98318
14	1.41297	0.70773	0.06054	0.08554	16.51895	11.69091
15	1.44830	0.69047	0.05577	0.08077	17.93193	12.38138
16	1.48451	0.67362	0.05160	0.07660	19.38022	13.05500
17	1.52162	0.65720	0.04793	0.07293	20.86473	13.71220
18	1.55966	0.64117	0.04467	0.06967	22.38635	14.35336
19	1.59865	0.62553	0.04176	0.06676	23.94601	14.97889
20	1.63862	0.61027	0.03915	0.06415	25.54466	15.58916
21	1.67958	0.59539	0.03679	0.06179	27.18327	16.18455
22	1.72157	0.58086	0.03465	0.05965	28.86286	16.76541
23	1.76461	0.56670	0.03270	0.05770	30.58443	17.33211
24	1.80873	0.55288	0.03091	0.05591	32.34904	17.88499
25	1.85394	0.53939	0.02928	0.05428	34.15776	18.42438
26	1.90029	0.52623	0.02777	0.05277	36.01171	18.95061
27	1.94780	0.51340	0.02638	0.05138	37.91200	19.46401
28	1.99650	0.50088	0.02509	0.05009	39.85980	19.96489
29	2.04641	0.48866	0.02389	0.04889	41.85630	20.45355
30	2.09757	0.47674	0.02278	0.04778	43.90270	20.93029
35	2.37321	0.42137	0.01821	0.04321	54.92821	23.14516
40	2.68506	0.37243	0.01484	0.03984	67.40255	25.10278
45	3.03790	0.32917	0.01227	0.03727	81.51613	26.83302
50	3.43711	0.29094	0.01026	0.03526	97.48435	28.36231
60	4.39979	0.22728	0.00735	0.03235	135.99159	30.90866
70	5.63210	0.17755	0.00540	0.03040	185.28411	32.89786
80	7.20957	0.13870	0.00403	0.02903	248.38271	34.45182
90	9.22886	0.10836	0.00304	0.02804	329.15425	35.66577
100	11.81372	0.08465	0.00231	0.02731	432.54865	36.61411

Interest rate: 2.50%

n	Arithmetic grad	
	A/G	P/G
1	0.00000	0.00000
2	0.49383	0.95181
3	0.98354	2.80901
4	1.46914	5.52687

(continued)

(continued)

n	Arithmetic grad	
	A/G	P/G
5	1.95063	9.06228
6	2.42801	13.37377
7	2.90128	18.42136
8	3.37045	24.16658
9	3.83552	30.57241
10	4.29649	37.60320
11	4.75338	45.22464
12	5.20618	53.40376
13	5.65490	62.10880
14	6.09955	71.30926
15	6.54013	80.97577
16	6.97665	91.08015
17	7.40912	101.59527
18	7.83754	112.49509
19	8.26193	123.75459
20	8.68230	135.34974
21	9.09865	147.25746
22	9.51099	159.45562
23	9.91933	171.92296
24	10.32369	184.63909
25	10.72408	197.58447
26	11.12050	210.74034
27	11.51298	224.08873
28	11.90152	237.61243
29	12.28613	251.29494
30	12.66683	265.12048
35	14.51218	335.88680
40	16.26203	408.22200
45	17.91848	480.80703
50	19.48389	552.60805
60	22.35185	690.86565
70	24.88807	818.76427
80	27.11666	934.21807
90	29.06288	1,036.54990
100	30.75249	1,125.97474

Interest rate: 3.00%

	Single payment		Uniform payment series			
n	F/P	P/F	A/F	A/P	F/A	P/A
1	1.03000	0.97087	1.00000	1.03000	1.00000	0.97087
2	1.06090	0.94260	0.49261	0.52261	2.03000	1.91347
3	1.09273	0.91514	0.32353	0.35353	3.09090	2.82861
4	1.12551	0.88849	0.23903	0.26903	4.18363	3.71710
5	1.15927	0.86261	0.18835	0.21835	5.30914	4.57971
6	1.19405	0.83748	0.15460	0.18460	6.46841	5.41719
7	1.22987	0.81309	0.13051	0.16051	7.66246	6.23028
8	1.26677	0.78941	0.11246	0.14246	8.89234	7.01969
9	1.30477	0.76642	0.09843	0.12843	10.15911	7.78611
10	1.34392	0.74409	0.08723	0.11723	11.46388	8.53020
11	1.38423	0.72242	0.07808	0.10808	12.80780	9.25262
12	1.42576	0.70138	0.07046	0.10046	14.19203	9.95400
13	1.46853	0.68095	0.06403	0.09403	15.61779	10.63496
14	1.51259	0.66112	0.05853	0.08853	17.08632	11.29607
15	1.55797	0.64186	0.05377	0.08377	18.59891	11.93794
16	1.60471	0.62317	0.04961	0.07961	20.15688	12.56110
17	1.65285	0.60502	0.04595	0.07595	21.76159	13.16612
18	1.70243	0.58739	0.04271	0.07271	23.41444	13.75351
19	1.75351	0.57029	0.03981	0.06981	25.11687	14.32380
20	1.80611	0.55368	0.03722	0.06722	26.87037	14.87747
21	1.86029	0.53755	0.03487	0.06487	28.67649	15.41502
22	1.91610	0.52189	0.03275	0.06275	30.53678	15.93692
23	1.97359	0.50669	0.03081	0.06081	32.45288	16.44361
24	2.03279	0.49193	0.02905	0.05905	34.42647	16.93554
25	2.09378	0.47761	0.02743	0.05743	36.45926	17.41315
26	2.15659	0.46369	0.02594	0.05594	38.55304	17.87684
27	2.22129	0.45019	0.02456	0.05456	40.70963	18.32703
28	2.28793	0.43708	0.02329	0.05329	42.93092	18.76411
29	2.35657	0.42435	0.02211	0.05211	45.21885	19.18845
30	2.42726	0.41199	0.02102	0.05102	47.57542	19.60044
35	2.81386	0.35538	0.01654	0.04654	60.46208	21.48722
40	3.26204	0.30656	0.01326	0.04326	75.40126	23.11477
45	3.78160	0.26444	0.01079	0.04079	92.71986	24.51871
50	4.38391	0.22811	0.00887	0.03887	112.79687	25.72976
60	5.89160	0.16973	0.00613	0.03613	163.05344	27.67556
70	7.91782	0.12630	0.00434	0.03434	230.59406	29.12342

(continued)

(continued)

n	Single payment		Uniform payment series			
	F/P	P/F	A/F	A/P	F/A	P/A
80	10.64089	0.09398	0.00311	0.03311	321.36302	30.20076
90	14.30047	0.06993	0.00226	0.03226	443.34890	31.00241
100	19.21863	0.05203	0.00165	0.03165	607.28773	31.59891

Interest rate: 3.00%

n	Arithmetic grad	
	A/G	P/G
1	0.00000	0.00000
2	0.49261	0.94260
3	0.98030	2.77288
4	1.46306	5.43834
5	1.94090	8.88878
6	2.41383	13.07620
7	2.88185	17.95475
8	3.34496	23.48061
9	3.80318	29.61194
10	4.25650	36.30879
11	4.70494	43.53300
12	5.14850	51.24818
13	5.58720	59.41960
14	6.02104	68.01413
15	6.45004	77.00020
16	6.87421	86.34770
17	7.29357	96.02796
18	7.70812	106.01367
19	8.11788	116.27882
20	8.52286	126.79866
21	8.92309	137.54964
22	9.31858	148.50939
23	9.70934	159.65661
24	10.09540	170.97108
25	10.47677	182.43362
26	10.85348	194.02598
27	11.22554	205.73090
28	11.59298	217.53197

(continued)

(continued)

n	Arithmetic grad	
	A/G	P/G
29	11.95582	229.41367
30	12.31407	241.36129
35	14.03749	301.62670
40	15.65016	361.74994
45	17.15557	420.63248
50	18.55751	477.48033
60	21.06742	583.05261
70	23.21454	676.08687
80	25.03534	756.08652
90	26.56665	823.63021
100	27.84445	879.85405

Interest rate: 4.00%

n	Single payment		Uniform payment series			
	F/P	P/F	A/F	A/P	F/A	P/A
1	1.04000	0.96154	1.00000	1.04000	1.00000	0.96154
2	1.08160	0.92456	0.49020	0.53020	2.04000	1.88609
3	1.12486	0.88900	0.32035	0.36035	3.12160	2.77509
4	1.16986	0.85480	0.23549	0.27549	4.24646	3.62990
5	1.21665	0.82193	0.18463	0.22463	5.41632	4.45182
6	1.26532	0.79031	0.15076	0.19076	6.63298	5.24214
7	1.31593	0.75992	0.12661	0.16661	7.89829	6.00205
8	1.36857	0.73069	0.10853	0.14853	9.21423	6.73274
9	1.42331	0.70259	0.09449	0.13449	10.58280	7.43533
10	1.48024	0.67556	0.08329	0.12329	12.00611	8.11090
11	1.53945	0.64958	0.07415	0.11415	13.48635	8.76048
12	1.60103	0.62460	0.06655	0.10655	15.02581	9.38507
13	1.66507	0.60057	0.06014	0.10014	16.62684	9.98565
14	1.73168	0.57748	0.05467	0.09467	18.29191	10.56312
15	1.80094	0.55526	0.04994	0.08994	20.02359	11.11839
16	1.87298	0.53391	0.04582	0.08582	21.82453	11.65230
17	1.94790	0.51337	0.04220	0.08220	23.69751	12.16567
18	2.02582	0.49363	0.03899	0.07899	25.64541	12.65930
19	2.10685	0.47464	0.03614	0.07614	27.67123	13.13394
20	2.19112	0.45639	0.03358	0.07358	29.77808	13.59033

(continued)

(continued)

| n | Single payment | | Uniform payment series | | | |
	F/P	P/F	A/F	A/P	F/A	P/A
21	2.27877	0.43883	0.03128	0.07128	31.96920	14.02916
22	2.36992	0.42196	0.02920	0.06920	34.24797	14.45112
23	2.46472	0.40573	0.02731	0.06731	36.61789	14.85684
24	2.56330	0.39012	0.02559	0.06559	39.08260	15.24696
25	2.66584	0.37512	0.02401	0.06401	41.64591	15.62208
26	2.77247	0.36069	0.02257	0.06257	44.31174	15.98277
27	2.88337	0.34682	0.02124	0.06124	47.08421	16.32959
28	2.99870	0.33348	0.02001	0.06001	49.96758	16.66306
29	3.11865	0.32065	0.01888	0.05888	52.96629	16.98371
30	3.24340	0.30832	0.01783	0.05783	56.08494	17.29203
35	3.94609	0.25342	0.01358	0.05358	73.65222	18.66461
40	4.80102	0.20829	0.01052	0.05052	95.02552	19.79277
45	5.84118	0.17120	0.00826	0.04826	121.02939	20.72004
50	7.10668	0.14071	0.00655	0.04655	152.66708	21.48218
60	10.51963	0.09506	0.00420	0.04420	237.99069	22.62349
70	15.57162	0.06422	0.00275	0.04275	364.29046	23.39451
80	23.04980	0.04338	0.00181	0.04181	551.24498	23.91539
90	34.11933	0.02931	0.00121	0.04121	827.98333	24.26728
100	50.50495	0.01980	0.00081	0.04081	1,237.62370	24.50500

Interest rate: 4.00%

| n | Arithmetic grad | |
	A/G	P/G
1	0.00000	0.00000
2	0.49020	0.92456
3	0.97386	2.70255
4	1.45100	5.26696
5	1.92161	8.55467
6	2.38571	12.50624
7	2.84332	17.06575
8	3.29443	22.18058
9	3.73908	27.80127
10	4.17726	33.88135
11	4.60901	40.37716
12	5.03435	47.24773

(continued)

(continued)

	Arithmetic grad	
n	A/G	P/G
13	5.45329	54.45462
14	5.86586	61.96179
15	6.27209	69.73550
16	6.67200	77.74412
17	7.06563	85.95809
18	7.45300	94.34977
19	7.83416	102.89333
20	8.20912	111.56469
21	8.57794	120.34136
22	8.94065	129.20242
23	9.29729	138.12840
24	9.64790	147.10119
25	9.99252	156.10400
26	10.33120	165.12123
27	10.66399	174.13846
28	10.99092	183.14235
29	11.31205	192.12059
30	11.62743	201.06183
35	13.11984	244.87679
40	14.47651	286.53030
45	15.70474	325.40278
50	16.81225	361.16385
60	18.69723	422.99665
70	20.19614	472.47892
80	21.37185	511.11614
90	22.28255	540.73692
100	22.98000	563.12487

Interest rate: 5.00%

	Single payment		Uniform payment series			
n	F/P	P/F	A/F	A/P	F/A	P/A
1	1.05000	0.95238	1.00000	1.05000	1.00000	0.95238
2	1.10250	0.90703	0.48780	0.53780	2.05000	1.85941
3	1.15763	0.86384	0.31721	0.36721	3.15250	2.72325
4	1.21551	0.82270	0.23201	0.28201	4.31013	3.54595

(continued)

(continued)

n	Single payment		Uniform payment series			
	F/P	P/F	A/F	A/P	F/A	P/A
5	1.27628	0.78353	0.18097	0.23097	5.52563	4.32948
6	1.34010	0.74622	0.14702	0.19702	6.80191	5.07569
7	1.40710	0.71068	0.12282	0.17282	8.14201	5.78637
8	1.47746	0.67684	0.10472	0.15472	9.54911	6.46321
9	1.55133	0.64461	0.09069	0.14069	11.02656	7.10782
10	1.62889	0.61391	0.07950	0.12950	12.57789	7.72173
11	1.71034	0.58468	0.07039	0.12039	14.20679	8.30641
12	1.79586	0.55684	0.06283	0.11283	15.91713	8.86325
13	1.88565	0.53032	0.05646	0.10646	17.71298	9.39357
14	1.97993	0.50507	0.05102	0.10102	19.59863	9.89864
15	2.07893	0.48102	0.04634	0.09634	21.57856	10.37966
16	2.18287	0.45811	0.04227	0.09227	23.65749	10.83777
17	2.29202	0.43630	0.03870	0.08870	25.84037	11.27407
18	2.40662	0.41552	0.03555	0.08555	28.13238	11.68959
19	2.52695	0.39573	0.03275	0.08275	30.53900	12.08532
20	2.65330	0.37689	0.03024	0.08024	33.06595	12.46221
21	2.78596	0.35894	0.02800	0.07800	35.71925	12.82115
22	2.92526	0.34185	0.02597	0.07597	38.50521	13.16300
23	3.07152	0.32557	0.02414	0.07414	41.43048	13.48857
24	3.22510	0.31007	0.02247	0.07247	44.50200	13.79864
25	3.38635	0.29530	0.02095	0.07095	47.72710	14.09394
26	3.55567	0.28124	0.01956	0.06956	51.11345	14.37519
27	3.73346	0.26785	0.01829	0.06829	54.66913	14.64303
28	3.92013	0.25509	0.01712	0.06712	58.40258	14.89813
29	4.11614	0.24295	0.01605	0.06605	62.32271	15.14107
30	4.32194	0.23138	0.01505	0.06505	66.43885	15.37245
35	5.51602	0.18129	0.01107	0.06107	90.32031	16.37419
40	7.03999	0.14205	0.00828	0.05828	120.79977	17.15909
45	8.98501	0.11130	0.00626	0.05626	159.70016	17.77407
50	11.46740	0.08720	0.00478	0.05478	209.34800	18.25593
60	18.67919	0.05354	0.00283	0.05283	353.58372	18.92929
70	30.42643	0.03287	0.00170	0.05170	588.52851	19.34268
80	49.56144	0.02018	0.00103	0.05103	971.22882	19.59646
90	80.73037	0.01239	0.00063	0.05063	1,594.60730	19.75226
100	131.50126	0.00760	0.00038	0.05038	2,610.02516	19.84791

Interest rate: 5.00%

n	Arithmetic grad	
	A/G	P/G
1	0.00000	0.00000
2	0.48780	0.90703
3	0.96749	2.63470
4	1.43905	5.10281
5	1.90252	8.23692
6	2.35790	11.96799
7	2.80523	16.23208
8	3.24451	20.96996
9	3.67579	26.12683
10	4.09909	31.65205
11	4.51444	37.49884
12	4.92190	43.62405
13	5.32150	49.98791
14	5.71329	56.55379
15	6.09731	63.28803
16	6.47363	70.15970
17	6.84229	77.14045
18	7.20336	84.20430
19	7.55690	91.32751
20	7.90297	98.48841
21	8.24164	105.66726
22	8.57298	112.84611
23	8.89706	120.00868
24	9.21397	127.14024
25	9.52377	134.22751
26	9.82655	141.25852
27	10.12240	148.22258
28	10.41138	155.11011
29	10.69360	161.91261
30	10.96914	168.62255
35	12.24980	200.58069
40	13.37747	229.54518
45	14.36444	255.31454
50	15.22326	277.91478
60	16.60618	314.34316
70	17.62119	340.84090

(continued)

(continued)

| n | Arithmetic grad | |
	A/G	P/G
80	18.35260	359.64605
90	18.87120	372.74879
100	19.23372	381.74922

Interest rate: 6.00%

| n | Single payment | | Uniform payment series | | | |
	F/P	P/F	A/F	A/P	F/A	P/A
1	1.06000	0.94340	1.00000	1.06000	1.00000	0.94340
2	1.12360	0.89000	0.48544	0.54544	2.06000	1.83339
3	1.19102	0.83962	0.31411	0.37411	3.18360	2.67301
4	1.26248	0.79209	0.22859	0.28859	4.37462	3.46511
5	1.33823	0.74726	0.17740	0.23740	5.63709	4.21236
6	1.41852	0.70496	0.14336	0.20336	6.97532	4.91732
7	1.50363	0.66506	0.11914	0.17914	8.39384	5.58238
8	1.59385	0.62741	0.10104	0.16104	9.89747	6.20979
9	1.68948	0.59190	0.08702	0.14702	11.49132	6.80169
10	1.79085	0.55839	0.07587	0.13587	13.18079	7.36009
11	1.89830	0.52679	0.06679	0.12679	14.97164	7.88687
12	2.01220	0.49697	0.05928	0.11928	16.86994	8.38384
13	2.13293	0.46884	0.05296	0.11296	18.88214	8.85268
14	2.26090	0.44230	0.04758	0.10758	21.01507	9.29498
15	2.39656	0.41727	0.04296	0.10296	23.27597	9.71225
16	2.54035	0.39365	0.03895	0.09895	25.67253	10.10590
17	2.69277	0.37136	0.03544	0.09544	28.21288	10.47726
18	2.85434	0.35034	0.03236	0.09236	30.90565	10.82760
19	3.02560	0.33051	0.02962	0.08962	33.75999	11.15812
20	3.20714	0.31180	0.02718	0.08718	36.78559	11.46992
21	3.39956	0.29416	0.02500	0.08500	39.99273	11.76408
22	3.60354	0.27751	0.02305	0.08305	43.39229	12.04158
23	3.81975	0.26180	0.02128	0.08128	46.99583	12.30338
24	4.04893	0.24698	0.01968	0.07968	50.81558	12.55036
25	4.29187	0.23300	0.01823	0.07823	54.86451	12.78336
26	4.54938	0.21981	0.01690	0.07690	59.15638	13.00317
27	4.82235	0.20737	0.01570	0.07570	63.70577	13.21053
28	5.11169	0.19563	0.01459	0.07459	68.52811	13.40616

(continued)

(continued)

n	Single payment		Uniform payment series			
	F/P	P/F	A/F	A/P	F/A	P/A
29	5.41839	0.18456	0.01358	0.07358	73.63980	13.59072
30	5.74349	0.17411	0.01265	0.07265	79.05819	13.76483
35	7.68609	0.13011	0.00897	0.06897	111.43478	14.49825
40	10.28572	0.09722	0.00646	0.06646	154.76197	15.04630
45	13.76461	0.07265	0.00470	0.06470	212.74351	15.45583
50	18.42015	0.05429	0.00344	0.06344	290.33590	15.76186
60	32.98769	0.03031	0.00188	0.06188	533.12818	16.16143
70	59.07593	0.01693	0.00103	0.06103	967.93217	16.38454
80	105.79599	0.00945	0.00057	0.06057	1,746.59989	16.50913
90	189.46451	0.00528	0.00032	0.06032	3,141.07519	16.57870
100	339.30208	0.00295	0.00018	0.06018	5,638.36806	16.61755

Interest rate: 6.00%

n	Arithmetic grad	
	A/G	P/G
1	0.00000	0.00000
2	0.48544	0.89000
3	0.96118	2.56924
4	1.42723	4.94552
5	1.88363	7.93455
6	2.33040	11.45935
7	2.76758	15.44969
8	3.19521	19.84158
9	3.61333	24.57677
10	4.02201	29.60232
11	4.42129	34.87020
12	4.81126	40.33686
13	5.19198	45.96293
14	5.56352	51.71284
15	5.92598	57.55455
16	6.27943	63.45925
17	6.62397	69.40108
18	6.95970	75.35692
19	7.28673	81.30615
20	7.60515	87.23044

(continued)

(continued)

| n | Arithmetic grad | |
	A/G	P/G
21	7.91508	93.11355
22	8.21662	98.94116
23	8.50991	104.70070
24	8.79506	110.38121
25	9.07220	115.97317
26	9.34145	121.46842
27	9.60294	126.85999
28	9.85681	132.14200
29	10.10319	137.30959
30	10.34221	142.35879
35	11.43192	165.74273
40	12.35898	185.95682
45	13.14129	203.10965
50	13.79643	217.45738
60	14.79095	239.04279
70	15.46135	253.32714
80	15.90328	262.54931
90	16.18912	268.39461
100	16.37107	272.04706

Interest rate: 7.00%

| n | Single payment | | Uniform payment series | | | |
	F/P	P/F	A/F	A/P	F/A	P/A
1	1.07000	0.93458	1.00000	1.07000	1.00000	0.93458
2	1.14490	0.87344	0.48309	0.55309	2.07000	1.80802
3	1.22504	0.81630	0.31105	0.38105	3.21490	2.62432
4	1.31080	0.76290	0.22523	0.29523	4.43994	3.38721
5	1.40255	0.71299	0.17389	0.24389	5.75074	4.10020
6	1.50073	0.66634	0.13980	0.20980	7.15329	4.76654
7	1.60578	0.62275	0.11555	0.18555	8.65402	5.38929
8	1.71819	0.58201	0.09747	0.16747	10.25980	5.97130
9	1.83846	0.54393	0.08349	0.15349	11.97799	6.51523
10	1.96715	0.50835	0.07238	0.14238	13.81645	7.02358
11	2.10485	0.47509	0.06336	0.13336	15.78360	7.49867
12	2.25219	0.44401	0.05590	0.12590	17.88845	7.94269

(continued)

(continued)

n	Single payment		Uniform payment series			
	F/P	P/F	A/F	A/P	F/A	P/A
13	2.40985	0.41496	0.04965	0.11965	20.14064	8.35765
14	2.57853	0.38782	0.04434	0.11434	22.55049	8.74547
15	2.75903	0.36245	0.03979	0.10979	25.12902	9.10791
16	2.95216	0.33873	0.03586	0.10586	27.88805	9.44665
17	3.15882	0.31657	0.03243	0.10243	30.84022	9.76322
18	3.37993	0.29586	0.02941	0.09941	33.99903	10.05909
19	3.61653	0.27651	0.02675	0.09675	37.37896	10.33560
20	3.86968	0.25842	0.02439	0.09439	40.99549	10.59401
21	4.14056	0.24151	0.02229	0.09229	44.86518	10.83553
22	4.43040	0.22571	0.02041	0.09041	49.00574	11.06124
23	4.74053	0.21095	0.01871	0.08871	53.43614	11.27219
24	5.07237	0.19715	0.01719	0.08719	58.17667	11.46933
25	5.42743	0.18425	0.01581	0.08581	63.24904	11.65358
26	5.80735	0.17220	0.01456	0.08456	68.67647	11.82578
27	6.21387	0.16093	0.01343	0.08343	74.48382	11.98671
28	6.64884	0.15040	0.01239	0.08239	80.69769	12.13711
29	7.11426	0.14056	0.01145	0.08145	87.34653	12.27767
30	7.61226	0.13137	0.01059	0.08059	94.46079	12.40904
35	10.67658	0.09366	0.00723	0.07723	138.23688	12.94767
40	14.97446	0.06678	0.00501	0.07501	199.63511	13.33171
45	21.00245	0.04761	0.00350	0.07350	285.74931	13.60552
50	29.45703	0.03395	0.00246	0.07246	406.52893	13.80075
60	57.94643	0.01726	0.00123	0.07123	813.52038	14.03918
70	113.98939	0.00877	0.00062	0.07062	1,614.13417	14.16039
80	224.23439	0.00446	0.00031	0.07031	3,189.06268	14.22201
90	441.10298	0.00227	0.00016	0.07016	6,287.18543	14.25333
100	867.71633	0.00115	0.00008	0.07008	12,381.66179	14.26925

Interest rate: 7.00%

n	Arithmetic grad	
	A/G	P/G
1	0.00000	0.00000
2	0.48309	0.87344
3	0.95493	2.50603
4	1.41554	4.79472

(continued)

(continued)

n	Arithmetic grad	
	A/G	P/G
5	1.86495	7.64666
6	2.30322	10.97838
7	2.73039	14.71487
8	3.14654	18.78894
9	3.55174	23.14041
10	3.94607	27.71555
11	4.32963	32.46648
12	4.70252	37.35061
13	5.06484	42.33018
14	5.41673	47.37181
15	5.75829	52.44605
16	6.08968	57.52707
17	6.41102	62.59226
18	6.72247	67.62195
19	7.02418	72.59910
20	7.31631	77.50906
21	7.59901	82.33932
22	7.87247	87.07930
23	8.13685	91.72013
24	8.39234	96.25450
25	8.63910	100.67648
26	8.87733	104.98137
27	9.10722	109.16556
28	9.32894	113.22642
29	9.54270	117.16218
30	9.74868	120.97182
35	10.66873	138.13528
40	11.42335	152.29277
45	12.03599	163.75592
50	12.52868	172.90512
60	13.23209	185.76774
70	13.66619	193.51853
80	13.92735	198.07480
90	14.08122	200.70420
100	14.17034	202.20008

Interest rate: 8.00%

n	Single payment		Uniform payment series			
	F/P	P/F	A/F	A/P	F/A	P/A
1	1.08000	0.92593	1.00000	1.08000	1.00000	0.92593
2	1.16640	0.85734	0.48077	0.56077	2.08000	1.78326
3	1.25971	0.79383	0.30803	0.38803	3.24640	2.57710
4	1.36049	0.73503	0.22192	0.30192	4.50611	3.31213
5	1.46933	0.68058	0.17046	0.25046	5.86660	3.99271
6	1.58687	0.63017	0.13632	0.21632	7.33593	4.62288
7	1.71382	0.58349	0.11207	0.19207	8.92280	5.20637
8	1.85093	0.54027	0.09401	0.17401	10.63663	5.74664
9	1.99900	0.50025	0.08008	0.16008	12.48756	6.24689
10	2.15892	0.46319	0.06903	0.14903	14.48656	6.71008
11	2.33164	0.42888	0.06008	0.14008	16.64549	7.13896
12	2.51817	0.39711	0.05270	0.13270	18.97713	7.53608
13	2.71962	0.36770	0.04652	0.12652	21.49530	7.90378
14	2.93719	0.34046	0.04130	0.12130	24.21492	8.24424
15	3.17217	0.31524	0.03683	0.11683	27.15211	8.55948
16	3.42594	0.29189	0.03298	0.11298	30.32428	8.85137
17	3.70002	0.27027	0.02963	0.10963	33.75023	9.12164
18	3.99602	0.25025	0.02670	0.10670	37.45024	9.37189
19	4.31570	0.23171	0.02413	0.10413	41.44626	9.60360
20	4.66096	0.21455	0.02185	0.10185	45.76196	9.81815
21	5.03383	0.19866	0.01983	0.09983	50.42292	10.01680
22	5.43654	0.18394	0.01803	0.09803	55.45676	10.20074
23	5.87146	0.17032	0.01642	0.09642	60.89330	10.37106
24	6.34118	0.15770	0.01498	0.09498	66.76476	10.52876
25	6.84848	0.14602	0.01368	0.09368	73.10594	10.67478
26	7.39635	0.13520	0.01251	0.09251	79.95442	10.80998
27	7.98806	0.12519	0.01145	0.09145	87.35077	10.93516
28	8.62711	0.11591	0.01049	0.09049	95.33883	11.05108
29	9.31727	0.10733	0.00962	0.08962	103.96594	11.15841
30	10.06266	0.09938	0.00883	0.08883	113.28321	11.25778
35	14.78534	0.06763	0.00580	0.08580	172.31680	11.65457
40	21.72452	0.04603	0.00386	0.08386	259.05652	11.92461
45	31.92045	0.03133	0.00259	0.08259	386.50562	12.10840
50	46.90161	0.02132	0.00174	0.08174	573.77016	12.23348
60	101.25706	0.00988	0.00080	0.08080	1,253.21330	12.37655
70	218.60641	0.00457	0.00037	0.08037	2,720.08007	12.44282

(continued)

(continued)

	Single payment		Uniform payment series			
n	F/P	P/F	A/F	A/P	F/A	P/A
80	471.95483	0.00212	0.00017	0.08017	5,886.93543	12.47351
90	1,018.91509	0.00098	0.00008	0.08008	12,723.93862	12.48773
100	2,199.76126	0.00045	0.00004	0.08004	27,484.51570	12.49432

Interest rate: 8.00%

	Arithmetic grad	
n	A/G	P/G
1	0.00000	0.00000
2	0.48077	0.85734
3	0.94874	2.44500
4	1.40396	4.65009
5	1.84647	7.37243
6	2.27635	10.52327
7	2.69366	14.02422
8	3.09852	17.80610
9	3.49103	21.80809
10	3.87131	25.97683
11	4.23950	30.26566
12	4.59575	34.63391
13	4.94021	39.04629
14	5.27305	43.47228
15	5.59446	47.88566
16	5.90463	52.26402
17	6.20375	56.58832
18	6.49203	60.84256
19	6.76969	65.01337
20	7.03695	69.08979
21	7.29403	73.06291
22	7.54118	76.92566
23	7.77863	80.67259
24	8.00661	84.29968
25	8.22538	87.80411
26	8.43518	91.18415
27	8.63627	94.43901
28	8.82888	97.56868

(continued)

(continued)

n	Arithmetic grad	
	A/G	P/G
29	9.01328	100.57385
30	9.18971	103.45579
35	9.96107	116.09199
40	10.56992	126.04220
45	11.04465	133.73309
50	11.41071	139.59279
60	11.90154	147.30001
70	12.17832	151.53262
80	12.33013	153.80008
90	12.41158	154.99254
100	12.45452	155.61073

Interest rate: 9.00%

n	Single payment		Uniform payment series			
	F/P	P/F	A/F	A/P	F/A	P/A
1	1.09000	0.91743	1.00000	1.09000	1.00000	0.91743
2	1.18810	0.84168	0.47847	0.56847	2.09000	1.75911
3	1.29503	0.77218	0.30505	0.39505	3.27810	2.53129
4	1.41158	0.70843	0.21867	0.30867	4.57313	3.23972
5	1.53862	0.64993	0.16709	0.25709	5.98471	3.88965
6	1.67710	0.59627	0.13292	0.22292	7.52333	4.48592
7	1.82804	0.54703	0.10869	0.19869	9.20043	5.03295
8	1.99256	0.50187	0.09067	0.18067	11.02847	5.53482
9	2.17189	0.46043	0.07680	0.16680	13.02104	5.99525
10	2.36736	0.42241	0.06582	0.15582	15.19293	6.41766
11	2.58043	0.38753	0.05695	0.14695	17.56029	6.80519
12	2.81266	0.35553	0.04965	0.13965	20.14072	7.16073
13	3.06580	0.32618	0.04357	0.13357	22.95338	7.48690
14	3.34173	0.29925	0.03843	0.12843	26.01919	7.78615
15	3.64248	0.27454	0.03406	0.12406	29.36092	8.06069
16	3.97031	0.25187	0.03030	0.12030	33.00340	8.31256
17	4.32763	0.23107	0.02705	0.11705	36.97370	8.54363
18	4.71712	0.21199	0.02421	0.11421	41.30134	8.75563
19	5.14166	0.19449	0.02173	0.11173	46.01846	8.95011
20	5.60441	0.17843	0.01955	0.10955	51.16012	9.12855

(continued)

(continued)

	Single payment		Uniform payment series			
n	F/P	P/F	A/F	A/P	F/A	P/A
21	6.10881	0.16370	0.01762	0.10762	56.76453	9.29224
22	6.65860	0.15018	0.01590	0.10590	62.87334	9.44243
23	7.25787	0.13778	0.01438	0.10438	69.53194	9.58021
24	7.91108	0.12640	0.01302	0.10302	76.78981	9.70661
25	8.62308	0.11597	0.01181	0.10181	84.70090	9.82258
26	9.39916	0.10639	0.01072	0.10072	93.32398	9.92897
27	10.24508	0.09761	0.00973	0.09973	102.72313	10.02658
28	11.16714	0.08955	0.00885	0.09885	112.96822	10.11613
29	12.17218	0.08215	0.00806	0.09806	124.13536	10.19828
30	13.26768	0.07537	0.00734	0.09734	136.30754	10.27365
35	20.41397	0.04899	0.00464	0.09464	215.71075	10.56682
40	31.40942	0.03184	0.00296	0.09296	337.88245	10.75736
45	48.32729	0.02069	0.00190	0.09190	525.85873	10.88120
50	74.35752	0.01345	0.00123	0.09123	815.08356	10.96168
60	176.03129	0.00568	0.00051	0.09051	1,944.79213	11.04799
70	416.73009	0.00240	0.00022	0.09022	4,619.22318	11.08445
80	986.55167	0.00101	0.00009	0.09009	10,950.57409	11.09985
90	2,335.52658	0.00043	0.00004	0.09004	25,939.18425	11.10635
100	5,529.04079	0.00018	0.00002	0.09002	61,422.67546	11.10910

Interest rate: 9.00%

	Arithmetic grad	
n	A/G	P/G
1	0.00000	0.00000
2	0.47847	0.84168
3	0.94262	2.38605
4	1.39250	4.51132
5	1.82820	7.11105
6	2.24979	10.09238
7	2.65740	13.37459
8	3.05117	16.88765
9	3.43123	20.57108
10	3.79777	24.37277
11	4.15096	28.24810
12	4.49102	32.15898

(continued)

(continued)

n	Arithmetic grad A/G	P/G
13	4.81816	36.07313
14	5.13262	39.96333
15	5.43463	43.80686
16	5.72446	47.58491
17	6.00238	51.28208
18	6.26865	54.88598
19	6.52358	58.38679
20	6.76745	61.77698
21	7.00056	65.05094
22	7.22322	68.20475
23	7.43574	71.23594
24	7.63843	74.14326
25	7.83160	76.92649
26	8.01556	79.58630
27	8.19064	82.12410
28	8.35714	84.54191
29	8.51538	86.84224
30	8.66566	89.02800
35	9.30829	98.35899
40	9.79573	105.37619
45	10.16029	110.55607
50	10.42952	114.32507
60	10.76832	118.96825
70	10.94273	121.29416
80	11.02994	122.43064
90	11.07256	122.97576
100	11.09302	123.23350

Interest rate: 10.00%

n	Single payment F/P	P/F	Uniform payment series A/F	A/P	F/A	P/A
1	1.10000	0.90909	1.00000	1.10000	1.00000	0.90909
2	1.21000	0.82645	0.47619	0.57619	2.10000	1.73554
3	1.33100	0.75131	0.30211	0.40211	3.31000	2.48685
4	1.46410	0.68301	0.21547	0.31547	4.64100	3.16987

(continued)

(continued)

	Single payment		Uniform payment series			
n	F/P	P/F	A/F	A/P	F/A	P/A
5	1.61051	0.62092	0.16380	0.26380	6.10510	3.79079
6	1.77156	0.56447	0.12961	0.22961	7.71561	4.35526
7	1.94872	0.51316	0.10541	0.20541	9.48717	4.86842
8	2.14359	0.46651	0.08744	0.18744	11.43589	5.33493
9	2.35795	0.42410	0.07364	0.17364	13.57948	5.75902
10	2.59374	0.38554	0.06275	0.16275	15.93742	6.14457
11	2.85312	0.35049	0.05396	0.15396	18.53117	6.49506
12	3.13843	0.31863	0.04676	0.14676	21.38428	6.81369
13	3.45227	0.28966	0.04078	0.14078	24.52271	7.10336
14	3.79750	0.26333	0.03575	0.13575	27.97498	7.36669
15	4.17725	0.23939	0.03147	0.13147	31.77248	7.60608
16	4.59497	0.21763	0.02782	0.12782	35.94973	7.82371
17	5.05447	0.19784	0.02466	0.12466	40.54470	8.02155
18	5.55992	0.17986	0.02193	0.12193	45.59917	8.20141
19	6.11591	0.16351	0.01955	0.11955	51.15909	8.36492
20	6.72750	0.14864	0.01746	0.11746	57.27500	8.51356
21	7.40025	0.13513	0.01562	0.11562	64.00250	8.64869
22	8.14027	0.12285	0.01401	0.11401	71.40275	8.77154
23	8.95430	0.11168	0.01257	0.11257	79.54302	8.88322
24	9.84973	0.10153	0.01130	0.11130	88.49733	8.98474
25	10.83471	0.09230	0.01017	0.11017	98.34706	9.07704
26	11.91818	0.08391	0.00916	0.10916	109.18177	9.16095
27	13.10999	0.07628	0.00826	0.10826	121.09994	9.23722
28	14.42099	0.06934	0.00745	0.10745	134.20994	9.30657
29	15.86309	0.06304	0.00673	0.10673	148.63093	9.36961
30	17.44940	0.05731	0.00608	0.10608	164.49402	9.42691
35	28.10244	0.03558	0.00369	0.10369	271.02437	9.64416
40	45.25926	0.02209	0.00226	0.10226	442.59256	9.77905
45	72.89048	0.01372	0.00139	0.10139	718.90484	9.86281
50	117.39085	0.00852	0.00086	0.10086	1,163.90853	9.91481
60	304.48164	0.00328	0.00033	0.10033	3,034.81640	9.96716
70	789.74696	0.00127	0.00013	0.10013	7,887.46957	9.98734
80	2,048.40021	0.00049	0.00005	0.10005	20,474.00215	9.99512
90	5,313.02261	0.00019	0.00002	0.10002	53,120.22612	9.99812
100	13,780.6123	0.00007	0.00001	0.10001	137,796.12340	9.99927

Interest rate: 10.00%

| n | Arithmetic grad | |
	A/G	P/G
1	0.00000	0.00000
2	0.47619	0.82645
3	0.93656	2.32908
4	1.38117	4.37812
5	1.81013	6.86180
6	2.22356	9.68417
7	2.62162	12.76312
8	3.00448	16.02867
9	3.37235	19.42145
10	3.72546	22.89134
11	4.06405	26.39628
12	4.38840	29.90122
13	4.69879	33.37719
14	4.99553	36.80050
15	5.27893	40.15199
16	5.54934	43.41642
17	5.80710	46.58194
18	6.05256	49.63954
19	6.28610	52.38268
20	6.50808	55.40691
21	6.71888	58.10952
22	6.91889	60.68929
23	7.10848	63.14621
24	7.28805	65.48130
25	7.45798	67.69640
26	7.61865	69.79404
27	7.77044	71.77726
28	7.91372	73.64953
29	8.04886	75.41463
30	8.17623	77.07658
35	8.70860	83.98715
40	9.09623	88.95254
45	9.37405	92.45443
50	9.57041	94.88887
60	9.80229	97.70101
70	9.91125	98.98702

(continued)

(continued)

	Arithmetic grad	
n	A/G	P/G
80	9.96093	99.56063
90	9.98306	99.81178
100	9.99274	99.92018

Interest rate: 12.00%

	Single payment		Uniform payment series			
n	F/P	P/F	A/F	A/P	F/A	P/A
1	1.12000	0.89286	1.00000	1.12000	1.00000	0.89286
2	1.25440	0.79719	0.47170	0.59170	2.12000	1.69005
3	1.40493	0.71178	0.29635	0.41635	3.37440	2.40183
4	1.57352	0.63552	0.20923	0.32923	4.77933	3.03735
5	1.76234	0.56743	0.15741	0.27741	6.35285	3.60478
6	1.97382	0.50663	0.12323	0.24323	8.11519	4.11141
7	2.21068	0.45235	0.09912	0.21912	10.08901	4.56376
8	2.47596	0.40388	0.08130	0.20130	12.29969	4.96764
9	2.77308	0.36061	0.06768	0.18768	14.77566	5.32825
10	3.10585	0.32197	0.05698	0.17698	17.54874	5.65022
11	3.47855	0.28748	0.04842	0.16842	20.65458	5.93770
12	3.89598	0.25668	0.04144	0.16144	24.13313	6.19437
13	4.36349	0.22917	0.03568	0.15568	28.02911	6.42355
14	4.88711	0.20462	0.03087	0.15087	32.39260	6.62817
15	5.47357	0.18270	0.02682	0.14682	37.27971	6.81086
16	6.13039	0.16312	0.02339	0.14339	42.75328	6.97399
17	6.86604	0.14564	0.02046	0.14046	48.88367	7.11963
18	7.68997	0.13004	0.01794	0.13794	55.74971	7.24967
19	8.61276	0.11611	0.01576	0.13576	63.43968	7.36578
20	9.64629	0.10367	0.01388	0.13388	72.05244	7.46944
21	10.80385	0.09256	0.01224	0.13224	81.69874	7.56200
22	12.10031	0.08264	0.01081	0.13081	92.50258	7.64465
23	13.55235	0.07379	0.00956	0.12956	104.60289	7.71843
24	15.17863	0.06588	0.00846	0.12846	118.15524	7.78432
25	17.00006	0.05882	0.00750	0.12750	133.33387	7.84314
26	19.04007	0.05252	0.00665	0.12665	150.33393	7.89566
27	21.32488	0.04689	0.00590	0.12590	169.37401	7.94255
28	23.88387	0.04187	0.00524	0.12524	190.69889	7.98442

(continued)

(continued)

n	Single payment			Uniform payment series			
	F/P	P/F	A/F	A/P	F/A	P/A
29	26.74993	0.03738	0.00466	0.12466	214.58275	8.02181
30	29.95992	0.03338	0.00414	0.12414	241.33268	8.05518
35	52.79962	0.01894	0.00232	0.12232	431.66350	8.17550
40	93.05097	0.01075	0.00130	0.12130	767.09142	8.24378
45	163.98760	0.00610	0.00074	0.12074	1,358.23003	8.28252
50	289.00219	0.00346	0.00042	0.12042	2,400.01825	8.30450
60	897.59693	0.00111	0.00013	0.12013	7,471.64111	8.32405
70	2,787.79983	0.00036	0.00004	0.12004	23,223.33190	8.33034
80	8,658.48310	0.00012	0.00001	0.12001	72,145.69250	8.33237
90	26,891.9342	0.00004	0.00000	0.12000	224,091.11853	8.33302
100	83,522.2657	0.00001	0.00000	0.12000	696,010.54772	8.33323

Interest rate: 12.00%

n	Arithmetic grad	
	A/G	P/G
1	0.00000	0.00000
2	0.47170	0.79719
3	0.92461	2.22075
4	1.35885	4.12731
5	1.77459	6.39702
6	2.17205	8.93017
7	2.55147	11.64427
8	2.91314	14.47145
9	3.25742	17.35633
10	3.58465	20.25409
11	3.89525	23.12885
12	4.18965	25.95228
13	4.46830	28.70237
14	4.73169	31.36242
15	4.98030	33.92017
16	5.21466	36.36700
17	5.43530	38.69731
18	5.64274	40.90798
19	5.83752	42.99790
20	6.02020	44.96757

(continued)

(continued)

n	Arithmetic grad	
	A/G	P/G
21	6.19132	46.81876
22	6.35141	48.55425
23	6.50101	50.17759
24	6.64064	51.69288
25	6.77084	53.10464
26	6.89210	54.41766
27	7.00491	55.63689
28	7.10976	56.76736
29	7.20712	57.81409
30	7.29742	58.78205
35	7.65765	62.60517
40	7.89879	65.11587
45	8.05724	66.73421
50	8.15972	67.76241
60	8.26641	68.81003
70	8.30821	69.21029
80	8.32409	69.35943
90	8.32999	69.41397
100	8.33214	69.43364

Interest rate: 15.00%

n	Single payment		Uniform payment series			
	F/P	P/F	A/F	A/P	F/A	P/A
1	1.15000	0.86957	1.00000	1.15000	1.00000	0.86957
2	1.32250	0.75614	0.46512	0.61512	2.15000	1.62571
3	1.52088	0.65752	0.28798	0.43798	3.47250	2.28323
4	1.74901	0.57175	0.20027	0.35027	4.99338	2.85498
5	2.01136	0.49718	0.14832	0.29832	6.74238	3.35216
6	2.31306	0.43233	0.11424	0.26424	8.75374	3.78448
7	2.66002	0.37594	0.09036	0.24036	11.06680	4.16042
8	3.05902	0.32690	0.07285	0.22285	13.72682	4.48732
9	3.51788	0.28426	0.05957	0.20957	16.78584	4.77158
10	4.04556	0.24718	0.04925	0.19925	20.30372	5.01877
11	4.65239	0.21494	0.04107	0.19107	24.34928	5.23371
12	5.35025	0.18691	0.03448	0.18448	29.00167	5.42062

(continued)

(continued)

	Single payment		Uniform payment series			
n	F/P	P/F	A/F	A/P	F/A	P/A
13	6.15279	0.16253	0.02911	0.17911	34.35192	5.58315
14	7.07571	0.14133	0.02469	0.17469	40.50471	5.72448
15	8.13706	0.12289	0.02102	0.17102	47.58041	5.84737
16	9.35762	0.10686	0.01795	0.16795	55.71747	5.95423
17	10.76126	0.09293	0.01537	0.16537	65.07509	6.04716
18	12.37545	0.08081	0.01319	0.16319	75.83636	6.12797
19	14.23177	0.07027	0.01134	0.16134	88.21181	6.19823
20	16.36654	0.06110	0.00976	0.15976	102.44358	6.25933
21	18.82152	0.05313	0.00842	0.15842	118.81012	6.31246
22	21.64475	0.04620	0.00727	0.15727	137.63164	6.35866
23	24.89146	0.04017	0.00628	0.15628	159.27638	6.39884
24	28.62518	0.03493	0.00543	0.15543	184.16784	6.43377
25	32.91895	0.03038	0.00470	0.15470	212.79302	6.46415
26	37.85680	0.02642	0.00407	0.15407	245.71197	6.49056
27	43.53531	0.02297	0.00353	0.15353	283.56877	6.51353
28	50.06561	0.01997	0.00306	0.15306	327.10408	6.53351
29	57.57545	0.01737	0.00265	0.15265	377.16969	6.55088
30	66.21177	0.01510	0.00230	0.15230	434.74515	6.56598
35	133.17552	0.00751	0.00113	0.15113	881.17016	6.61661
40	267.86355	0.00373	0.00056	0.15056	1,779.09031	6.64178
45	538.76927	0.00186	0.00028	0.15028	3,585.12846	6.65429
50	1,083.65744	0.00092	0.00014	0.15014	7,217.71628	6.66051
60	4,383.99875	0.00023	0.00003	0.15003	29,219.99164	6.66515
70	17,735.7200	0.00006	0.00001	0.15001	118,231.46693	6.66629
80	71,750.8794	0.00001	0.00000	0.15000	478,332.52934	6.66657
90	290,272.325	0.0000	0.0000	0.15000	1,935,142.168	6.6666
100	1,174,313.4	0.0000	0.0000	0.1500	7,828,749.671	6.6666

Interest rate: 15.00%

	Arithmetic grad	
n	A/G	P/G
1	0.00000	0.00000
2	0.46512	0.75614
3	0.90713	2.07118
4	1.32626	3.78644

(continued)

(continued)

n	Arithmetic grad	
	A/G	P/G
5	1.72281	5.77514
6	2.09719	7.93678
7	2.44985	10.19240
8	2.78133	12.48072
9	3.09223	14.75481
10	3.38320	16.97948
11	3.65494	19.12891
12	3.90820	21.18489
13	4.14376	23.13522
14	4.36241	24.97250
15	4.56496	26.69302
16	4.75225	28.29599
17	4.92509	29.78280
18	5.08431	31.15649
19	5.23073	32.42127
20	5.36514	33.58217
21	5.48832	34.64479
22	5.60102	35.61500
23	5.70398	36.49884
24	5.79789	37.30232
25	5.88343	38.03139
26	5.96123	38.69177
27	6.03190	39.28899
28	6.09600	39.82828
29	6.15408	40.31460
30	6.20663	40.75259
35	6.40187	42.35864
40	6.51678	43.28299
45	6.58299	43.80513
50	6.62048	44.09583
60	6.65298	44.34307
70	6.66272	44.41563
80	6.66555	44.43639
90	6.66636	44.44222
100	6.66658	44.44384

Interest rate: 18.00%

n	Single payment		Uniform payment series			
	F/P	P/F	A/F	A/P	F/A	P/A
1	1.18000	0.84746	1.00000	1.18000	1.00000	0.84746
2	1.39240	0.71818	0.45872	0.63872	2.18000	1.56564
3	1.64303	0.60863	0.27992	0.45992	3.57240	2.17427
4	1.93878	0.51579	0.19174	0.37174	5.21543	2.69006
5	2.28776	0.43711	0.13978	0.31978	7.15421	3.12717
6	2.69955	0.37043	0.10591	0.28591	9.44197	3.49760
7	3.18547	0.31393	0.08236	0.26236	12.14152	3.81153
8	3.75886	0.26604	0.06524	0.24524	15.32700	4.07757
9	4.43545	0.22546	0.05239	0.23239	19.08585	4.30302
10	5.23384	0.19106	0.04251	0.22251	23.52131	4.49409
11	6.17593	0.16192	0.03478	0.21478	28.75514	4.65601
12	7.28759	0.13722	0.02863	0.20863	34.93107	4.79322
13	8.59936	0.11629	0.02369	0.20369	42.21866	4.90951
14	10.14724	0.09855	0.01968	0.19968	50.81802	5.00806
15	11.97375	0.08352	0.01640	0.19640	60.96527	5.09158
16	14.12902	0.07078	0.01371	0.19371	72.93901	5.16235
17	16.67225	0.05998	0.01149	0.19149	87.06804	5.22233
18	19.67325	0.05083	0.00964	0.18964	103.74028	5.27316
19	23.21444	0.04308	0.00810	0.18810	123.41353	5.31624
20	27.39303	0.03651	0.00682	0.18682	146.62797	5.35275
21	32.32378	0.03094	0.00575	0.18575	174.02100	5.38368
22	38.14206	0.02622	0.00485	0.18485	206.34479	5.40990
23	45.00763	0.02222	0.00409	0.18409	244.48685	5.43212
24	53.10901	0.01883	0.00345	0.18345	289.49448	5.45095
25	62.66863	0.01596	0.00292	0.18292	342.60349	5.46691
26	73.94898	0.01352	0.00247	0.18247	405.27211	5.48043
27	87.25980	0.01146	0.00209	0.18209	479.22109	5.49189
28	102.96656	0.00971	0.00177	0.18177	566.48089	5.50160
29	121.50054	0.00823	0.00149	0.18149	669.44745	5.50983
30	143.37064	0.00697	0.00126	0.18126	790.94799	5.51681
35	327.99729	0.00305	0.00055	0.18055	1,816.65161	5.53862
40	750.37834	0.00133	0.00024	0.18024	4,163.21303	5.54815
45	1,716.68388	0.00058	0.00010	0.18010	9,531.57711	5.55232
50	3,927.35686	0.00025	0.00005	0.18005	21,813.09367	5.55414
60	20,555.1399	0.00005	0.00001	0.18001	114,189.66648	5.55529
70	107,582.222	0.0000	0.0000	0.1800	597,673.4576	5.5555

(continued)

(continued)

n	Single payment		Uniform payment series			
	F/P	P/F	A/F	A/P	F/A	P/A
80	563,067.660	0.0000	0.0000	0.1800	3,128,148.113	5.5555
90	2,947,003.5	0.0000	0.0000	0.1800	16,372,236.33	5.5555
100	15,424,131	0.0000	0.0000	0.1800	85,689,616.14	5.5555

Interest rate: 18.00%

	Arithmetic grad	
n	A/G	P/G
1	0.00000	0.00000
2	0.45872	0.71818
3	0.89016	1.93545
4	1.29470	3.48281
5	1.67284	5.23125
6	2.02522	7.08341
7	2.35259	8.96696
8	2.65581	10.82922
9	2.93581	12.63287
10	3.19363	14.35245
11	3.43033	15.97164
12	3.64703	17.48106
13	3.84489	18.87651
14	4.02504	20.15765
15	4.18866	21.32687
16	4.33688	22.38852
17	4.47084	23.34820
18	4.59161	24.21231
19	4.70026	24.98769
20	4.79778	25.68130
21	4.88514	26.30004
22	4.96324	26.85061
23	5.03292	27.33942
24	5.09498	27.77249
25	5.15016	28.15546
26	5.19914	28.49353
27	5.24255	28.79149
28	5.28096	29.05371

(continued)

(continued)

	Arithmetic grad	
n	A/G	P/G
29	5.31489	29.28416
30	5.34484	29.48643
35	5.44852	30.17728
40	5.50218	30.52692
45	5.52933	30.70059
50	5.54282	30.78561
60	5.55264	30.84648
70	5.55490	30.86030
80	5.55541	30.86335
90	5.55553	30.86402
100	5.55555	30.86416

Interest rate: 20.00%

	Single payment		Uniform payment series			
n	F/P	P/F	A/F	A/P	F/A	P/A
1	1.20000	0.83333	1.00000	1.20000	1.00000	0.83333
2	1.44000	0.69444	0.45455	0.65455	2.20000	1.52778
3	1.72800	0.57870	0.27473	0.47473	3.64000	2.10648
4	2.07360	0.48225	0.18629	0.38629	5.36800	2.58873
5	2.48832	0.40188	0.13438	0.33438	7.44160	2.99061
6	2.98598	0.33490	0.10071	0.30071	9.92992	3.32551
7	3.58318	0.27908	0.07742	0.27742	12.91590	3.60459
8	4.29982	0.23257	0.06061	0.26061	16.49908	3.83716
9	5.15978	0.19381	0.04808	0.24808	20.79890	4.03097
10	6.19174	0.16151	0.03852	0.23852	25.95868	4.19247
11	7.43008	0.13459	0.03110	0.23110	32.15042	4.32706
12	8.91610	0.11216	0.02526	0.22526	39.58050	4.43922
13	10.69932	0.09346	0.02062	0.22062	48.49660	4.53268
14	12.83918	0.07789	0.01689	0.21689	59.19592	4.61057
15	15.40702	0.06491	0.01388	0.21388	72.03511	4.67547
16	18.48843	0.05409	0.01144	0.21144	87.44213	4.72956
17	22.18611	0.04507	0.00944	0.20944	105.93056	4.77463
18	26.62333	0.03756	0.00781	0.20781	128.11667	4.81219
19	31.94800	0.03130	0.00646	0.20646	154.74000	4.84350
20	38.33760	0.02608	0.00536	0.20536	186.68800	4.86958

(continued)

(continued)

	Single payment		Uniform payment series			
n	F/P	P/F	A/F	A/P	F/A	P/A
21	46.00512	0.02174	0.00444	0.20444	225.02560	4.89132
22	55.20614	0.01811	0.00369	0.20369	271.03072	4.90943
23	66.24737	0.01509	0.00307	0.20307	326.23686	4.92453
24	79.49685	0.01258	0.00255	0.20255	392.48424	4.93710
25	95.39622	0.01048	0.00212	0.20212	471.98108	4.94759
26	114.47546	0.00874	0.00176	0.20176	567.37730	4.95632
27	137.37055	0.00728	0.00147	0.20147	681.85276	4.96360
28	164.84466	0.00607	0.00122	0.20122	819.22331	4.96967
29	197.81359	0.00506	0.00102	0.20102	984.06797	4.97472
30	237.37631	0.00421	0.00085	0.20085	1,181.88157	4.97894
35	590.66823	0.00169	0.00034	0.20034	2,948.34115	4.99154
40	1,469.77157	0.00068	0.00014	0.20014	7,343.85784	4.99660
45	3,657.26199	0.00027	0.00005	0.20005	18,281.30994	4.99863
50	9,100.43815	0.00011	0.00002	0.20002	45,497.19075	4.99945
60	56,347.51435	0.00002	0.00000	0.20000	281,732.5717	4.9999
70	348,888.956	0.0000	0.0000	0.2000	1,744,439.78	4.9999
80	2,160,228.4	0.0000	0.0000	0.2000	10,801,137.3	5.0000
90	13,375,565.2	0.0000	0.0000	0.2000	66,877,821.2	5.0000
100	82,817,974.5	0.0000	0.0000	0.2000	414,089,867	5.0000

Interest rate: 20.00%

	Arithmetic grad	
n	A/G	P/G
1	0.00000	0.00000
2	0.45455	0.69444
3	0.87912	1.85185
4	1.27422	3.29861
5	1.64051	4.90612
6	1.97883	6.58061
7	2.29016	8.25510
8	2.57562	9.88308
9	2.83642	11.43353
10	3.07386	12.88708
11	3.28929	14.23296
12	3.48410	15.46668

(continued)

(continued)

n	Arithmetic grad	
	A/G	P/G
13	3.65970	16.58825
14	3.81749	17.60078
15	3.95884	18.50945
16	4.08511	19.32077
17	4.19759	20.04194
18	4.29752	20.68048
19	4.38607	21.24390
20	4.46435	21.73949
21	4.53339	22.17423
22	4.59414	22.55462
23	4.64750	22.88671
24	4.69426	23.17603
25	4.73516	23.42761
26	4.77088	23.64600
27	4.80201	23.83527
28	4.82911	23.99906
29	4.85265	24.14061
30	4.87308	24.26277
35	4.94064	24.66140
40	4.97277	24.84691
45	4.98769	24.93164
50	4.99451	24.96978
60	4.99894	24.99423
70	4.99980	24.99893
80	4.99996	24.99980
90	4.99999	24.99996
100	5.00000	24.99999

Interest rate: 25.00%

n	Single payment		Uniform payment series			
	F/P	P/F	A/F	A/P	F/A	P/A
1	1.25000	0.80000	1.00000	1.25000	1.00000	0.80000
2	1.56250	0.64000	0.44444	0.69444	2.25000	1.44000
3	1.95313	0.51200	0.26230	0.51230	3.81250	1.95200
4	2.44141	0.40960	0.17344	0.42344	5.76563	2.36160

(continued)

(continued)

n	Single payment		Uniform payment series			
	F/P	P/F	A/F	A/P	F/A	P/A
5	3.05176	0.32768	0.12185	0.37185	8.20703	2.68928
6	3.81470	0.26214	0.08882	0.33882	11.25879	2.95142
7	4.76837	0.20972	0.06634	0.31634	15.07349	3.16114
8	5.96046	0.16777	0.05040	0.30040	19.84186	3.32891
9	7.45058	0.13422	0.03876	0.28876	25.80232	3.46313
10	9.31323	0.10737	0.03007	0.28007	33.25290	3.57050
11	11.64153	0.08590	0.02349	0.27349	42.56613	3.65640
12	14.55192	0.06872	0.01845	0.26845	54.20766	3.72512
13	18.18989	0.05498	0.01454	0.26454	68.75958	3.78010
14	22.73737	0.04398	0.01150	0.26150	86.94947	3.82408
15	28.42171	0.03518	0.00912	0.25912	109.68684	3.85926
16	35.52714	0.02815	0.00724	0.25724	138.10855	3.88741
17	44.40892	0.02252	0.00576	0.25576	173.63568	3.90993
18	55.51115	0.01801	0.00459	0.25459	218.04460	3.92794
19	69.38894	0.01441	0.00366	0.25366	273.55576	3.94235
20	86.73617	0.01153	0.00292	0.25292	342.94470	3.95388
21	108.42022	0.00922	0.00233	0.25233	429.68087	3.96311
22	135.52527	0.00738	0.00186	0.25186	538.10109	3.97049
23	169.40659	0.00590	0.00148	0.25148	673.62636	3.97639
24	211.75824	0.00472	0.00119	0.25119	843.03295	3.98111
25	264.69780	0.00378	0.00095	0.25095	1,054.79118	3.98489
26	330.87225	0.00302	0.00076	0.25076	1,319.48898	3.98791
27	413.59031	0.00242	0.00061	0.25061	1,650.36123	3.99033
28	516.98788	0.00193	0.00048	0.25048	2,063.95153	3.99226
29	646.23485	0.00155	0.00039	0.25039	2,580.93941	3.99381
30	807.79357	0.00124	0.00031	0.25031	3,227.17427	3.99505
35	2,465.19033	0.00041	0.00010	0.25010	9,856.76132	3.99838
40	7,523.16385	0.00013	0.00003	0.25003	30,088.65538	3.99947
45	22,958.87404	0.00004	0.00001	0.25001	91,831.49616	3.99983
50	70,064.9232	0.0000	0.0000	0.2500	280,255.6928	3.99994
60	652,530.4468	0.0000	0.0000	0.2500	2,610,117.78	3.9999
70	6,077,163.35	0.0000	0.0000	0.2500	24,308,649.4	4.0000
80	56,597,994.2	0.0000	0.0000	0.2500	226,391,972	4.0000
90	527,109,897	0.000	0.000	0.250	2,108,439,58	4.000

Interest rate: 25.00%

| n | Arithmetic grad | |
	A/G	P/G
1	0.00000	0.00000
2	0.44444	0.64000
3	0.85246	1.66400
4	1.22493	2.89280
5	1.56307	4.20352
6	1.86833	5.51424
7	2.14243	6.77253
8	2.38725	7.94694
9	2.60478	9.02068
10	2.79710	9.98705
11	2.96631	10.84604
12	3.11452	11.60195
13	3.24374	12.26166
14	3.35595	12.83341
15	3.45299	13.32599
16	3.53660	13.74820
17	3.60838	14.10849
18	3.66979	14.41473
19	3.72218	14.67414
20	3.76673	14.89320
21	3.80451	15.07766
22	3.83646	15.23262
23	3.86343	15.36248
24	3.88613	15.47109
25	3.90519	15.56176
26	3.92118	15.63732
27	3.93456	15.70019
28	3.94574	15.75241
29	3.95506	15.79574
30	3.96282	15.83164
35	3.98580	15.93672
40	3.99468	15.97661
45	3.99804	15.99146
50	3.99929	15.99692
60	3.99991	15.99961
70	3.99999	15.99995

(continued)

(continued)

n	Arithmetic grad	
	A/G	P/G
80	4.00000	15.99999
90	4.00000	16.00000

Appendix C
Selected Discrete Payment Compound Interest Factors with Geometric Gradient Percentages

Interest rate: 1.0%; gradient rates: 0.25%, 0.50%

	Single payment		Single payment	
n	P/A, i, 0.25%	F/A, i, 0.25%	P/A, i, 0.50%	F/A, i, 0.50%
1	0.99010	1.00000	0.99010	1.00000
2	1.97285	2.01250	1.97530	2.01500
3	2.94829	3.03763	2.95562	3.04518
4	3.91650	4.07553	3.93108	4.09070
5	4.87752	5.12632	4.90172	5.15176
6	5.83140	6.19015	5.86756	6.22853
7	6.77819	7.26714	6.82861	7.32119
8	7.71796	8.35744	7.78490	8.42993
9	8.65075	9.46119	8.73646	9.55494
10	9.57661	10.57853	9.68331	10.69640
11	10.49559	11.70960	10.62547	11.85450
12	11.40775	12.85454	11.56297	13.02944
13	12.31314	14.01350	12.49583	14.22142
14	13.21181	15.18663	13.42406	15.43062
15	14.10380	16.37407	14.34771	16.65724
16	14.98917	17.57598	15.26678	17.90150
17	15.86796	18.79250	16.18130	19.16358
18	16.74023	20.02378	17.09129	20.44371
19	17.60602	21.26999	17.99678	21.74207
20	18.46538	22.53127	18.89779	23.05889
21	19.31836	23.80779	19.79433	24.39438

(continued)

© The Editor(s) (if applicable) and The Author(s), under exclusive license to Springer Nature Switzerland AG 2022
T. S. Cotter, *Engineering Managerial Economic Decision and Risk Analysis*, Topics in Safety, Risk, Reliability and Quality 39,
https://doi.org/10.1007/978-3-030-87767-5

(continued)

n	Single payment		Single payment	
	P/A, i, 0.25%	F/A, i, 0.25%	P/A, i, 0.50%	F/A, i, 0.50%
22	20.16500	25.09970	20.68644	25.74874
23	21.00536	26.40717	21.57413	27.12220
24	21.83948	27.73035	22.45743	28.51497
25	22.66741	29.06941	23.33635	29.92728
26	23.48918	30.42451	24.21092	31.35935
27	24.30486	31.79583	25.08117	32.81141
28	25.11448	33.18353	25.94710	34.28367
29	25.91808	34.58778	26.80875	35.77638
30	26.71572	36.00875	27.66613	37.28977
35	30.61594	43.37062	31.88980	45.17517
40	34.37348	51.17743	36.00994	53.61390
45	37.99357	59.45274	40.02911	62.63799
50	41.48123	68.22135	43.94978	72.28120
60	48.07847	87.34399	51.50519	93.56931
70	54.20186	108.77030	58.69480	117.78657
80	59.88544	132.74897	65.53631	145.27533
90	65.16080	159.55487	72.04658	176.41560
100	70.05726	189.49186	78.24164	211.62907

Interest rate: 2.0%; gradient rates: 0.50%, 1.25%

n	Single payment		Single payment	
	P/A, i, 0.50%	F/A, i, 0.50%	P/A, i, 1.25%	F/A, i, 1.25%
1	0.98039	1.00000	0.98039	1.00000
2	1.94637	2.02500	1.95358	2.03250
3	2.89814	3.07553	2.91960	3.09831
4	3.83591	4.15211	3.87853	4.19824
5	4.75989	5.25530	4.83040	5.33315
6	5.67028	6.38566	5.77528	6.50390
7	6.56729	7.54375	6.71320	7.71136
8	7.45110	8.73016	7.64423	8.95644
9	8.32192	9.94547	8.56842	10.24005
10	9.17993	11.19029	9.48581	11.56315
11	10.02533	12.46523	10.39645	12.92668
12	10.85829	13.77093	11.30040	14.33164
13	11.67900	15.10803	12.19770	15.77902

(continued)

(continued)

n	Single payment		Single payment	
	P/A, i, 0.50%	F/A, i, 0.50%	P/A, i, 1.25%	F/A, i, 1.25%
14	12.48764	16.47718	13.08840	17.26987
15	13.28439	17.87904	13.97256	18.80522
16	14.06942	19.31430	14.85021	20.38615
17	14.84291	20.78366	15.72141	22.01377
18	15.60503	22.28782	16.58620	23.68918
19	16.35593	23.82751	17.44464	25.41354
20	17.09580	25.40345	18.29676	27.18802
21	17.82478	27.01642	19.14262	29.01382
22	18.54304	28.66717	19.98225	30.89216
23	19.25074	30.35648	20.81572	32.82429
24	19.94804	32.08516	21.64305	34.81149
25	20.63508	33.85403	22.46431	36.85507
26	21.31201	35.66390	23.27952	38.95637
27	21.97899	37.51564	24.08874	41.11674
28	22.63616	39.41011	24.89201	43.33759
29	23.28367	41.34818	25.68937	45.62033
30	23.92165	43.33077	26.48087	47.96643
35	26.97358	53.94418	30.35192	60.70049
40	29.80760	65.81636	34.08273	75.25603
45	32.43928	79.08223	37.67838	91.85441
50	34.88306	93.89081	41.14377	110.74209
60	39.25963	128.81204	47.70247	156.51326
70	43.03355	172.11518	53.79455	215.15443
80	46.28780	225.67337	59.45322	289.86056
90	49.09395	291.77190	64.70931	384.57607
100	51.51370	373.19851	69.59147	504.16558

Interest rate: 3.0%; gradient rates: 1.00%, 2.00%

n	Single payment		Single payment	
	P/A, i, 1.00%	F/A, i, 1.00%	P/A, i, 2.00%	F/A, i, 2.00%
1	0.97087	1.00000	0.97087	1.00000
2	1.92290	2.04000	1.93232	2.05000
3	2.85643	3.12130	2.88443	3.15190
4	3.77184	4.24524	3.82730	4.30766
5	4.66947	5.41320	4.76102	5.51933

(continued)

(continued)

n	Single payment		Single payment	
	P/A, i, 1.00%	F/A, i, 1.00%	P/A, i, 2.00%	F/A, i, 2.00%
6	5.54968	6.62661	5.68567	6.78899
7	6.41279	7.88693	6.60134	8.11882
8	7.25915	9.19567	7.50813	9.51107
9	8.08907	10.55440	8.40611	10.96806
10	8.90287	11.96471	9.29537	12.49220
11	9.70087	13.42828	10.17599	14.08596
12	10.48338	14.94679	11.04807	15.75191
13	11.25069	16.52202	11.91168	17.49271
14	12.00311	18.15578	12.76691	19.31110
15	12.74091	19.84992	13.61383	21.20991
16	13.46439	21.60639	14.45253	23.19207
17	14.17382	23.42716	15.28309	25.26062
18	14.86947	25.31428	16.10559	27.41868
19	15.55162	27.26986	16.92009	29.66949
20	16.22052	29.29606	17.72670	32.01638
21	16.87643	31.39513	18.52547	34.46282
22	17.51961	33.56938	19.31648	37.01237
23	18.15029	35.82117	20.09982	39.66872
24	18.76873	38.15297	20.87555	42.43569
25	19.37517	40.56730	21.64374	45.31719
26	19.96982	43.06675	22.40448	48.31732
27	20.55293	45.65401	23.15784	51.44025
28	21.12472	48.33184	23.90388	54.69035
29	21.68541	51.10308	24.64268	58.07208
30	22.23520	53.97068	25.37430	61.59009
35	24.82814	69.86298	28.92725	81.39729
40	27.17893	88.65870	32.31103	105.39981
45	29.31018	110.83925	35.53372	134.37416
50	31.24239	136.96371	38.60297	169.23180
60	34.58232	203.74532	44.31005	261.05723
70	37.32755	295.55293	49.48664	391.82637
80	39.58398	421.20877	54.18204	576.54514
90	41.43863	592.59172	58.44099	835.73340
100	42.96304	825.69091	62.30405	1,197.39859

Interest rate: 5.0%; gradient rates: 2.00%, 4.00%

n	Single payment		Single payment	
	P/A, i, 2.00%	F/A, i, 2.00%	P/A, i , 4.00%	F/A, i, 4.00%
1	0.95238	1.00000	0.95238	1.00000
2	1.87755	2.07000	1.89569	2.09000
3	2.77629	3.21390	2.83002	3.27610
4	3.64935	4.43580	3.75545	4.56477
5	4.49746	5.74003	4.67206	5.96287
6	5.32134	7.13111	5.57995	7.47766
7	6.12168	8.61383	6.47919	9.11686
8	6.89916	10.19320	7.36986	10.88864
9	7.65442	11.87452	8.25205	12.80164
10	8.38811	13.66334	9.12584	14.86503
11	9.10083	15.56550	9.99131	17.08853
12	9.79318	17.58715	10.84854	19.48241
13	10.46576	19.73475	11.69760	22.05756
14	11.11912	22.01509	12.53857	24.82552
15	11.75381	24.43533	13.37154	27.79847
16	12.37037	27.00296	14.19657	30.98933
17	12.96931	29.72590	15.01375	34.41178
18	13.55114	32.61243	15.82314	38.08027
19	14.11635	35.67130	16.62482	42.01010
20	14.66540	38.91168	17.41887	46.21746
21	15.19877	42.34321	18.20536	50.71945
22	15.71690	45.97603	18.98436	55.53419
23	16.22023	49.82082	19.75593	60.68082
24	16.70917	53.88876	20.52016	66.17958
25	17.18415	58.19163	21.27711	72.05186
26	17.64556	62.74182	22.02686	78.32029
27	18.09378	67.55233	22.76946	85.00877
28	18.52919	72.63683	23.50499	92.14258
29	18.95217	78.00970	24.23351	99.74841
30	19.36306	83.68603	24.95510	107.85449
35	21.24798	117.20419	28.46124	156.99264
40	22.87858	161.06497	31.80357	223.89681
45	24.28918	218.23845	34.98975	314.38321
50	25.50945	292.52706	38.02707	436.07164
60	27.47828	513.27184	43.68262	815.95585

(continued)

(continued)

n	Single payment		Single payment	
	P/A, i, 2.00%	F/A, i, 2.00%	P/A, i , 4.00%	F/A, i, 4.00%
70	28.95166	880.89558	48.82206	1,485.48072
80	30.05428	1,489.53340	53.49248	2,651.16420
90	30.87943	2,492.90773	57.73668	4,661.10317
100	31.49694	4,141.88706	61.59356	8,099.63097

Interest rate: 7.0%; gradient rates: 3.00%, 5.00%

n	Single payment		Single payment	
	P/A, i, 3.00%	F/A, i, 3.00%	P/A, i, 5.00%	F/A, i, 5.00%
1	0.93458	1.00000	0.93458	1.00000
2	1.83422	2.10000	1.85169	2.12000
3	2.70023	3.30790	2.75166	3.37090
4	3.53387	4.63218	3.63481	4.76449
5	4.33634	6.08194	4.50144	6.31351
6	5.10881	7.66695	5.35188	8.03174
7	5.85241	9.39769	6.18643	9.93405
8	6.56821	11.28540	7.00537	12.03654
9	7.25725	13.34215	7.80901	14.35655
10	7.92053	15.58087	8.59763	16.91284
11	8.55901	18.01545	9.37150	19.72563
12	9.17363	20.66077	10.13092	22.81676
13	9.76527	23.53278	10.87613	26.20979
14	10.33479	26.64861	11.60742	29.93013
15	10.88302	30.02660	12.32504	34.00517
16	11.41076	33.68643	13.02924	38.46446
17	11.91877	37.64919	13.72028	43.33984
18	12.40779	41.93748	14.39841	48.66565
19	12.87852	46.57554	15.06386	54.47887
20	13.33166	51.58933	15.71687	60.81934
21	13.76786	57.00670	16.35768	67.72999
22	14.18776	62.85746	16.98651	75.25705
23	14.59195	69.17358	17.60358	83.45031
24	14.98104	75.98932	18.20912	92.36335
25	15.35558	83.34137	18.80334	102.05388
26	15.71612	91.26904	19.38646	112.58401
27	16.06318	99.81447	19.95868	124.02057

(continued)

(continued)

| | Single payment | | Single payment | |
n	P/A, i, 3.00%	F/A, i, 3.00%	P/A, i, 5.00%	F/A, i, 5.00%
28	16.39727	109.02277	20.52020	136.43546
29	16.71886	118.94229	21.07122	149.90607
30	17.02844	129.62481	21.61194	164.51563
35	18.41113	196.56798	24.16769	258.02831
40	19.55400	292.81050	26.49334	396.72346
45	20.49863	430.52140	28.60962	600.87220
50	21.27941	626.82798	30.53537	899.48126
60	22.45817	1,301.37059	33.88237	1,963.36205
70	23.26347	2,651.78926	36.65383	4,178.14833
80	23.81364	5,339.83743	38.94874	8,733.64733
90	24.18951	10,670.06282	40.84903	18,018.63074
100	24.44629	21,212.44234	42.42257	36,810.75339

Interest rate: 10.0%; gradient rates: 4.00%, 8.00%

| | Single payment | | Single payment | |
n	P/A, i, 4.00%	F/A, i, 4.00%	P/A, i, 8.00%	F/A, i, 8.00%
1	0.90909	1.00000	0.90909	1.00000
2	1.76860	2.14000	1.80165	2.18000
3	2.58122	3.43560	2.67799	3.56440
4	3.34951	4.90402	3.53839	5.18055
5	4.07590	6.56428	4.38314	7.05910
6	4.76267	8.43737	5.21254	9.23433
7	5.41198	10.54642	6.02686	11.74464
8	6.02587	12.91700	6.82637	14.63293
9	6.60628	15.57726	7.61134	17.94715
10	7.15503	18.55830	8.38205	21.74087
11	7.67385	21.89438	9.13874	26.07389
12	8.16436	25.62327	9.88167	31.01291
13	8.62813	29.78663	10.61109	36.63237
14	9.06659	34.43036	11.32726	43.01524
15	9.48114	39.60508	12.03040	50.25395
16	9.87308	45.36653	12.72075	58.45152
17	10.24364	51.77616	13.39856	67.72261
18	10.59398	58.90168	14.06404	78.19489
19	10.92522	66.81766	14.71742	90.01040

(continued)

(continued)

n	Single payment		Single payment	
	P/A, i, 4.00%	F/A, i, 4.00%	P/A, i, 8.00%	F/A, i, 8.00%
20	11.23839	75.60628	15.35892	103.32714
21	11.53448	85.35803	15.98876	118.32081
22	11.81442	96.17260	16.60715	135.18673
23	12.07909	108.15978	17.21429	154.14194
24	12.32932	121.44048	17.81039	175.42760
25	12.56590	136.14783	18.39566	199.31154
26	12.78958	152.42845	18.97028	226.09117
27	13.00106	170.44376	19.53446	256.09664
28	13.20100	190.37150	20.08838	289.69436
29	13.39003	212.40736	20.63223	327.29090
30	13.56876	236.76675	21.16618	369.33727
35	14.32637	402.60580	23.69384	665.85463
40	14.89870	674.30392	25.99991	1,176.73670
45	15.33106	1,117.48847	28.10383	2,048.50171
50	15.65769	1,838.06949	30.02331	3,524.46202
60	16.09085	4,899.36687	33.37222	10,161.22879
70	16.33805	12,902.92231	36.15972	28,557.02754
80	16.47912	33,755.84026	38.47992	78,822.26902
90	16.55964	87,981.72131	40.41115	214,705.37613
100	16.60558	228,835.12319	42.01864	579,042.55417

Appendix D
Cumulative Standard Normal Distribution

Z	0.00	0.01	0.02	0.03	0.04
−3.9	0.00005	0.00005	0.00004	0.00004	0.00004
−3.8	0.00007	0.00007	0.00007	0.00006	0.00006
−3.7	0.00011	0.00010	0.00010	0.00010	0.00009
−3.6	0.00016	0.00015	0.00015	0.00014	0.00014
−3.5	0.00023	0.00022	0.00022	0.00021	0.00020
3.4	0.00034	0.00032	0.00031	0.00030	0.00029
−3.3	0.00048	0.00047	0.00045	0.00043	0.00042
−3.2	0.00069	0.00066	0.00064	0.00062	0.00060
−3.1	0.00097	0.00094	0.00090	0.00087	0.00084
−3.0	0.00135	0.00131	0.00126	0.00122	0.00118
−2.9	0.00187	0.00181	0.00175	0.00169	0.00164
−2.8	0.00256	0.00248	0.00240	0.00233	0.00226
−2.7	0.00347	0.00336	0.00326	0.00317	0.00307
−2.6	0.00466	0.00453	0.00440	0.00427	0.00415
−2.5	0.00621	0.00604	0.00587	0.00570	0.00554
−2.4	0.00820	0.00798	0.00776	0.00755	0.00734
−2.3	0.01072	0.01044	0.01017	0.00990	0.00964
−2.2	0.01390	0.01355	0.01321	0.01287	0.01255
−2.1	0.01786	0.01743	0.01700	0.01659	0.01618
−2.0	0.02275	0.02222	0.02169	0.02118	0.02068
−1.9	0.02872	0.02807	0.02743	0.02680	0.02619
−1.8	0.03593	0.03515	0.03438	0.03362	0.03288

(continued)

T. S. Cotter, *Engineering Managerial Economic Decision and Risk Analysis*,
Topics in Safety, Risk, Reliability and Quality 39,
https://doi.org/10.1007/978-3-030-87767-5

(continued)

Z	0.00	0.01	0.02	0.03	0.04
−1.7	0.04457	0.04363	0.04272	0.04182	0.04093
−1.6	0.05480	0.05370	0.05262	0.05155	0.05050
−1.5	0.06681	0.06552	0.06426	0.06301	0.06178
−1.4	0.08076	0.07927	0.07780	0.07636	0.07493
−1.3	0.09680	0.09510	0.09342	0.09176	0.09012
−1.2	0.11507	0.11314	0.11123	0.10935	0.10749
−1.1	0.13567	0.13350	0.13136	0.12924	0.12714
−1.0	0.15866	0.15625	0.15386	0.15151	0.14917
−0.9	0.18406	0.18141	0.17879	0.17619	0.17361
−0.8	0.21186	0.20897	0.20611	0.20327	0.20045
−0.7	0.24196	0.23885	0.23576	0.23270	0.22965
−0.6	0.27425	0.27093	0.26763	0.26435	0.26109
−0.5	0.30854	0.30503	0.30153	0.29806	0.29460
−0.4	0.34458	0.34090	0.33724	0.33360	0.32997
−0.3	0.38209	0.37828	0.37448	0.37070	0.36693
−0.2	0.42074	0.41683	0.41294	0.40905	0.40517
−0.1	0.46017	0.45620	0.45224	0.44828	0.44433
0.0	0.50000	0.50399	0.50798	0.51197	0.51595

Cumulative Standard Normal Distribution

Z	0.05	0.06	0.07	0.08	0.09
−3.9	0.00004	0.00004	0.00004	0.00003	0.00003
−3.8	0.00006	0.00006	0.00005	0.00005	0.00005
−3.7	0.00009	0.00008	0.00008	0.00008	0.00008
−3.6	0.00013	0.00013	0.00012	0.00012	0.00011
−3.5	0.00019	0.00019	0.00018	0.00017	0.00017
−3.4	0.00028	0.00027	0.00026	0.00025	0.00024
−3.3	0.00040	0.00039	0.00038	0.00036	0.00035
−3.2	0.00058	0.00056	0.00054	0.00052	0.00050
−3.1	0.00082	0.00079	0.00076	0.00074	0.00071
−3.0	0.00114	0.00111	0.00107	0.00104	0.00100
−2.9	0.00159	0.00154	0.00149	0.00144	0.00139
−2.8	0.00219	0.00212	0.00205	0.00199	0.00193
−2.7	0.00298	0.00289	0.00280	0.00272	0.00264
−2.6	0.00402	0.00391	0.00379	0.00368	0.00357
−2.5	0.00539	0.00523	0.00508	0.00494	0.00480

(continued)

(continued)

Z	0.05	0.06	0.07	0.08	0.09
−2.4	0.00714	0.00695	0.00676	0.00657	0.00639
−2.3	0.00939	0.00914	0.00889	0.00866	0.00842
−2.2	0.01222	0.01191	0.01160	0.01130	0.01101
−2.1	0.01578	0.01539	0.01500	0.01463	0.01426
−2.0	0.02018	0.01970	0.01923	0.01876	0.01831
−1.9	0.02559	0.02500	0.02442	0.02385	0.02330
−1.8	0.03216	0.03144	0.03074	0.03005	0.02938
−1.7	0.04006	0.03920	0.03836	0.03754	0.03673
−1.6	0.04947	0.04846	0.04746	0.04648	0.04551
−1.5	0.06057	0.05938	0.05821	0.05705	0.05592
−1.4	0.07353	0.07215	0.07078	0.06944	0.06811
−1.3	0.08851	0.08691	0.08534	0.08379	0.08226
−1.2	0.10565	0.10383	0.10204	0.10027	0.09853
−1.1	0.12507	0.12302	0.12100	0.11900	0.11702
−1.0	0.14686	0.14457	0.14231	0.14007	0.13786
−0.9	0.17106	0.16853	0.16602	0.16354	0.16109
−0.8	0.19766	0.19489	0.19215	0.18943	0.18673
−0.7	0.22663	0.22363	0.22065	0.21770	0.21476
−0.6	0.25785	0.25463	0.25143	0.24825	0.24510
−0.5	0.29116	0.28774	0.28434	0.28096	0.27760
−0.4	0.32636	0.32276	0.31918	0.31561	0.31207
−0.3	0.36317	0.35942	0.35569	0.35197	0.34827
−0.2	0.40129	0.39743	0.39358	0.38974	0.38591
−0.1	0.44038	0.43644	0.43251	0.42858	0.42465
0.0	0.51994	0.52392	0.52790	0.53188	0.53586

Cumulative Standard Normal Distribution

Z	0.00	0.01	0.02	0.03	0.04
0.0	0.50000	0.50399	0.50798	0.51197	0.51595
0.1	0.53983	0.54380	0.54776	0.55172	0.55567
0.2	0.57926	0.58317	0.58706	0.59095	0.59483
0.3	0.61791	0.62172	0.62552	0.62930	0.63307
0.4	0.65542	0.65910	0.66276	0.66640	0.67003
0.5	0.69146	0.69497	0.69847	0.70194	0.70540
0.6	0.72575	0.72907	0.73237	0.73565	0.73891
0.7	0.75804	0.76115	0.76424	0.76730	0.77035

(continued)

(continued)

Z	0.00	0.01	0.02	0.03	0.04
0.8	0.78814	0.79103	0.79389	0.79673	0.79955
0.9	0.81594	0.81859	0.82121	0.82381	0.82639
1.0	0.84134	0.84375	0.84614	0.84849	0.85083
1.1	0.86433	0.86650	0.86864	0.87076	0.87286
1.2	0.88493	0.88686	0.88877	0.89065	0.89251
1.3	0.90320	0.90490	0.90658	0.90824	0.90988
1.4	0.91924	0.92073	0.92220	0.92364	0.92507
1.5	0.93319	0.93448	0.93574	0.93699	0.93822
1.6	0.94520	0.94630	0.94738	0.94845	0.94950
1.7	0.95543	0.95637	0.95728	0.95818	0.95907
1.8	0.96407	0.96485	0.96562	0.96638	0.96712
1.9	0.97128	0.97193	0.97257	0.97320	0.97381
2.0	0.97725	0.97778	0.97831	0.97882	0.97932
2.1	0.98214	0.98257	0.98300	0.98341	0.98382
2.2	0.98610	0.98645	0.98679	0.98713	0.98745
2.3	0.98928	0.98956	0.98983	0.99010	0.99036
2.4	0.99180	0.99202	0.99224	0.99245	0.99266
2.5	0.99379	0.99396	0.99413	0.99430	0.99446
2.6	0.99534	0.99547	0.99560	0.99573	0.99585
2.7	0.99653	0.99664	0.99674	0.99683	0.99693
2.8	0.99744	0.99752	0.99760	0.99767	0.99774
2.9	0.99813	0.99819	0.99825	0.99831	0.99836
3.0	0.99865	0.99869	0.99874	0.99878	0.99882
3.1	0.99903	0.99906	0.99910	0.99913	0.99916
3.2	0.99931	0.99934	0.99936	0.99938	0.99940
3.3	0.99952	0.99953	0.99955	0.99957	0.99958
3.4	0.99966	0.99968	0.99969	0.99970	0.99971
3.5	0.99977	0.99978	0.99978	0.99979	0.99980
3.6	0.99984	0.99985	0.99985	0.99986	0.99986
3.7	0.99989	0.99990	0.99990	0.99990	0.99991
3.8	0.99993	0.99993	0.99993	0.99994	0.99994
3.9	0.99995	0.99995	0.99996	0.99996	0.99996

Cumulative Standard Normal Distribution

Z	0.05	0.06	0.07	0.08	0.09
0.0	0.51994	0.52392	0.52790	0.53188	0.53586

(continued)

(continued)

Z	0.05	0.06	0.07	0.08	0.09
0.1	0.55962	0.56356	0.56749	0.57142	0.57535
0.2	0.59871	0.60257	0.60642	0.61026	0.61409
0.3	0.63683	0.64058	0.64431	0.64803	0.65173
0.4	0.67364	0.67724	0.68082	0.68439	0.68793
0.5	0.70884	0.71226	0.71566	0.71904	0.72240
0.6	0.74215	0.74537	0.74857	0.75175	0.75490
0.7	0.77337	0.77637	0.77935	0.78230	0.78524
0.8	0.80234	0.80511	0.80785	0.81057	0.81327
0.9	0.82894	0.83147	0.83398	0.83646	0.83891
1.0	0.85314	0.85543	0.85769	0.85993	0.86214
1.1	0.87493	0.87698	0.87900	0.88100	0.88298
1.2	0.89435	0.89617	0.89796	0.89973	0.90147
1.3	0.91149	0.91309	0.91466	0.91621	0.91774
1.4	0.92647	0.92785	0.92922	0.93056	0.93189
1.5	0.93943	0.94062	0.94179	0.94295	0.94408
1.6	0.95053	0.95154	0.95254	0.95352	0.95449
1.7	0.95994	0.96080	0.96164	0.96246	0.96327
1.8	0.96784	0.96856	0.96926	0.96995	0.97062
1.9	0.97441	0.97500	0.97558	0.97615	0.97670
2.0	0.97982	0.98030	0.98077	0.98124	0.98169
2.1	0.98422	0.98461	0.98500	0.98537	0.98574
2.2	0.98778	0.98809	0.98840	0.98870	0.98899
2.3	0.99061	0.99086	0.99111	0.99134	0.99158
2.4	0.99286	0.99305	0.99324	0.99343	0.99361
2.5	0.99461	0.99477	0.99492	0.99506	0.99520
2.6	0.99598	0.99609	0.99621	0.99632	0.99643
2.7	0.99702	0.99711	0.99720	0.99728	0.99736
2.8	0.99781	0.99788	0.99795	0.99801	0.99807
2.9	0.99841	0.99846	0.99851	0.99856	0.99861
3.0	0.99886	0.99889	0.99893	0.99896	0.99900
3.1	0.99918	0.99921	0.99924	0.99926	0.99929
3.2	0.99942	0.99944	0.99946	0.99948	0.99950
3.3	0.99960	0.99961	0.99962	0.99964	0.99965
3.4	0.99972	0.99973	0.99974	0.99975	0.99976
3.5	0.99981	0.99981	0.99982	0.99983	0.99983
3.6	0.99987	0.99987	0.99988	0.99988	0.99989
3.7	0.99991	0.99992	0.99992	0.99992	0.99992

(continued)

(continued)

Z	0.05	0.06	0.07	0.08	0.09
3.8	0.99994	0.99994	0.99995	0.99995	0.99995
3.9	0.99996	0.99996	0.99996	0.99997	0.99997

Appendix E
Fundamentals of Engineering (FE) Examination Example Questions

1. A product has the following annual manufacturing expenses.

Direct labor	$187,000
Direct materials	$220,000
Fixed overhead	$ 70,000
G&A and sale overhead	$ 60,000
Annual demand	20,000 units

   The manufacturing firm adds 20% profit margin to its product cost to the retailer. On average, retailers add 10% profit margin to their carrying cost to the final customer. How much will it cost for retailers to purchase the product in 100-unit shipments?

   a. $4907
   b. $3183
   c. $3222
   d. $5570

2. For the product in problem 1, what is the per-unit cost to the final customer?

   a. $35.45
   b. $53.98
   c. $61.27
   d. $35.01

3. An organization has total annual fixed costs of $500,000 and variable cost of $5.10 per unit produced. The organization prices its products at $10.60 per unit. What is the breakeven production quantity?

   a. 54,535
   b. 98,530
   c. 132,223

T. S. Cotter, *Engineering Managerial Economic Decision and Risk Analysis*, Topics in Safety, Risk, Reliability and Quality 39, https://doi.org/10.1007/978-3-030-87767-5

   d.   90,909

4.   A one-metric-ton/hour fluid bed jet mill costs $270,000 installed 10 years ago. What will a two-metric-ton/hour fluid bed jet mill cost today if the power sizing index exponent is 0.40 and the cost index for fluid bed jet mills increased from 108 to 116 over the last 10 years?

   a. $355,545
   b. $382,657
   c. $322,238
   d.   390,428

5.   If 370 labor hours were required to produce the first unit in the production run and 112 labor hours were required to produce the 16th unit, what is the learning curve rate for this product?

   a.   66%
   b.   81%
   c.   74%
   d.   79%

6.   An organization's balance sheet shows the following account information

Cash	$405,970	Accounts payable	$831,000
Accounts receivable	$762,590	Wages payable	$200,000
Inventory	$387,530	Salaries payable	$20,000
		Dividend payable (short term)	$73,000

   What are the organization's current ratio and acid-test ratios?

   a.   Current = 1.26; acid-test = 1.08
   b.   Current = 1.31; acid-test = 1.09
   c.   Current = 1.15; acid-test = 0.99
   d.   Current = 1.38; acid-test = 1.04

7.   An organization's accrual income statement shows the following account information

Sales	$12,000,000
Returns and allowances	$350,000
Cost of goods sold	$8,780,000
Depreciation	$329,250
Selling expenses	$658,500
G&A expenses	$1,566,650

   If the organization's combined federal and state income taxes were $94,680, what was the organization's net profit.

   a. $220,920
   b. $331,100
   c. $215,990

    d. $268,100

8.   If $1000 is invested at an annual interest rate of 6% per year, its future worth at the end of 20 years will be:

    a. $2,920
    b. $3,310
    c. $3,156
    d. $3,207

9.   A maintenance manager needs $37,200 in 5 years to pay for an upgrade to some equipment. If a fund earns 9% annually, how much must he invest today for the amount to grow to the required $37,200?

    a. $23,290
    b. $24,827
    c. $23,651
    d. $20,730

10.   An account pays 6.0% nominal interest compounded monthly. If $5,000 was deposited at the beginning of the current year, how much will be in the account at the end of four years?

    a. $6208
    b. $5872
    c. $6512
    d. $6352

11.   The loan shark credit card only charges 2.5% per month interest. What is the effective annual interest rate?

    a.   26.08%
    b.   28.27%
    c.   34.48%
    d.   36.23%

12.   A machine must be replaced at a cost of $47,000. A local bank will finance the cost of replacement with a loan that charges 9% nominal annual interest rate with monthly payments for three years. What is the uniform monthly payment?

    a. $1680
    b. $2082
    c. $1844
    d. $1495

13.   An engineer deposits $1000 into a retirement account today and deposits $100 at the end of every month starting with the first month. The account earns a guaranteed 0.25% per month if the account remains open and active. How much will the engineering have in the retirement account at the end of the engineer's 40-year career?

   a. $91,887
   b. $95,922
   c. $101,484
   d. $94,195

14.   An asset returns the following annual cash flows: year 1—$73,000, year 2—
      $78,000, year 3—$83,000, year 4—$88,000, year 5—$93,000, and year 6—
      $98,000. Given MARR = 12%, what is the asset's present worth?

   a. $419,787
   b. $359,292
   c. $344,753
   d. $299,451

15.   A project has an initial cost of $90,200, generates $60,200 additional revenues
      per year, requires $36,000 annual operating and maintenance expenses, and
      has a salvage value of $24,000 at the end of its 4-year life. If the organization's
      MARR = 10% for this project, what is its net present worth?

   a. $2906
   b. $2290
   c. $3443
   d. $2995

16.   What is the equivalent uniform annual cash flow for the project in question
      15?

   a. $962
   b. $892
   c. $1034
   d. $917

17.   An investment costed $50,000 and provided positive net annual cash flows of
      $7790. What was the investment's internal rate of return?

   a.   9.0%
   b.   8.0%
   c.   10.0%
   d.   7.0%

18.   Consider the following two mutually exclusive investment choices. What is
      the incremental rate of return?

Year	Alternative 1	Alternative 2
0	($20,000)	($30,000)
1–10	$7500	$9492

   a.   15.0%

    b.   18.0%

    c.   12.0%

    d.   8.0%

19. A MACRS GDS 5-year property asset was purchased and installed for $100,000 and has a salvage value of $15,000 at the end of its 10-year useful life. The asset's MACRS depreciation charge for years 1 and 2 of service is:

    a. $33,330 and $44,450

    b. $17,490 and $12,490

    c. $20,000 and $32,000

    d. $10,000 and $18,000

20. A one-metric-ton unit of raw material cost $1000 6 years ago. If the raw material prices have inflated at 3.0% per year for the last 6 years, what is the per-metric-ton cost this year?

    a. $1194

    b. $1160

    c. $1229

    d. $116

21. An investor is considering purchasing $10,000 stock in a new risky venture but only if he can double his money in one year and sell the stock. Based on his research of other similar risky ventures, he believes the following probabilities of return apply. What is his expected return in dollars?

Return	Probability
$25,000	0.1
$20,000	0.2
$15,000	0.2
$10,000	0.3
$5000	0.1
$0	0.1

    a. $9000

    b. $13,000

    c. $18,000

    d. $20,000

22. During the next year, the cost of a commodity is expected to vary as a uniform distribution U($3.20, $4.00) per unit. What is the commodity's expected cost and standard deviation?

    a. $3.40 and 0.038

    b. $3.60 and $0.053

    c. $3.70 and 0.068

    d. $3.80 and 0.079

23.   Construction time for an addition to a manufacturing facility depends on the weather. The time distribution can be described by a triangular distribution with minimum = 5 months, mode = 6 months, and maximum = 10 months. What is the expected completion time in months?

    a.   8 months

    b.   6.5 months

    c.   7 months

    d.   7.5 months

24.   What is the standard deviation in months of construction time in problem 22?

    a.   1.08 month

    b.   2.55 months

    c.   0.73 month

    d.   1.75 month

25.   Compute the sample mean and standard deviation of the following data: 90, 96, 102, 104, 108.

    a.   100 and 6.32

    b.   99.3 and 5.98

    c.   101.2 and 6.48

    d.   102.0 and 7.08.

Bibliography

1. Baasel W (1974) Preliminary chemical engineering plant design. Elsevier, New York
2. Bussey L (1978) The economic analysis of industrial projects. Prentice Hall, Englewood Cliffs, NJ
3. Fabrycky W, Blanchard B (1991) Life-cycle cost and economic analysis. Prentice Hall, Englewood Cliffs, NJ
4. Hodson W (ed) (2001) Maynard's industrial engineering handbook. McGraw-Hill, New York
5. Killough L, Leininger W (1987) Cost accounting: concepts and techniques for management. West Publishing Company, St. Paul, MN
6. Lori J, Savage L (1955) Three problems in rationing capital. J Bus 229–239
7. Meyers F (1999) Motion and time study for lean manufacturing. Prentice Hall, Englewood Cliffs, NJ
8. NATO (2007) Methods and models for life cycle costing. RTO Technical Report TR-SAS-054
9. Newman D, Eschenbach T, Lavelle J, Lewis N (2019) Engineering economic analysis. Oxford University Press, New York
10. Ostwald P (1992) Engineering cost estimating. Prentice Hall, Englewood Cliffs, NJ
11. Park C, Sharp-Bette G (1990) Advanced engineering economics. Wiley, New York
12. Park C (2007) Contemporary engineering economics. Pearson, Upper Saddle River, NJ
13. Perry R (ed) (1997) Perry's chemical engineering handbook. McGraw-Hill, New York
14. Peters M, Timmerhaus K (1991) Plant design and economics for chemical engineers. McGraw-Hill, New York
15. Thuesen G, Fabrycky W (2000) Engineering economy. Pearson, Upper Saddle River, NJ
16. Zandin K (1990) MOST work measurement systems. Marcel Dekker, New York

Index

Printed in the United States
by Baker & Taylor Publisher Services